Introduction to Data E
and Its Uses

INTRODUCTION TO DATA ENVELOPMENT ANALYSIS AND ITS USES

With DEA-Solver Software and References

WILLIAM W. COOPER
University of Texas at Austin, U.S.A.

LAWRENCE M. SEIFORD
University of Michigan, U.S.A.

KAORU TONE
National Graduate Institute for Policy Studies, Japan

Springer

William W. Cooper
University of Texas, USA

Lawrence M. Seiford
University of Michigan, USA

Kaoru Tone
National Graduate Institute for Policy Studies, Japan

Library of Congress Cataloging-in-Publication Data

ISBN-10: 0-387-28580-6 (SC)
ISBN-10: 0-387-29122-9 (e-book)
ISBN-13: 978-0387-28580-1 (SC)
ISBN-13: 978-0387-29122-2 (e-book)

Printed in the United States of America.

Printed on acid-free paper.

9 8 7 6 5 4 3 2 1

SPIN 11542650

springeronline.com

To

Ruth, Bev and Michiko

Contents

List of Tables

List of Figures

Preface

1. Introduction

We earlier coauthored a text, published in 2000, that is described in the following "Preface to Data Envelopment Analysis: A Comprehensive Text with Models, Applications, References and DEA Solver Software." The present text modifies and revises that book in order to incorporate important new developments. These developments have appeared in papers, books, monographs, etc., that now number in the thousands and address a wide range of different problems by many different authors in many different countries. See the publications that appear in the disk that accompanies this text. See also Tavares (2003)[1] "A Bibliography of Data Envelopment Analysis (1978-2001)" which lists more than 3200 papers, books, monographs, etc. addressing a great variety of problems by more than 1600 authors in 42 countries.

To accommodate all of these developments is now nearly impossible. We have therefore directed our attention to an "introduction" — hence the title of this book. This means that we have focused on approaches that have been most widely used or are most likely to be encountered by users of DEA and we hope to include others in a follow-on book.

2. What is Data Envelopment Analysis?

Data Envelopment Analysis (DEA) was accorded this name because of the way it "envelops" observations in order to identify a "frontier" that is used to evaluate observations representing the performances of all of the entities that are to be evaluated. Uses of DEA have involved a wide range of different kinds of entities that include not only business firms but also government and non-profit agencies including schools, hospitals, military units, police forces and court and criminal justice systems as well as countries, regions, etc. The term "Decision Making Unit" (DMU) was therefore introduced to cover, in a flexible manner, any such entity, with each such entity to be evaluated as part of a collection that utilizes similar inputs to produce similar outputs. These evaluations result in a performance score that ranges between zero and unity and represents the "degree of efficiency" obtained by the thus evaluated entity. In arriving at these scores, DEA also identifies the sources and amounts of inefficiency in each input and output for every DMU. It also identifies the DMUs (located on the "efficiency frontier") that entered actively in arriving at these results. These evaluating entities are all efficient DMUs and hence can serve as benchmarks en route to effecting improvements in future performances of the thus evaluated DMUs.

The different types of efficiency covered in this text range from "allocative," or "price," efficiency, and extend through "scale" and "technical" efficiency, as well as "mix" and other kinds of efficiencies. Technical inefficiency, which represents "waste," is the one we focus on in this Preface because it requires the least information, makes the fewest assumptions, and is the one most likely to be agreed upon as to what is meant by the term "inefficiency." Uses of DEA to effect these evaluations are almost entirely "data dependent" and do not require explicit characterizations of relations like "linearity," "nonlinearity," etc., which are customarily used in statistical regressions and related approaches where they are assumed to connect inputs to outputs, etc.

3. Engineering-Science Definitions of Efficiency

To illustrate these uses we start with the usual "output-to-input ratio" definitions (and measures) of efficiency used in engineering and science. Consider, for instance, the following definition (quoted from the *Encyclopedia Americana* (1966)[2]) where it is illustrated by an example from the field of combustion engineering: "Efficiency is the ratio of the amount of heat liberated in a given device to the maximum amount which could be liberated by the fuel [being used]." In symbols, $E = y_r/y_R$, where y_R= Maximum heat that can be obtained from a given input of fuel and y_r = Heat obtained from this same fuel input by the device being rated.

In "pure science" the rating would probably be obtained from theoretical considerations, such as are available from the theory of thermodynamics, to obtain a value $0 \le E \le 1$, with a value of unity not being obtainable. In applied science or engineering, alternative approaches are likely be used. For instance, the rating of a furnace that is being developed for sale would be evaluated by comparing it with a "reference furnace" with a value of unity being obtained when the heat obtained is equal to the heat from the reference unit.

Other approaches are used when ratings for $n(> 1)$ devices are to be developed in order to determine relative efficiencies. See Bulla *et al.* (2000)[3] for an example involving relative evaluations of 29 jet aircraft engines in which DEA is compared with the customarily employed engineering measure of efficiency. This engineering measure of efficiency consists of the ratio of "work rate output" to "fuel energy input." Using this same input and output, DEA produced the same efficiency rankings as the engineering measure. However, other factors (including cost) also need to be considered *en route* to a decision on which engine to choose. To accommodate these other measures, DEA was expanded to three inputs and two outputs and produced rankings that were markedly different from the engineering measure.

To accomplish this kind of result, DEA expands the ratio definition of efficiency to accommodate multiple outputs and multiple inputs. This expansion to multiple inputs and outputs is accomplished without using preassigned weights or other devices such as are commonly employed in index number (or like) constructions and no *a priori* identification of reference units, etc., are required. Proceeding in this manner DEA identifies a set of "best performers" from a collection of DMUs and uses them to generate "efficiency scores" obtained by evaluating each of the n DMUs. This is accomplished by identifying a point on this frontier that assigns a value of $0 \leq E \leq 1$ to each of the n DMUs. This score is obtained by comparing the performance of the DMU to be evaluated relative to the performances of *all* DMUs.

As obtained by DEA, this score reflects a series of weights as determined from the data by DEA — one for each output and one for each input - that produces the highest efficiency score that the data will allow relative to the values $0 \leq E_j \leq 1$ for each of the several DMU_j $j = 1, \ldots, n$, being considered. Moreover, these weights (called "multipliers" to distinguish them from customary weighting processes) have additional uses. A multiplier can be used, for instance, to evaluate changes in the efficiency score for a DMU that will accompany a change in the corresponding input or output. Since these multipliers are applicable to all DMUs, they can also be used to conduct "sensitivity analyses" to determine changes in *all* data (considered simultaneously) that will result in a change of classification from "efficient" to "inefficient," or *vice versa*, for any DMU, and also identify the ranges of data variation that can be allowed before such a change occurs. Unlike statistical and mathematical programming analyses, in which sensitivity analyses apply to only one data point at a time, these sensitivity analysis can allow simultaneous variation in *all* of the data. How this is done is described in this book. Also described in this book are methods for undertaking a one-data-point-at-a-time analysis (as in statistics, say) that can also be used to determine the allowable range of stability for the data changes associated with any individual DMU.

4. Different Definitions of Efficiency—and Duality

The above discussion is based on generalizations (and additional uses) of engineering-science definitions of efficiency. There are, however, other definitions of efficiency that are used in other disciplines. One such is the very general definition of efficiency, called "Pareto optimality," as used in "welfare economics," which was formulated by the Swiss-Italian economist, Vilfredo-Pareto. The definition, as given on page 319 in Pearce (1986)[4] is as follows: "A Pareto optimum is a welfare maximum defined as a position [in an economy] from which it is impossible to improve anyone's welfare by altering production or exchange without impairing someone else's welfare." This definition forms a basis for "welfare economics" in which proposed changes in policy (e.g., enactment of a tariff) are to be judged according to whether the proposed change can increase the welfare of some without worsening the welfare of others. This approach, we might note, avoids the need for making interpersonal comparisons such as would take the form of weights, or similar devices, that assign relative values to different individuals in making such decisions.

Tjalling Koopmans (1951), [5] a Dutch-American economist, subsequently extended this definition to "production economics" by introducing "efficiency prices" (=multipliers)[6] to guide production and exchange to positions that are similar to a Pareto optimum. This approach is further extended and exploited in DEA by reference to what is referred to as the "Pareto-Koopmans" definition of efficiency.

As shown in the present book, these seemingly different definitions involved in engineering-science and Pareto-Koopmans definitions of efficiency are mathematically related to each other by the "duality theorem" of linear programming. See Appendix A in this book for a description and discussion of duality and other parts of linear programming. To state this differently, there is a problem that is dual to the one used to implement the engineering definition of efficiency which employs the same data in a different manner to arrive at a measure of Pareto-Koopmans efficiency. At their corresponding minimum and maximum values the Pareto-Koopmans and engineering science scores are then equal.

This dual problem utilizes the following variant of a "welfare optimum" in defining efficiency:

Definition 1 (*Pareto-Koopmans Definition of Efficiency*) *The performance of a DMU is efficient if and only if it is not possible to improve any input or output without worsening any other input or output.*

From this we also have

Definition 2 (*Definition of Inefficiency*) *The performance of a DMU is inefficient if and only if it is possible to improve some input or output without worsening some other input and output.*

Thus a characterization of being inefficient holds if and only if the observation — i.e., the input-output coordinates — for the DMU being evaluated is not on the efficiency frontier of the set of production possibilities that can be generated by DEA from the data on all observations. Conversely, the geometric points associated with the performance of the DMU to be evaluated by DEA will be found to be efficient if and only if it is on the efficiency frontier.

5. Extensions and Further Uses of DEA

The above generalization of the engineering-science definition of efficiency involves maximizing a ratio of inputs to outputs subject to a collection of n ratios that also take the form of inequality constraints with a bound of unity on their possible values. However, as shown in this book, this nonlinear (and non-convex) problem can be replaced by a simple (ordinary) linear programming problem that makes the solution easily obtainable from any of numerous computer codes that are used to solve linear programming problems. It also opens other possibilities for use from now available computer codes that are specifically designed for use in DEA. See, for instance, the DEA-Solver code that is described in Appendix B of this book.

Using the duality relation of linear programming to comprehend these two different definitions of efficiency and giving them implementable form greatly extends the ability of DEA to deal with a wide range of problems. For instance, the study recommending a relocation of the Japanese capital to a new site (now being considered by the Japanese Diet) had to deal with "inputs" like "susceptibility to earthquakes" and with "outputs" like "ability to recover from earthquakes." These are not "inputs" or "outputs" in the ordinary senses of these terms and are therefore referred to as "goods" or "bads" according to whether they should be placed in the numerator or denominator of an engineering-science type of "efficiency ratio" in accordance with whether an increase in this item should result in an increase or a decrease in the performance score. This greatly simplifies matters that would otherwise be difficult to accomplish en route to identifying inputs as distinguished from outputs (in borderline cases) that could be difficult to do with the Pareto-Koopmans definition of efficiency. This illustrates one way that the "conceptual power" of the engineering-science definition of efficiency can be used to obtain characterizations that also satisfy the Pareto-Koopmans definition. The thus identified inputs and outputs that had to be considered in the Japan relocation study were then incorporated in a DEA model that was directed to "effectiveness" rather than "efficiency" in order to determine the best location (two "equally best" locations were identified). As defined in this book, these two concepts differ as follows:

$$\text{Effectiveness: } \begin{cases} \text{Ability to state goals} \\ \text{Ability to achieve goals.} \end{cases}$$

$$\text{Efficiency: } \begin{cases} \text{Benefits secured} \\ \text{Resources used.} \end{cases}$$

Thus effectiveness refers to goal achievement and, in contrast to efficiency, effects its evaluations without reference to the resources used. (See Chapter 6 in this book for further discussion of how the committee appointed to conduct this study developed its reports to produce recommendations that were submitted to the Prime Minister who has transmitted them to the Japanese Diet for consideration).

These and the other developments in the use of DEA are all given implementable form in this book. Technical developments are accompanied by numerical examples and illustrated by applications with accompanying interpretations and descriptions that are intended to provide new possibilities for uses of DEA. The way these topics are developed is described in the following "Preface to Data Envelopment Analysis, a Comprehensive Text with Models, Applications, References and DEA-Solver Software" — to which this new volume adds much new material. This includes a new chapter 10, for instance, that deals with a concept called "super-efficiency" that extends the efficiency measures to allow values of $E > 1$. This extension makes it possible, for example, to ascertain the consequences (in the form of the increases in inputs that are needed to avoid reducing any output) that will accompany the elimination of an entire DMU — especially an efficient DMU — from the production possibility set.

Acknowledgments

The authors of this text have all been actively involved in applications and uses of DEA that could not have been undertaken except as part of a team effort. We hope that the citations made in this text will serve as acknowledgments to the other team members who helped to make these applications possible, and who also helped to point up the subsequent advances in DEA that are described in this text.

Acknowledgment is also due to Michiko Tone in Tokyo, Japan, and Bisheng Gu in Austin, Texas. In great measure this text could not have been completed without their extraordinary efforts in typing, correcting and compiling the many versions this manuscript passed through en route to its present form. All of the authors benefited from their numerous suggestions as well as their efforts in coordinating these activities, which spanned an ocean and a continent.

Finally, thanks are also due to Gary Folven of Kluwer Publishers (now Springer) who encouraged, cajoled and importuned the authors in manners that a lawyer might say, "aided and abetted them in the commission of this text."

We are very much benefited by readers of our previous DEA textbook who kindly pointed out mistakes and typos. Further mistakes should be attributed only to the authors and we will be grateful if they are notified to us.

WILLIAM W. COOPER
The Red McCombs School of Business
University of Texas at Austin
Austin, Texas 78712-1175, USA
cooperw@mail.utexas.edu

LAWRENCE M. SEIFORD
Department of Industrial and
Operations Engineering
University of Michigan
1205 Beal Avenue Ann Arbor
MI 48109-2117, USA
seiford@umich.edu

KAORU TONE
National Graduate Institute for Policy Studies
7-22-1 Roppongi, Minato-ku
Tokyo 106-8677, Japan
tone@grips.ac.jp

June, 2005

Kaoru Tone would like to express his heartfelt thanks to his colleagues and students: B. K. Sahoo, K. Nakabayashi, M. Tsutsui, J. Liu, S. Iwasaki, K. Igarashi and H. Shimatsuji for their help in writing this book. W.W. Cooper wants to express his appreciation to the IC^2 Institute of the University of Texas at Austin for their continuing support of his research. Larry Seiford would like to acknowledge the many DEA researchers worldwide whose collective efforts provide the backdrop for this book, and he especially wants to thank K. Tone and W.W. Cooper for their dedicated efforts in producing this text.

Notes

1. G. Tavares (2003), "A Bibliography of Data Envelopment Analysis (1978-2001)," (Pascatuwy, N.J.: Rutcor, Rutgers University), gtavares@rutcor.rutgers.edu.
2. *Encyclopedia Americana* (1966). New York: Encyclopedia Americana Corporation.
3. S. Bulla, W.W. Cooper, K.S. Park and D. Wilson (2000), "Evaluating Efficiencies of Turbo-Fan Jet Engines: A Data Envelopment Analysis Approach," *Journal of Propulsion and Power*, 16, 431-439.
4. D.W. Pearce (1986), *The MIT Dictionary of Modern Economics,* 3rd edition (Cambridge, Mass: The MIT Press).
5. T.C. Koopmans (1951), "Analysis of Production as an Efficient Combination of Activities" in, T.C. Koopmans, ed., *Activity Analysis of Production and Allocation* (New York: John Wiley & Sons).
6. See Chapter IX in A. Charnes and W.W. Cooper (1962), *Management Models and Industrial Applications of Linear Programming* (New York: John Wiley & Sons, Inc.) for a proof.

Preface to
DATA ENVELOPMENT ANALYSIS
A Comprehensive Text with
Models, Applications, References
and DEA Solver Software
(Kluwer Academic Publishers, 2000)

1. Introduction

Recent years have seen a great variety of applications of DEA (Data Envelopment Analysis) for use in evaluating the performances of many different kinds of entities engaged in many different activities in many different contexts in many different countries. One reason is that DEA has opened up possibilities for use in cases which have been resistant to other approaches because of the complex (often unknown) nature of the relations between the multiple inputs and multiple outputs involved in many of these activities (which are often reported in non-commeasurable units). Examples include the maintenance activities of U.S. Air Force bases in different geographic locations, or police forces in England and Wales as well as performances of branch banks in Cyprus and Canada and the efficiency of universities in performing their education and research functions in the U.S., England and France. These kinds of applications extend to evaluating the performances of cities, regions and countries with many different kinds of inputs and outputs that include "social" and "safety-net" expenditures as inputs and various "quality-of-life" dimensions as outputs.

DEA has also been used to supply new insights into activities (and entities) that have previously been evaluated by other methods. For instance, studies of benchmarking practices with DEA have identified numerous sources of inefficiency in some of the most profitable firms — firms that served as benchmarks by reference to their (profitability) criterion. DEA studies of the efficiency of different legal-organizations forms as in "stock" vs. "mutual" insurance companies have shown that previous studies have fallen short in their attempts to evaluate the potentials of these different forms of organizations. (See below.) Similarly, a use of DEA has suggested reconsideration of previous studies of the efficiency with which pre- and post-merger activities have been conducted in banks that were studied by DEA.

The study of insurance company organization forms referenced above can be used to show not only how new results can be secured but also to show some of the new methods of data exploitation that DEA makes available. In order to study the efficiency of these organization forms it is, of course, necessary to remove other sources of inefficiency from the observations — unless one wants to *assume* that all such inefficiencies are absent.

In the study referenced in Chapter 7, this removal is accomplished by first developing separate efficiency frontiers for each of the two forms — stock vs. mutual. Each firm is then brought onto its respective frontier by removing its inefficiencies. A third frontier is then erected by reference to the thus adjusted data for each firm with deviations from this overall frontier then representing organization inefficiency. Statistical tests using the Mann-Whitney rank order statistic are then used in a manner that takes into account the full array of firms in each of the two categories. This led to the conclusion that stock companies were uniformly more efficient in all of the dimensions that were studied.

In Chapter 7 this study is contrasted with the earlier studies by E. Fama and M. Jensen.[1] Using elements from "agency theory" to study the relative efficiency of these two forms of organization, Fama and Jensen concluded that each type of organization was relatively most efficient in supplying its special brand of services.

Lacking access to methods like those described above, Fama and Jensen utilized agency theory constructs which, like other parts of micro-economics, assume that performances of all firms occur on efficiency frontiers. This assumption can now be called into question by the hundreds of studies in DEA and related approaches[2] which have shown it to be seriously deficient for use with empirical data. Moreover, it is no longer necessary to make this assumption since the requisite conditions for fully efficient performances can be fulfilled by using concepts and procedures like those we have just described — and which are given implementable form in the chapters that follows.[3]

Fama and Jensen also base their conclusions on relatively few summary ratios which they reenforce by observing the long standing co-existence of mutual and stock companies. Left unattended, however, is what Cummins and Zi (1998, p.132)[4] refer to as a "wave of conversions of insurers from mutual to stock ownership which is accelerating as competition increases from non-traditional services such as banks, mutual funds and security banks."

See also the discussion in Chapter 7 which notes that this movement has generally taken the form of "demutualizations" with no movement in the opposite direction.

In contrast to the relatively few summary ratios used by Fama and Jensen, DEA supplies a wealth of information in the form of estimates of inefficiencies in both inputs and outputs for every DMU (= Decision Making Unit). It also identifies the peer (comparison) group of efficient firms (stock and mutual) used to obtain these estimates and effect these evaluations. Moreover, the two-stage analysis described above provides additional information. Stage two identifies inefficiencies in each input and output which are attributable to the organization form used. Stage 1 identifies inefficiencies in each input and output attributable to the way each of these organization forms is managed. (P.L. Brockett and B. Golany (1996)[5] suggest a use of the MDI (Minimum Discrimination Information) Statistic to determine whether crossovers occur with the possibility that one form of organization may be more efficient in some areas and less efficient in others — and hence automatically supplies a test of the Fama and Jensen hypothesis in a way that identifies where the crossovers, with accompanying advantages, occur.)

Turning to methodology, we emphasize that the linear (and mathematical) programming models and methods used in DEA effect their evaluations from observed (i.e., already executed) performances and hence reverse the usual manner in which programming models are used. This has led to new results as well as new insights for use in other parts of the programming literature. This use of linear programming has also led to new principles of inference from empirical data which are directed to obtaining best estimates for *each* of the observed entities (or activities) in a collection of entities. This is in contrast to the usual *averaging* over all of the observations which are the characteristic approaches used in statistics and accounting. These estimates thus take the form of identifying the sources and amounts of inefficiency in each input and each output for every entity while also providing an overall measure of efficiency for *each* entity or activity that may be of interest.

2. Motivation

The great amount of activity involved in these uses of DEA have been accompanied by research directed to expanding its methods and concepts is evidenced by the hundreds of publications which are referenced in this book and elsewhere in the literature. See, for instance, A. Charnes, W.W. Cooper, A.Y. Lewin and L.M. Seiford (1994) *Data Envelopment Analysis: Theory, Methodology and Applications* (Norwell, Mass: Kluwer Academic Publishers). The references, which appear in the attached disk (at the end of this text) are comprehensive but not exhaustive. Indeed, they are directed only to published materials and do not even attempt to cover the still larger (and increasing) numbers of unpublished studies and reports which are now in circulation which range from Ph.D. theses to brief notes and summaries. As a case in point we can report that more than 50 papers are presently scheduled for presentation in sessions devoted to DEA which the authors of this text helped to organize for the Decision Sciences Institute Meetings to be held in Athens, Greece, from July 4-7, 1999. Additional references may be secured from various DEA web sites such as the ones at the University of Massachusetts–Amherst: http://www.ecs.umass.edu/dea/ or the Business School of the University of Warwick in England: http://www.csv.warwick.ac.uk/~bsrlu/. See also The Productivity Analysis Research Network (PARN), Odense University, Denmark: PARN@SAM.SDU.DK. In addition the developments in DEA have been (and are being) reported in two different literatures: (1) the literature of Operations Research/Management Science and (2) the literature of Economics.

Some or all of this activity is reflected in the texts that are presently available — as referenced at suitable points in the volume. However, these texts tend to be relatively advanced and directed to audiences of either economists or operations researchers and management scientists in both the methods and concepts they use. The present text is directed to a wider audience and proceeds on more elementary levels. For instance, efficiency considerations which are central to the DEA evaluations of interest are introduced by using the familiar and very simple ratio definition of "output divided by input." This ratio formulation is then extended to multiple outputs and multiple inputs in a manner that makes contact with more complex formulations. This includes the Pareto concept of "welfare efficiency" used in economics which is here referred to as "Pareto-Koopmans efficiency" in recognition of the adaptation of this "welfare economics" concept by T.C. Koopmans for use in "production economics."

3. Methodology

The methodology used in this text is based mainly on linear algebra (including matrices and vectors) rather than the less familiar set-theoretic and axiomatic formulations used in other texts on DEA. An additional advantage of this approach is that it very naturally and easily makes contacts with the linear programming methods and concepts on which the developments in this book rest.

The power of this programming approach is greatly enhanced, as is well known, by virtue of mathematical duality relations which linear programming provides access to in unusually simple forms. This, in turn, provides opportunities for extending results and simplifying proofs which are not available from approaches used in texts that have tended to slight these linear-programming-duality relations. One reason for such slighting may be attributed to a lack of familiarity (or appreciation) of these concepts by the audiences to which these texts were directed. To comprehend this possibility we have included an appendix — Appendix A — which provides a summary of linear programming and its duality relations and extends this to more advanced concepts such as the strong form of the "complementary slackness" principle.

Proofs are supplied for results which are critical but not obvious. Readers not interested in proof details can, if they wish, take the statements in the theorems on faith or belief. They can also check results and perhaps firm up their understanding by working the many small numerical examples which are included to illustrate theorems and their potential uses. In addition many examples are supplied in the form of miniaturized numerical illustrations.

To facilitate the use of this material a "DEA-Solver" disk is supplied with accompanying instructions which are introduced early in the text. In addition more comprehensive instructions for use of "DEA-Solver" are provided in Appendix B. This Solver was developed using VBA (Visual Basic for Applications) and Excel Macros in Microsoft Office 97 (a trademark of Microsoft Corporation) and is completely compatible with Excel data sheets. It can read a data set directly from an Excel worksheet and returns the results of computation to an Excel workbook. The code works on Excel 2000 as well. The results provide both primal (envelopment form) and dual (multiplier form) solutions as well as slacks, projections onto efficient frontiers, graphs, etc. The linear programming code was originally designed and implemented for DEA, taking advantage of special features of DEA formulations. Readers are encouraged to use this Solver for solving examples and problems in this text and for deepening their understanding of the models and their solutions. Although the attached Solver is primarily for learning purposes and can deal only with relatively small sized problems within a limited number of models, a more advanced "Professional version" can be found at the web site http://www.saitech-inc.com/.

Numerous specific references are also supplied for those who want to follow up some of the developments and example applications provided in this text. Still further references may be secured from the bibliography appended to the disk.

4. Strategy of Presentation

To serve the wider audience that is our objective, we have supplied numerous problems and suggested answers at the end of each chapter. The latter are referred to as "suggested responses" in order to indicate that other answers are possible in many cases and interested readers are encouraged to examine such possibilities.

In order to serve practitioner as well as classroom purposes we have presented these suggested responses immediately after the problems to which they refer. We have also provided appendices in cases where a deeper understanding or a broadened view of possibilities is desirable. We have thus tried to make this volume relatively self contained.

A wealth of developments in theory and methodology, computer codes, etc., have been (are being) stimulated by the widespread use of DEA on many different problems in many different contexts. It is not possible to cover them all in a single text so, as already noted, we supply references for interested readers. Most of these references, including the more advanced texts that are also referenced, will require some knowledge of DEA. Supplying this requisite background is the objective of this text.

The first 5 chapters should suffice for persons seeking an introduction to DEA. In addition to different DEA models and their uses, these chapters cover different kinds of efficiency — from "technical" to "mix" and "returns-to-scale" inefficiencies. These topics are addressed in a manner that reflects the "minimal assumption" approach of DEA in that information (= data) requirements are also minimal. Additional information such as a knowledge of unit prices and costs or other weights such as weights stated in "utils" is not needed. Treatments of these topics are undertaken in Chapter 8 which deals with allocation (= allocative) efficiency on the assumption that access to the requisite information in the form of an exact knowledge of prices, costs, etc., is available.

Relaxation of the need for such "exact" information can be replaced with bounds on the values of the variables to be used. This topic is addressed in Chapter 6 because it has wider applicability than its potential for use in treating allocative inefficiency.

Chapter 7 treats variables the values of which cannot be completely controlled by users. Referred to as being "non-discretionary," such variables are illustrated by weather conditions that affect the "sortie rates" that are reported by different air force bases or the unemployment rates that affect the recruitment potentials in different offices (and regions) of the U.S. Army Recruiting Command.[6]

Categorical (= classificatory) variables are also treated in Chapter 7. As examplified by outlets which do or do not have drive-through capabilities in a chain of fast-food outlets, these variables are also non-discretionary but need to be treated in a different manner. Also, as might be expected, different degrees of discretion may need to be considered so extensions to the standard treatments of these topics are also included in Chapter 7.

It is characteristic of DEA that still further progress has been made in treating data which are imperfectly known while this text was in its final stages. Such imperfect knowledge may take a form in which the data can only be treated in the form of ordinal relations such as "less" or "more than." See, for example, Ali, Cook and Seiford (1991).[7] It can also take the form of knowing only that the values of the data lie within limits prescribed by certain upper and lower bounds. See Cooper, Park and Yu (1999)[8] which unifies all of these approaches. See also Yu, Wei and Brockett (1996)[9] for an approach to unification of DEA models and methods in a manner which relates them to multiple objective programming. Nor is this all, as witness recent articles which join DEA to "multiple objective programming" and "fuzzy sets." See Joro, Korhonen and Wallenius (1998)[10] and K. Triantis and O. Girod (1998).[11]

As is apparent, much is going on with DEA in both research and uses. This kind of activity is likely to increase. Indeed, we hope that this volume will help to stimulate further such developments. The last chapter — *viz.*, Chapter 9 — is therefore presented in this spirit of continuing research in joining DEA with other approaches. Thus, the possibility of joining statistical regression approaches to efficiency evaluation with DEA is there treated in the form of a two-stage process in the following manner. Stage one: Use DEA to determine which observations are associated with efficient and which observations are associated with inefficient performances. Stage 2: Incorporate these results in the form of "dummy variables" in a statistical regression. As described in Chapter 9, the resulting regression was found to perform satisfactorily using the same form of regression relation and the same data that had previously yielded unsatisfactory results in a study undertaken to expand possible use of regression relations in response to a legislative mandate to provide improved methods to evaluate performances of secondary school in Texas. Moreover, a subsequent use of a simulation study confirmed these "satisfactory" results by showing that (1) the two-stage process *always* gave "correct" parameter estimates for all of the regression forms that were studied while (2) the ordinary one-stage process with this same regression *never* gave correct estimates — i.e., the estimates were always significantly different from the true parameter values.

There is, of course, more to be done and the discussions in Chapter 9 point up some of these possibilities for research. This is the same spirit in which the other topics are treated in Chapter 9.

Another direction for research is provided by modifying the full (100%) efficiency evaluations of DEA by invoking the "satisficing" concepts of H.A. Simon,[12] as derived from the literatures of management and psychology. Using "chance-constrained programming" approaches, a formulation is presented in Chapter 9 which replaces the usual deterministic requirements of 100% efficiency as a basis for evaluation in DEA with a less demanding objective of achieving a desired level of efficiency with a sufficiently high probability. This formulation is then expanded to consider possible behaviors when a manager aspires to one level and superior echelons of management prescribe other (higher) levels, including levels which are probablistically impossible of achievement.

Sensitivity of the DEA evaluations to data variations are topics that are also treated in Chapter 9. Methods for effecting such studies range from studies that consider the ranges over which a single data point may be varied without altering the ratings and extend to methods for examining effects on evaluations when all data are varied simultaneously — e.g., by worsening the output-input data for efficient performers and improving the output-input data for inefficient performers.

Such sensitivity analyses are not restricted to varying the components of given observations. Sensitivity analyses extend to eliminating some observations entirely and/or introducing additional observations. This topic is studied in Chapter 9 by reference to a technique referred to as "window analysis." Originally viewed as a method for studying trends, window analysis is reminiscent of the use of "moving averages" in statistical time series. Here, however, it is also treated as a method for studying the stability of DEA results because such window analyses involve the removal of entire sets of observations and their replacement by other (previously not considered) observations.

Comments and suggestions for further research are offered as each topic is covered in Chapter 9. These comments identify shortcomings that need to be addressed as well as extensions that might be made. The idea is to provide readers with possible directions and "head starts" in pursuing any of these topics. Other uses are also kept in mind, however, as possibilities for present use are covered along with openings for further advances.

W.W. COOPER L.M. SEIFORD K. TONE

June, 1999

Notes

1. E.F. Fama and M.C. Jensen (1983), "Separation of Ownership and Control," *Journal of Law and Economics* 26, pp.301-325. See also, in this same issue, Fama and Jensen "Agency Problems and Residual Claims," pp.327-349.
2. E.g., the stochastic frontier regression approaches for which descriptions and references are supplied in Chapter 9.
3. See the discussion following (3.22) and (3.33) in Chapter 3.
4. J.D. Cummins and H. Zi (1998), "Comparisons of Frontier Effecting Methods: An Application to the U.S. Life Insurance Industry," *Journal of Productivity Analysis* 10, pp.131-152.
5. P.L. Brockett and B. Golany (1996), "Using Rank Statistics for Determining Programmatic Efficiency Differences in Data Envelopment Analysis," *Management Science* 42, pp.466-472.
6. See D.A. Thomas (1990), "Data Envelopment Analysis Methods in the Management of Personnel Recruitment under Competition in the Context of U.S. Army Recruiting," Ph.D. Thesis (Austin, Texas: The University of Texas Graduate School of Business. Also available from University Micro-Films, Inc., Ann Arbor, Michigan.
7. A.I. Ali, W.D. Cook and L.M. Seiford (1991), "Strict vs Weak Ordinal Relations for Multipliers in Data Envelopment Analysis," *Management Science* 37, pp.733-738.
8. W.W. Cooper, K.S. Park and G. Yu (1999), "IDEA and AR: IDEA: Models for Dealing with Imprecise Data in DEA," *Management Science* 45, pp.597-607.
9. G. Yu, Q. Wei and P. Brockett (1996), "A Generalized Data Envelopment Model: A Unification and Extension Method for Efficiency Analysis of Decision Making Units," *Annals of Operations Research* 66, pp.47-92.
10. T. Joro, P. Korhonen and J. Wallenius, (1998) "Structural Comparisons of Data Envelopment Analysis and Multiple Objective Programming," *Management Science* 44, pp.962-970.
11. K. Triantis and O. Girod (1998), "A Mathematical Programming Approach for Measuring Technical Efficiency in a Fuzzy Environment," *Journal of Productivity Analysis* 10, pp.85-102.
12. H.A. Simon (1957), *Models of Man: Social and Rational* (New York: John Wiley & Sons, Inc.). See also his autobiography, H.A. Simon (1991), *Models of My Life* (New York: Basic Books).

Glossary of symbols

$\boldsymbol{x},\ \boldsymbol{y},\ \boldsymbol{\lambda}$	Small letters in bold face denote vectors.
$\boldsymbol{x} \in R^n$	$\boldsymbol{x}$ is a point in the n dimensional vector space R^n.
$A \in R^{m \times n}$	A is a matrix with m rows and n columns.
$\boldsymbol{x}^T,\ A^T$	The symbol ^{T} denotes transposition.
$\boldsymbol{e}$	$\boldsymbol{e}$ denotes a row vector in which all elements are equal to 1.
$\boldsymbol{e}_j$	$\boldsymbol{e}_j$ is the unit row vector with the j-th element 1 and others 0.
I	I is the identity matrix.
$\text{rank}(A)$	$\text{rank}(A)$ denotes the rank of the matrix A.
$\sum_{j=1}^{n} \lambda_j$	indicates the sum of $\lambda_1 + \lambda_2 + \cdots + \lambda_n$.
$\sum_{j=1,\ j \neq k}^{n} \lambda_j$	indicates the sum with λ_k omitted: $\lambda_1 + \cdots + \lambda_{k-1} + \lambda_{k+1} + \cdots + \lambda_n$.

1 GENERAL DISCUSSION

1.1 INTRODUCTION

This book is concerned with evaluations of performance and it is especially concerned with evaluating the activities of organizations such as business firms, government agencies, hospitals, educational institutions, etc. Such evaluations take a variety of forms in customary analyses. Examples include cost per unit, profit per unit, satisfaction per unit, and so on, which are measures stated in the form of a ratio like the following,

$$\frac{\text{Output}}{\text{Input}}. \tag{1.1}$$

This is a commonly used measure of efficiency. The usual measure of "productivity" also assumes a ratio form when used to evaluate worker or employee performance. "Output per worker hour" or "output per worker employed" are examples with sales, profit or other measures of output appearing in the numerator. Such measures are sometimes referred to as "partial productivity measures." This terminology is intended to distinguish them from "total factor productivity measures," because the latter attempt to obtain an output-to-input ratio value which takes account of *all* outputs and *all* inputs. Moving from partial to total factor productivity measures by combining all inputs and all outputs to obtain a single ratio helps to avoid imputing gains to one factor (or one output) that are really attributable to some other input (or output). For instance, a gain in output resulting from an increase in capital or improved management might be mistakenly attributed to labor (when a single output to input ratio is used) even though the performance of labor *deteriorated* during the period being considered. However, an attempt to move from partial to total factor productivity measures encounters difficulties such as choosing the inputs and outputs to be considered and the weights to be used in order to obtain a single-output-to-single-input ratio that reduces to a form like expression (1.1).

Other problems and limitations are also incurred in traditional attempts to evaluate productivity or efficiency when multiple outputs and multiple inputs need to be taken into account. Some of the problems that need to be addressed will be described as we proceed to deal in more detail with Data Envelopment Analysis (DEA), the topic of this book. The relatively new approach embodied in DEA does not require the user to prescribe weights to be attached to each input and output, as in the usual index number approaches, and it also does not require prescribing the functional forms that are needed in statistical regression approaches to these topics.

DEA utilizes techniques such as mathematical programming which can handle large numbers of variables and relations (constraints) and this relaxes the requirements that are often encountered when one is limited to choosing only a few inputs and outputs because the techniques employed will otherwise encounter difficulties. Relaxing conditions on the number of candidates to be used in calculating the desired evaluation measures makes it easier to deal with complex problems and to deal with other considerations that are likely to be confronted in many managerial and social policy contexts. Moreover, the extensive body of theory and methodology available from mathematical programming can be brought to bear in guiding analyses and interpretations. It can also be brought to bear in effecting computations because much of what is needed has already been developed and adapted for use in many prior applications of DEA. Much of this is now available in the literature on research in DEA and a lot of this has now been incorporated in commercially available computer codes that have been developed for use with DEA. This, too, is drawn upon in the present book and a CD with supporting DEA-Solver software and instructions, has been included to provide a start by applying it to some problems given in this book.

DEA provides a number of additional opportunities for use. This includes opportunities for collaboration between analysts and decision-makers, which extend from collaboration in choices of the inputs and outputs to be used and includes choosing the types of "what-if" questions to be addressed. Such collaborations extend to "benchmarking" of "what-if" behaviors of competitors and include identifying potential (new) competitors that may emerge for consideration in some of the scenarios that might be generated.

1.2 SINGLE INPUT AND SINGLE OUTPUT

To provide a start to our study of DEA and its uses, we return to the single output to single input case and apply formula (1.1) to the following simple example. Suppose there are 8 branch stores which we label A to H at the head of each column in Table 1.1.

Table 1.1. Single Input and Single Output Case

Store	A	B	C	D	E	F	G	H
Employee	2	3	3	4	5	5	6	8
Sale	1	3	2	3	4	2	3	5
Sale/Employee	0.5	1	0.667	0.75	0.8	0.4	0.5	0.625

The number of employees and sales (measured in 100,000 dollars) are as recorded in each column. The bottom line of Table 1.1 shows the sales per employee — a measure of "productivity" often used in management and investment analysis. As noted in the sentence following expression (1.1), this may also be treated in the more general context of "efficiency." Then, by this measure, we may identify B as the most efficient branch and F as least efficient.

Let us represent these data as in Figure 1.1 by plotting "number of employees" on the horizontal and "sales" on the vertical axis. The slope of the line connecting each point to the origin corresponds to the sales per employee and the highest such slope is attained by the line from the origin through B. This line is called the "efficient frontier." Notice that this frontier touches at least one point and all points are therefore on or below this line. The name Data Envelopment Analysis, as used in DEA, comes from this property because in mathematical parlance, such a frontier is said to "envelop" these points.

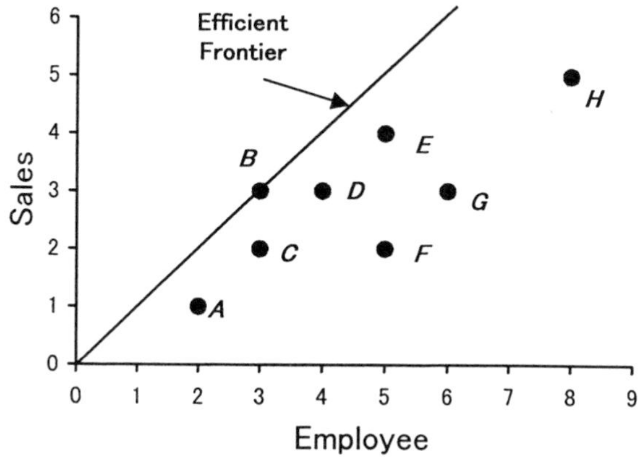

Figure 1.1. Comparisons of Branch Stores

Given these data, one might be tempted to draw a statistical regression line fitted to them. The dotted line in Figure 1.2 shows the regression line passing through the origin which, under the least squares principle, is expressed by $y = 0.622x$. This line, as normally determined in statistics, goes through the "middle" of these data points and so we could define the points above it as *excellent* and the points below it as *inferior* or *unsatisfactory*. One can measure the degree of excellence or inferiority of these data points by the magnitude of the deviation from the thus fitted line. On the other hand, the frontier line designates the performance of the best store (B) and measures the efficiency of other stores by deviations from it. There thus exists a fundamental difference between statistical approaches via regression analysis and DEA. The former reflects "average" or "central tendency" behavior of the observations while the latter deals with best performance and evaluates all performances by deviations from the frontier line. These two points of view can result in major differences when used as methods of evaluation. They can also result in different approaches to improvement. DEA identifies a point like B for future examination or to serve as a "benchmark" to use in seeking improvements. The statistical approach, on the other hand, averages B along with the other observations, including F as a basis for suggesting where improvements might be sought.

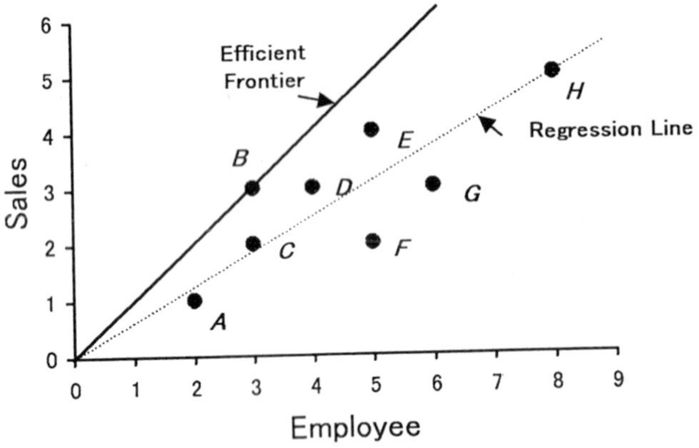

Figure 1.2. Regression Line vs. Frontier Line

Returning to the example above, it is not really reasonable to believe that the frontier line stretches to infinity with the same slope. We will analyze this problem later by using different DEA models. However, we assume that this line is effective in the range of interest and call it the *constant returns-to-scale* assumption.

Compared with the best store B, the others are inefficient. We can measure the efficiency of others relative to B by

$$0 \leq \frac{\text{Sales per employee of others}}{\text{Sales per employee of } B} \leq 1 \tag{1.2}$$

and arrange them in the following order by reference to the results shown in Table 1.2.

$$1 = B > E > D > C > H > A = G > F = 0.4.$$

Thus, the worst, F, attains $0.4 \times 100\% = 40\%$ of B's efficiency.

Table 1.2. Efficiency

Store	A	B	C	D	E	F	G	H
Efficiency	0.5	1	0.667	0.75	0.8	0.4	0.5	0.625

Now we observe the problem of how to make the inefficient stores efficient, i.e., how to move them up to the efficient frontier. For example, store A in Figure 1.3 can be improved in several ways. One is achieved by reducing the input (number of employees) to A_1 with coordinates (1, 1) on the efficient frontier. Another is achieved by raising the output (sales in \$100,000 units) up to $A_2(2, 2)$. Any point on the line segment $\overline{A_1 A_2}$ offers a chance to effect the *improvements* in a manner which assumes that the input should not be increased and the output should not be decreased in making the store efficient.

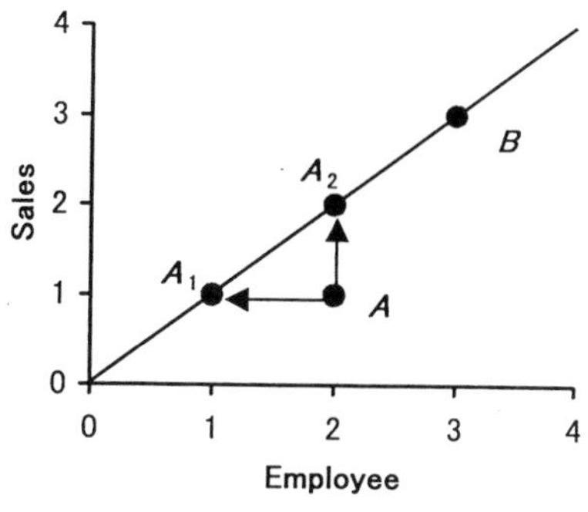

Figure 1.3. Improvement of Store A

This very simple example moves from the ratio in Table 1.1 to the "ratio of ratios" in Table 1.2, which brings to the fore an important point. The values in (1.1) depend on the units of measure used whereas this is not the case for (1.2). For instance, if sales were stated in units of \$10,000, the ratio for F would change from $2/5 = 0.4$ to $20/5 = 4.0$. However, the value of (1.2) would remain unchanged at $4/10 = 0.4$ and the *relative efficiency* score associated with F is not affected by this choice of a different unit of measure. This property, sometimes referred to as "units invariance" has long been recognized as important in engineering and science. Witness, for instance, the following example from the field of combustion engineering where ratings of furnace efficiency are obtained from the following formula,[1]

$$0 \leq E_r = \frac{y_r}{y_R} \leq 1 \tag{1.3}$$

where

y_r = Heat obtained from a given unit of fuel by the furnace being rated,

y_R = Maximum heat that can be obtained from this same fuel input.

The latter, i.e., the maximum heat can be calculated from thermodynamic principles by means of suitable chemical-physical analyses. The point to be emphasized here, however, is that x, the amount of fuel used must be the same so that, mathematically,

$$0 \leq \frac{y_r/x}{y_R/x} = \frac{y_r}{y_R} \leq 1 \tag{1.4}$$

Hence, (1.3) is obtained from a ratio of ratios that is "units invariant."

Returning to the ratios in Table 1.2, we might observe that these values are also bounded by zero and unity. However, the variations recorded in Table 1.2 may result from an excess amount of input or a deficiency in the output. Moreover, this situation is general in the business and social-policy (economics) applications which are of concern in this book. This is one reason we can make little use of formulas like (1.4). Furthermore, this formula is restricted to the case of a single output and input. Attempts to extend it to multiple inputs and multiple outputs encounter the troubles which were identified in our earlier discussion of "partial" and "total factor productivity" measures.

1.3 TWO INPUTS AND ONE OUTPUT CASE

To move to multiple inputs and outputs and their treatment, we turn to Table 1.3 which lists the performance of 9 supermarkets each with two inputs and one output. *Input* x_1 is the number of employees (unit: 10), *Input* x_2 the floor area (unit: $1000m^2$) and *Output* y the sales (unit: 100,000 dollars). However, notice that the sales are unitized to 1 under the constant returns-to-scale assumption. Hence, input values are normalized to values for getting 1 unit of sales. We plot the stores, taking *Input* x_1/*Output* y and *Input* x_2/*Output* y as axes which we may think of as "unitized axes" in Figure 1.4.

Table 1.3. Two Inputs and One Output Case

Store		A	B	C	D	E	F	G	H	I
Employee	x_1	4	7	8	4	2	5	6	5.5	6
Floor Area	x_2	3	3	1	2	4	2	4	2.5	2.5
Sale	y	1	1	1	1	1	1	1	1	1

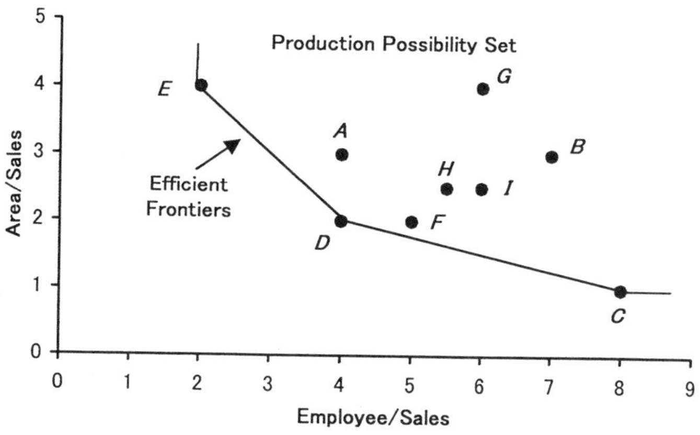

Figure 1.4. Two Inputs and One Output Case

From the efficiency point of view, it is natural to judge stores which use less inputs to get one unit output as more efficient. We therefore identify the line connecting C, D and E as the efficient frontier. We do not discuss the tradeoffs between these three stores but simply note here that no point on this frontier line can improve one of its input values without worsening the other. We can envelop all the data points within the region enclosed by the frontier line, the horizontal line passing through C and the vertical line through E. We call this region the *production possibility set.* (More accurately, it should be called the *piecewise linear* production possibility set assumption, since it is not guaranteed that the (true) boundary of this region is piecewise linear, i.e., formed of linear segments like the segment connecting E and D and the segment connecting D and C.) This means that the observed points are assumed to provide (empirical) evidence that production is possible at the rates specified by the coordinates of any point in this region.

The efficiency of stores not on the frontier line can be measured by referring to the frontier point as follows. For example, A is inefficient. To measure its inefficiency let $\overline{OA}$, the line from zero to A, cross the frontier line at P (see Figure 1.5). Then, the efficiency of A can be evaluated by

$$\frac{OP}{OA} = 0.8571.$$

This means that the inefficiency of A is to be evaluated by a combination of D and E because the point P is on the line connecting these two points. D and E are called the *reference set* for A. The reference set for an inefficient store may differ from store to store. For example, B has the reference set composed of C and D in Figure 1.4. We can also see that many stores come together around D and hence it can be said that D is an efficient store which is also "representative," while C and E are also efficient but also possess unique characteristics in their association with segments of the frontiers that are far removed from any observations.

Now we extend the analysis in Figure 1.3 to identify improvements by referring inefficient behaviors to the efficient frontier in this two inputs (and one output) case. For example, A can be effectively improved by movement to P with *Input* $x_1 = 3.4$ and *Input* $x_2 = 2.6$, because these are the coordinates of P, the point on the efficient frontier that we previously identified with the line segment $\overline{OA}$ in Figure 1.5. However, any point on the line segment $\overline{DA_1}$ may also be used as a candidate for improvement. D is attained by reducing *Input* x_2 (floor area), while A_1 is achieved by reducing *Input* x_1 (employees). Yet another possibility for improvement remains by increasing output and keeping the *status quo* for inputs. This will be discussed later.

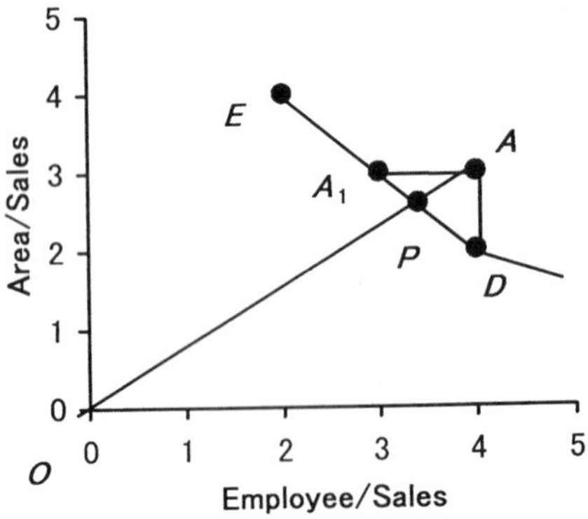

Figure 1.5. Improvement of Store A

1.4 ONE INPUT AND TWO OUTPUTS CASE

Table 1.4 shows the number of customers (unit=10) per salesman and the sales (unit=100,000 dollars) per salesman of 7 branch offices. To obtain a unitized frontier in this case, we divide by the number of employees (=salesmen) which is considered to be the only input of interest. The efficient frontier then consists of the lines connecting B, E, F and G as shown in Figure 1.6.

Table 1.4. One Input and Two Outputs Case

Store		A	B	C	D	E	F	G
Employees	x	1	1	1	1	1	1	1
Customers	y_1	1	2	3	4	4	5	6
Sales	y_2	5	7	4	3	6	5	2

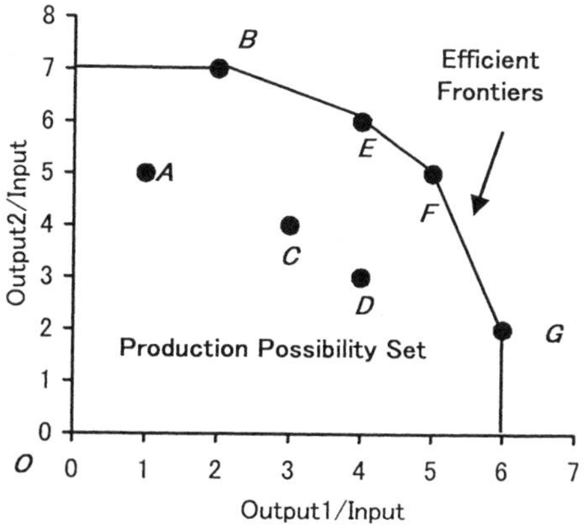

Figure 1.6. One Input and Two Outputs Case

The production possibility set is the region bounded by the axes and the frontier line. Branches A, C and D are inefficient and their efficiency can be evaluated by referring to the frontier lines. For example, from Figure 1.7, the efficiency of D is evaluated by

$$\frac{d(O,D)}{d(O,P)} = 0.75, \tag{1.5}$$

where $d(O,D)$ and $d(O,P)$ mean "distance from zero to D" and "distance from zero to P," respectively.

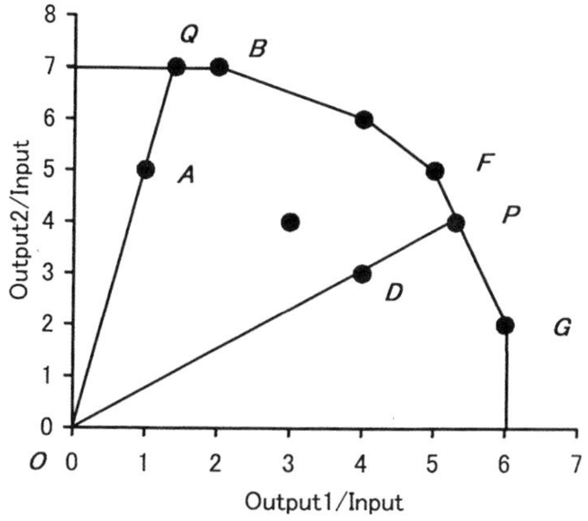

Figure 1.7. Improvement

The above ratio is referred to as a "radial measure" and can be interpreted as the ratio of two distance measures. The choice of distance measures is not unique so, [2] because of familiarity, we select the Euclidean measures given by

$$d(O,D) = \sqrt{4^2 + 3^2} = 5$$

$$d(O,P) = \sqrt{\left(\frac{16}{3}\right)^2 + 4^2} = \frac{20}{3},$$

where the terms under the radical sign are squares of the coordinates of D and P, respectively, as obtained from Table 1.4 for D and from the intersection of $y_2 = \frac{3}{4}y_1$ and $y_2 = 20 - 3y_1$ for P. As claimed, substitution in (1.5) then gives

$$5 \div \frac{20}{3} = \frac{15}{20} = 0.75.$$

This interpretation as a ratio of distances aligns the results with our preceding discussion of such ratios. Because the ratio is formed relative to the Euclidean distance from the origin over the production possibility set, we will always obtain a measure between zero and unity.

We can also interpret the results for managerial (or other) uses in a relatively straightforward manner. The value of the ratio in (1.5) will always have a value between zero and unity. Because we are concerned with output, however, it is easier to interpret (1.5) in terms of its reciprocal

$$\frac{d(O,P)}{d(O,D)} = \frac{20}{3} \div 5 = 1.33.$$

This result means that, to be efficient, D would have had to increase both of its outputs by 4/3. To confirm that this is so we simply apply this ratio to the coordinates of D and obtain

$$\frac{4}{3}(4,3) = \left(\frac{16}{3}, 4\right),$$

which would bring coincidence with the coordinates of P, the point on the efficient frontier used to evaluate D.

Returning to (1.5) we note that 0.75 refers to the proportion of the output that P shows was possible of achievement. It is important to note that this refers to the proportion of inefficiency present in *both* outputs by D. Thus, the shortfall in D's output can be repaired by increasing both outputs without changing their proportions — until P is attained.

As might be expected, this is only one of the various types of inefficiency that will be of concern in this book. This kind of inefficiency which can be eliminated without changing proportions is referred to as "technical inefficiency."

Another type of inefficiency occurs when only some (but not all) outputs (or inputs) are identified as exhibiting inefficient behavior. This kind of inefficiency is referred to as "mix inefficiency" because its elimination will alter the proportions in which outputs are produced (or inputs are utilized).[3]

We illustrated the case of "technical inefficiency" by using D and P in Figure 1.7. We can use Q and B to illustrate "mix inefficiency" or we can use A, Q and

B to illustrate both technical and mix inefficiency. Thus, using the latter case we identify the technical efficiency component in A's performance by means of the following radial measure,

$$\frac{d(O,A)}{d(O,Q)} = 0.714. \tag{1.6}$$

Using the reciprocal of this measure, as follows, and applying it to the coordinates of A at (1, 5) gives

$$\frac{1}{0.714}(1,5) = (1.4,7),$$

as the coordinates of Q.

We can now note that the thus adjusted outputs are in the ratio 1.4/7=1/5, which is the same as the ratio for A in Table 1.4 — viz., $y_1/y_2 = 1/5$. This augments both of the outputs of A without worsening its input and without altering the output proportions. This improvement in technical efficiency by movement to Q does not remove all of the inefficiencies. Even though Q is on the frontier it is not on an efficient part of the frontier. Comparison of Q with B shows a shortfall in output 1 (number of customers served) so a further increase in this output can be achieved by a lateral movement from Q to B. Thus this improvement can also be achieved without worsening the other output or the value of the input. Correcting output value, y_1, without altering y_2 will change their proportions, however, and so we can identify two sources of inefficiencies in the performance of A: first a technical inefficiency via the radial measure given in (1.6) followed by a mix inefficiency represented by the output shortfall that remains in y_1 *after* all of the technical inefficiencies are removed.

We now introduce the term "purely technical inefficiency" so that, in the interest of simplicity, we can use the term "technical inefficiency" to refer to all sources of waste — purely technical and mix — which can be eliminated without worsening any other input or output. This also has the advantage of conforming to usages that are now fixed in the literature. It will also simplify matters when we come to the discussion of prices, costs and other kinds of values or weights that may be assigned to the different sources of inefficiency.

Comment: The term "technical efficiency" is taken from the literature of economics where it is used to distinguish the "technological" aspects of production from other aspects, generally referred to as "economic efficiency" which are of interest to economists.[4] The latter involves recourse to information on prices, costs or other value considerations which we shall cover later in this text. Here, and in the next two chapters, we shall focus on purely technical and mix inefficiencies which represent "waste" that can be justifiably eliminated without requiring additional data such as prices and costs. It only requires assuring that the resulting improvements are worthwhile even when we do not specifically assign them a value.

As used here, the term mix inefficiency is taken from the accounting literatures where it is also given other names such as "physical variance" or "efficiency variance."[5] In this usage, the reference is to physical aspects of production which exceed a prescribed standard and hence represent excessive uses of labor, raw materials, etc.

1.5 FIXED AND VARIABLE WEIGHTS

The examples used to this point have been very limited in the number of inputs and outputs used. This made it possible to use simple graphic displays to clarify matters but, of course, this was at the expense of the realism needed to deal with the multiple inputs and multiple outputs that are commonly encountered in practice. The trick is to develop approaches that make it possible to deal with such applications without unduly burdening users with excessive analyses or computations and without requiring large numbers of (often arbitrary or questionable) assumptions.

Consider, for instance, the situation in Table 1.5 which records behavior intended to serve as a basis for evaluating the relative efficiency of 12 hospitals in terms of two inputs, number of doctors and number of nurses, and two outputs identified as number of outpatients and inpatients (each in units of 100 persons/month).

Table 1.5. Hospital Case

Hospital	*A*	*B*	*C*	*D*	*E*	*F*	*G*	*H*	*I*	*J*	*K*	*L*
Doctors	20	19	25	27	22	55	33	31	30	50	53	38
Nurses	151	131	160	168	158	255	235	206	244	268	306	284
Outpatients	100	150	160	180	94	230	220	152	190	250	260	250
Inpatients	90	50	55	72	66	90	88	80	100	100	147	120

One way to simplify matters would be to weight the various inputs and outputs by pre-selected (fixed) weights. The resulting ratio would then yield an index for evaluating efficiencies. For instance, the weight

$$v_1(\text{weight for doctor}) \;:\; v_2(\text{weight for nurse}) = 5:1$$

$$u_1(\text{weight for outpatient}) \;:\; u_2(\text{weight for inpatient}) = 1:3$$

would yield the results shown in the row labelled "Fixed" of Table 1.6. (Notice that these ratios are normalized so that the maximum becomes unity, i.e., by dividing by the ratio of *A*.) This simplifies matters for use, to be sure, but raises a host of other questions such as justifying the 5 to 1 ratio for doctor vs. nurse and/or the 3 to 1 ratio of the weights for inpatients and outpatients. Finally, and even more important, are problems that can arise with the results shown – since it is not clear how much of the efficiency ratings are due to the weights and how much inefficiency is associated with the observations.

Table 1.6. Comparisons of Fixed vs. Variable Weights

Hospital	*A*	*B*	*C*	*D*	*E*	*F*	*G*	*H*	*I*	*J*	*K*	*L*
Fixed	1	.90	.77	.89	.74	.64	.82	.74	.84	.72	.83	.87
CCR	1	1	.88	1	.76	.84	.90	.80	.96	.87	.96	.96

DEA, by contrast, uses variable weights. In particular, the weights are derived directly from the data with the result that the numerous *a priori* assumptions and computations involved in fixed weight choices are avoided. Moreover, the weights are chosen in a manner that assigns a best set of weights to each hospital. The term "best" is used here to mean that the resulting input-to-output ratio for each hospital is maximized relative to all other hospitals when these weights are assigned to these inputs and outputs for every hospital. The row labelled CCR in Table 1.6 shows results obtained from DEA using what is called the "CCR model"[6] in DEA. As can be seen, these efficiency values are always at least as great as the ratio values obtained from the previous fixed value weights. Moreover, this "best ratio" result is general, under the following conditions: (1) all data and all weights are positive (or at least nonnegative), (2) the resulting ratio must lie between zero and unity and (3) these same weights for the target entity (=hospital) are applied to all entities. Consequently, the entity being evaluated cannot choose a better set of weights for its evaluation (relative to the other entities). The meaning of these results is clear. In each case, the evaluation is effected from a point on the efficient frontier so that a value like .88 for hospital C means that it is 12% inefficient. That is, compared to members of an efficient reference set, it is possible to identify a purely technical inefficiency of 12%—and possible mix inefficiencies as well—even under the best set of weights that *each* of these hospitals could choose to evaluate its own inefficiencies.

As we shall later see, the sources of inefficiency, such as purely technical and mix inefficiencies are automatically identified for each entity by DEA and their amounts estimated. Moreover, the reference set used to benchmark these inefficiencies are also identified. Finally, as we shall also see, these results are obtained using only minimal *a priori* assumptions. In addition to avoiding a need for *a priori* choices of weights, DEA does not require specifying the form of the relation between inputs and outputs in, perhaps, an arbitrary manner and, even more important, it does not require these relations to be the same for each hospital.

1.6 SUMMARY AND CONCLUSION

We have now covered a variety of topics which will be refined and extended in this book. Employing commonly used output-to-input ratio measures we related them to topics such as measurements of productivity as well as the efficiency evaluation methods commonly used in economics, business and engineering. We will subsequently introduce other (non-ratio) approaches but will do so in ways that maintain contact with these ratio forms.

Extensions to multiple outputs and multiple inputs were examined in terms of fixed weights to be applied uniformly to the inputs and outputs of all entities to be evaluated, as in economic indices of "total factor productivity." This usage was then contrasted with the use of variable weights based on a best set being chosen for *each* entity to be evaluated, as in DEA. We also described interpretations and uses of the latter as derived from the efficient frontiers from which the evaluations were effected. This was then contrasted with the mixed, generally unclear, sources of inefficiencies that are implicit in the use of fixed weight approaches.

Additional advantages of DEA were also noted in terms of (*a*) its ability to identify sources and amounts of inefficiency in each input and each output for each entity (hospital, store, furnace, etc.) and (*b*) its ability to identify the benchmark members of the efficient set used to effect these evaluations and identify these sources (and amounts) of inefficiency.

All of the thus assessed entities were assumed to use the same inputs to produce the same outputs. Also, all data were assumed to be positive and weight choices were also restricted to positive values. These assumptions will be maintained in the immediately following chapters and then relaxed. Inputs and outputs were also assumed to be variable at the discretion of managers or designers. This assumption will also be maintained and then relaxed so that we will be able to distinguish between discretionary and non-discretionary inputs and outputs — to allow for differences in the circumstances under which different entities operate. Then we shall also introduce categorical variables to allow for further difference such as rural vs. urban categories, etc., to obtain more refined evaluations and insights.

The discussion in this chapter was confined to physical aspects of efficiency with distinctions between "purely technical" and "mix" inefficiencies. These were referred to as "waste" because they could be removed from any input or output without worsening any other input and output. Other types of inefficiency covered in this book will involve movement along efficient frontiers and hence also involve exchanges or substitutions. Such movements may be effected to achieve returns-to-scale economies or to improve cost and profit performances. All such movements along frontiers, however, imply an absence of technical inefficiencies (=purely technical plus mix). Hence this topic will be the center of attention in the immediately following chapters.

1.7 PROBLEM SUPPLEMENT FOR CHAPTER 1

Problem 1.1

To deal with multiple inputs and outputs, a ratio like the following may be used.

$$\frac{\sum_{r=1}^{s} u_r y_r}{\sum_{i=1}^{m} v_i x_i} = \frac{u_1 y_1 + u_2 y_2 + \cdots + u_s y_s}{v_1 x_1 + v_2 x_2 + \cdots + v_m x_m}$$

where

$$\begin{aligned} y_r &= \text{amount of output } r \\ u_r &= \text{weight assigned to output } r \\ x_i &= \text{amount of input } i \\ v_i &= \text{weight assigned to input } i. \end{aligned}$$

The weights may be (1) fixed in advance or (2) derived from the data. The former is sometimes referred as an *a priori* determination..

1. **Assignment 1**

   The weights given in the text for use in Table 1.5 are as follows:

   $$v_1 = 5, \quad v_2 = 1$$

   $$u_1 = 1, \quad u_2 = 3.$$

   Apply these results to the example of Table 1.5 and compare your answer to the first two rows of Table 1.6. Then, interpret your results.

   Suggestion: Be sure to normalize all results by dividing with the ratio for A, which is $370/251 \doteq 1.474$. Notice that this division cancels all units of measure. This is not the same as "units invariance," however, which means that a change in the unit of measure will not affect the solution, e.g., if the number of doctors were restated in units of "10 doctors" or any other unit, then resulting solution values would be the same if the solution is "units invariant."

2. **Assignment 2**

   The manager of Hospital B asks you to determine a set of weights which will improve its standing relative to A.

   Suggested Answer:

   - Using the data in Table 1.5 you could determine weights which bring B to a par with A by solving

   $$\frac{100u_1 + 90u_2}{20v_1 + 151v_2} = \frac{150u_1 + 50u_2}{19v_1 + 131v_2}. \tag{1.7}$$

   - An easier route would be to solve the following problem.

   $$\max \frac{150u_1 + 50u_2}{19v_1 + 131v_2}$$

subject to

$$\frac{150u_1 + 50u_2}{19v_1 + 131v_2} \leq 1$$

$$\frac{100u_1 + 90u_2}{20v_1 + 151v_2} \leq 1.$$

with $u_1, u_2, v_1, v_2 > 0$. The choice of u_1, u_2 and v_1, v_2 should maximize the ratio for Hospital B, so no better choice can be made from this manager's standpoint.

- As shown in the next chapter, this nonlinear programming problem can be replaced by the following linear program.

$$\max 150u_1 + 50u_2$$

subject to

$$150u_1 + 50u_2 \leq 19v_1 + 131v_2$$

$$100u_1 + 90u_2 \leq 20v_1 + 151v_2$$

$$19v_1 + 131v_2 = 1$$

and all variables are constrained to be positive. Note that the normalization in the last constraint ensures that the weights will be relative. Since they sum to unity, no further normalization is needed such as the one used in the answer to Assignment 1. Also the possibility of zero values for all weights is eliminated, even though it is one possible solution for the equality pair (1.7) stated Assignment 2 for Problem 1.1.

Suggested Answer : Solutions to the first 2 problems involve treating nonlinear problems so we focus on this last problem. By the attached software DEA-Solver, we then obtain

$$u_1^* \doteq 0.00463, \ u_2^* \doteq 0.00611, \ v_1^* \doteq 0.0275, \ v_2^* \doteq 0.00364,$$

where "$\doteq$" means "approximately equal to." This solution is also optimal for the preceding problem with

$$\frac{150(0.00463) + 50(0.00611)}{19(0.0275) + 131(0.00364)} \doteq 1$$

for Hospital B and

$$\frac{100(0.00463) + 90(0.00611)}{20(0.0275) + 151(0.00364)} \doteq 0.921$$

for Hospital A. This reversal of efficiency ratings might lead to A responding by calculating his own best weights, and similarly for the other hospitals. A regulatory agency might then respond by going beyond such pairwise comparisons, in which case we could use an extension of the above approaches – to be described in succeeding chapters – which effects such a "best rating" by considering all 12 hospitals simultaneously for each such evaluation. In fact this was the approach used to derive the evaluations in the row labelled "CCR" in Table 1.6. Note that Hospitals A and B are both rated as fully efficient when each is given its own "best" weights. With the exception of D, none of the other hospitals achieve full (=100%) DEA efficiency even when each is accorded its own "best" weights.

Suggestion for using the attached DEA-Solver

You can solve the best weights for the above two hospital problem using the supporting "DEA-Solver" software on the included CD. See Appendix B for installation instructions for a PC. Then follow the procedures below:

1. Create an Excel 97 file containing the data sheet as exhibited in Figure 1.8 where (I) and (O) indicate Input and Output, respectively.

	A	B	C	D	E	F
1	Hospital	(I)Doctor	(I)Nurse	(O)Outpatient	(O)Inpatient	
2	A	20	151	100	90	
3	B	19	131	150	50	
4						

Figure 1.8. Excel File "HospitalAB.xls"

2. Save the file with the file name "HospitalAB.xls" in an appropriate folder and close Excel.

3. Start DEA-Solver and follow the instructions on the display.

4. Choose "CCR-I" as DEA model.

5. Choose "HospitalAB.xls" as data file.

6. Choose "Hospital.xls" as the Workbook for saving results of the computation.

7. Click "Run."

8. After the computation is over, click the "Exit" button.

9. Open the sheet "Weight" which contains optimal weights obtained for each hospital. You will see results like in Table 1.7. From the table, we can see a set of optimal weights for Hospital A as given by

$$v_1^* = .025,\ v_2^* = 3.31E - 3 = 3.31 \times 10^{-3} = .00331$$

$$u_1^* = 3.74E{-}3 = 3.74{\times}10^{-3} = .00374,\ u_2^* = 6.96E{-}3 = 6.96{\times}10^{-3} = .00696$$

Table 1.7. Optimal Weights for Hospitals A and B

No.	DMU	Score	v(1)	v(2)	u(1)	u(2)
1	A	1	0.025	3.31E-03	3.74E-03	6.96E-03
2	B	1	2.75E-02	3.64E-03	4.63E-03	6.11E-03

and for Hospital B as

$v_1^* = 2.75E{-}2 = 2.75 \times 10^{-2} = .0275,\ v_2^* = 3.64E{-}3 = 3.64 \times 10^{-3} = .00364$

$u_1^* = 4.63E{-}3 = 4.63 \times 10^{-3} = .00463,\ u_2^* = 6.11E{-}3 = 6.11 \times 10^{-3} = .00611.$

These weights give the best ratio score 1 (100%) to each hospital.

However, notice that the best weights are not necessarily unique as you can see from the "Fixed" weight case in Table 1.6. Actually, the weights

$$v_1 : v_2 = 5 : 1, \quad u_1 : u_2 = 1 : 3$$

or more concretely,

$$v_1 = .02,\ v_2 = .004,\ u_1 = .0027,\ u_2 = .0081$$

are applied for this fixed weight evaluation and these also made Hospital A efficient.

Problem 1.2

The comptroller and treasurer of an industrial enterprise discuss whether the company's sales should be treated as an input or an output.

Suggested Resolution: Take the ratio of output ÷ input and ask whether an increase in sales should improve or worsen the company's efficiency rating in terms of its effects on the value of this ratio. Compare this with whether you could treat an increase in expenses as an output or an input in terms of its effects on the ratio.

Problem 1.3

The ratios in Table 1.6 are designed for use in evaluating the performance efficiency of each hospital. This means that entire hospitals are to be treated as Decision Making Units (DMUs) which convert inputs of Doctors and Nurses into outputs of Inpatients and Outpatients. Can you suggest other DMUs to evaluate the performance of hospital?

Suggested Answer : David Sherman used surgical units to evaluate performances of teaching hospitals for the rate-setting Commission of Massachusetts because outputs such as interns, residents and nurses to be trained in the performances of such services have proved difficult to treat with measures, such as cost/patient, which the Commission had been using.[7]

Problem 1.4

Suppose that a manager of a chain-store is trying to evaluate performance of each store. He selected factors for evaluation as follows: (1) the annual average salary per employee as input, (2) the number of employees as input, and (3) the annual average sales per employee as output. Criticize this selection.

Suggested Answer : Let p_i be the number of employees of store i, c_i be the total annual salary paid by store i, and d_i be the total annual sales of store i. Then the weighted ratio scale which expresses the manager's criterion would be

$$\frac{u_1 d_i/p_i}{v_1 c_i/p_i + v_2 p_i},$$

where u_1 = weight for output d_i/p_i (the average sales per employee), v_1 = weight for input c_i/p_i (the average salary per employee) and v_2 = weight for input p_i (the number of employees).

The above ratio can be transformed into

$$\frac{u_1 d_i}{v_1 c_i + v_2 p_i^2}.$$

This evaluation thus puts emphasis on the number of employees by squaring this value, while other factors are evaluated in a linear manner. If such uneven treatment has no special justification, it may be better to use a ratio form such as,

$$\frac{u_1 d_i}{v_1 c_i + v_2 p_i}.$$

This choice would be (1) the total salary paid by the store as input, (2) the number of employees as input, and (3) the total annual sales of the store as output.

You should be careful in dealing with processed data, e.g., value per head, average, percentile, and raw data at the same time.

Problem 1.5

Enumerate typical inputs and outputs for performance measurement of the following organizations: (1) airlines, (2) railways, (3) car manufacturers, (4) universities, (5) farms, and (6) baseball teams.

Notes

1. This example is taken from A. Charnes, W.W. Cooper and E. Rhodes, "Measuring the Efficiency of Decision Making Units," *European Journal of Operational Research* 2, 1978, pp.429-444.

2. Our measures of distance and their uses are related to each other and discussed in W.W. Cooper, L.M. Seiford, K. Tone and J. Zhu "DEA: Past Accomplishments and Future Prospects," *Journal of Productivity Analysis* (submitted, 2005).

3. The latter is referred to as an input mix and the former as an output mix inefficiency.

4. See, for example, pp.15-18 in H.R. Varian *Microeconomic Analysis* 2nd ed. (New York. W.W. Norton & Co., 1984.)

5. Vide p.192 in W.W. Cooper and Y. Ijiri, eds., *Kohler's Dictionary For Accountants, 6th Edition* (Englewood Cliffs, N.J., Prentice-Hall, Inc., 1981.)

6. After Charnes, Cooper and Rhodes (1978) above.

7. See H.D. Sherman, "Measurement of Hospital Technical Efficiency: A Comparative Evaluation of Data Envelopment Analysis and Other Techniques for Measuring and Locating Efficiency in Health Care Organizations," Ph.D. Thesis (Boston: Harvard University Graduate School of Business, 1981.) Also available from University Microfilms, Inc., Ann Arbor, Michigan.

2 THE BASIC CCR MODEL

2.1 INTRODUCTION

This chapter deals with one of the most basic DEA models, the *CCR model*, which was initially proposed by Charnes, Cooper and Rhodes in 1978. Tools and ideas commonly used in DEA are also introduced and the concepts developed in Chapter 1 are extended. There, for each DMU, we formed the virtual input and output by (yet unknown) weights (v_i) and (u_r):

$$\text{Virtual input} = v_1 x_{1o} + \cdots + v_m x_{mo}$$
$$\text{Virtual output} = u_1 y_{1o} + \cdots + u_s y_{so}.$$

Then we tried to determine the weight, using linear programming so as to maximize the ratio

$$\frac{\text{virtual output}}{\text{virtual input}}.$$

The optimal weights may (and generally will) vary from one DMU to another DMU. Thus, the "weights" in DEA are derived from the data instead of being fixed in advance. Each DMU is assigned a best set of weights with values that may vary from one DMU to another. Additional details and the algorithms used to implement these concepts are explained in succeeding chapters.

2.2 DATA

In DEA, the organization under study is called a *DMU* (Decision Making Unit). The definition of DMU is rather loose to allow flexibility in its use over a wide range of possible applications. Generically a DMU is regarded as the entity responsible for converting inputs into outputs and whose performances are to be evaluated. In managerial applications, DMUs may include banks, department stores and supermarkets, and extend to car makers, hospitals, schools, public libraries and so forth. In engineering, DMUs may take such forms as airplanes or their components such as jet engines. For the purpose of securing *relative* comparisons, a group of DMUs is used to evaluate each other with each DMU having a certain degree of managerial freedom in decision making.

Suppose there are n DMUs: DMU_1, DMU_2,..., and DMU_n. Some common input and output items for each of these $j = 1, ..., n$ DMUs are selected as follows:

1. Numerical data are available for each input and output, with the data assumed to be positive[1] for all DMUs.

2. The items (inputs, outputs and choice of DMUs) should reflect an analyst's or a manager's interest in the components that will enter into the relative efficiency evaluations of the DMUs.

3. In principle, smaller input amounts are preferable and larger output amounts are preferable so the efficiency scores should reflect these principles.

4. The measurement units of the different inputs and outputs need not be congruent. Some may involve number of persons, or areas of floor space, money expended, etc.

Suppose m input items and s output items are selected with the properties noted in 1 and 2. Let the input and output data for DMU_j be $(x_{1j}, x_{2j}, ..., x_{mj})$ and $(y_{1j}, y_{2j}, ..., y_{sj})$, respectively. The input data matrix X and the output data matrix Y can be arranged as follows,

$$X = \begin{pmatrix} x_{11} & x_{12} & \cdots & x_{1n} \\ x_{21} & x_{22} & \cdots & x_{2n} \\ \cdot & \cdot & \cdots & \cdot \\ \cdot & \cdot & \cdots & \cdot \\ x_{m1} & x_{m2} & \cdots & x_{mn} \end{pmatrix} \tag{2.1}$$

$$Y = \begin{pmatrix} y_{11} & y_{12} & \cdots & y_{1n} \\ y_{21} & y_{22} & \cdots & y_{2n} \\ \cdot & \cdot & \cdots & \cdot \\ \cdot & \cdot & \cdots & \cdot \\ y_{s1} & y_{s2} & \cdots & y_{sn} \end{pmatrix} \tag{2.2}$$

where X is an $(m \times n)$ matrix and Y an $(s \times n)$ matrix. For example, the hospital case in Section 1.5 has the data matrices:

$$X = \begin{pmatrix} 20 & 19 & 25 & 27 & 22 & 55 & 33 & 31 & 30 & 50 & 53 & 38 \\ 151 & 131 & 160 & 168 & 158 & 255 & 235 & 206 & 244 & 268 & 306 & 284 \end{pmatrix}$$

$$Y = \begin{pmatrix} 100 & 150 & 160 & 180 & 94 & 230 & 220 & 152 & 190 & 250 & 260 & 250 \\ 90 & 50 & 55 & 72 & 66 & 90 & 88 & 80 & 100 & 100 & 147 & 120 \end{pmatrix}$$

so x_{1j} = number of doctors and x_{2j} = number of nurses used by hospital j in servicing (= producing) y_{1j} = number of outpatients and y_{2j} = number of inpatients.

2.3 THE CCR MODEL

Given the data, we measure the efficiency of each DMU once and hence need n optimizations, one for each DMU_j to be evaluated. Let the DMU_j to be evaluated on any trial be designated as DMU_o where o ranges over 1, 2,..., n. We solve the following fractional programming problem to obtain values for the input "weights" (v_i) $(i = 1, ..., m)$ and the output "weights" (u_r) $(r = 1, ..., s)$ as variables.

$$(FP_o) \qquad \max_{v,u} \ \theta = \frac{u_1 y_{1o} + u_2 y_{2o} + \cdots + u_s y_{so}}{v_1 x_{1o} + v_2 x_{2o} + \cdots + v_m x_{mo}} \tag{2.3}$$

$$\text{subject to} \qquad \frac{u_1 y_{1j} + \cdots + u_s y_{sj}}{v_1 x_{1j} + \cdots + v_m x_{mj}} \leq 1 \ \ (j = 1, \ldots, n) \tag{2.4}$$

$$v_1, v_2, \ldots, v_m \geq 0 \tag{2.5}$$

$$u_1, u_2, \ldots, u_s \geq 0. \tag{2.6}$$

The constraints mean that the ratio of "virtual output" vs. "virtual input" should not exceed 1 for every DMU. The objective is to obtain weights (v_i) and (u_r) that maximize the ratio of DMU_o, the DMU being evaluated. By virtue of the constraints, the optimal objective value θ^* is at most 1. Mathematically, the nonnegativity constraint (2.5) is not sufficient for the fractional terms in (2.4) to have a positive value. We do not treat this assumption in explicit mathematical form at this time. Instead we put this in managerial terms by assuming that all outputs and inputs have some nonzero worth and this is to be reflected in the weights u_r and v_i being assigned some positive value.

2.4 FROM A FRACTIONAL TO A LINEAR PROGRAM

We now replace the above fractional program (FP_o) by the following linear program (LP_o),

$$(LP_o) \qquad \max_{\mu,\nu} \ \theta = \mu_1 y_{1o} + \cdots + \mu_s y_{so} \tag{2.7}$$

$$\text{subject to} \quad \nu_1 x_{1o} + \cdots + \nu_m x_{mo} = 1 \tag{2.8}$$

$$\mu_1 y_{1j} + \cdots + \mu_s y_{sj} \leq \nu_1 x_{1j} + \cdots + \nu_m x_{mj} \quad (j = 1, \ldots, n) \tag{2.9}$$

$$\nu_1, \nu_2, \ldots, \nu_m \geq 0 \tag{2.10}$$

$$\mu_1, \mu_2, \ldots, \mu_s \geq 0. \tag{2.11}$$

Theorem 2.1 *The fractional program* (FP_o) *is equivalent to* (LP_o).

Proof. Under the nonzero assumption of $\boldsymbol{v}$ and $X > 0$, the denominator of the constraint of (FP_o) is positive for every j, and hence we obtain (2.3) by multiplying both sides of (2.4) by the denominator. Next, we note that a fractional number is invariant under multiplication of both numerator and denominator by the same nonzero number. After making this multiplication, we set the denominator of (2.3) equal to 1, move it to a constraint, as is done in (2.8), and maximize the numerator, resulting in (LP_o). Let an optimal solution of (LP_o) be $(\boldsymbol{\nu} = \boldsymbol{\nu}^*,\ \boldsymbol{\mu} = \boldsymbol{\mu}^*)$ and the optimal objective value θ^*. The solution $(\boldsymbol{v} = \boldsymbol{\nu}^*, \boldsymbol{u} = \boldsymbol{\mu}^*)$ is also optimal for (FP_o), since the above transformation is reversible under the assumptions above. (FP_o) and (LP_o) therefore have the same optimal objective value θ^*. □

We also note that the measures of efficiency we have presented are "units invariant" — i.e., they are independent of the units of measurement used in the sense that multiplication of each input by a constant $\delta_i > 0$, $i = 1, \ldots, m$, and each output by a constant $p_r > 0$, $r = 1, \ldots, s$, does not change the obtained solution. Stated in precise form we have

Theorem 2.2 (Units Invariance Theorem) *The optimal values of max* $\theta = \theta^*$ *in (2.3) and (2.7) are independent of the units in which the inputs and outputs are measured provided these units are the same for every DMU.*

Thus, one person can measure outputs in miles and inputs in gallons of gasoline and quarts of oil while another measures these same outputs and inputs in kilometers and liters. They will nevertheless obtain the same efficiency value from (2.3) or (2.7) when evaluating the same collection of automobiles, say. See Note 2 for proof.[2]

Before proceeding we note that (LP_o) can be solved by the simplex method of linear programming. The optimal solution can be more easily obtained by dealing with the dual side of (LP_o), however, as will be explained in detail in Chapter 3.

In any case let us suppose we have an optimal solution of (LP_o) which we represent by $(\theta^*,\ \boldsymbol{v}^*,\ \boldsymbol{u}^*)$[3] where $\boldsymbol{v}^*$ and $\boldsymbol{u}^*$ are values with constraints given in (2.10) and (2.11). We can then identify whether *CCR-efficiency* has been achieved as follows:

Definition 2.1 (CCR-Efficiency)
1. *DMU_o is* CCR-efficient *if* $\theta^* = 1$ *and there exists at least one optimal* $(\boldsymbol{v}^*, \boldsymbol{u}^*)$, *with* $\boldsymbol{v}^* > \mathbf{0}$ *and* $\boldsymbol{u}^* > \mathbf{0}$.
2. *Otherwise, DMU_o is* CCR-inefficient.

Thus, CCR-inefficiency means that either (i) $\theta^* < 1$ or (ii) $\theta^* = 1$ and at least one element of $(\boldsymbol{v}^*, \boldsymbol{u}^*)$ is zero for every optimal solution of (LP_o). We will explain the latter case using an example in Section 2.6.2, and a detailed description of CCR-efficiency will be given in Chapter 3.

Now we observe the case where DMU_o has $\theta^* < 1$ (CCR-inefficient). Then there must be at least one constraint (or DMU) in (2.9) for which the weight $(\boldsymbol{v}^*, \boldsymbol{u}^*)$ produces equality between the left and right hand sides since, otherwise, θ^* could be enlarged. Let the set of such $j \in \{1, \ldots, n\}$ be

$$E'_o = \{\, j : \sum_{r=1}^{s} u_r^* y_{rj} = \sum_{i=1}^{m} v_i^* x_{ij} \}. \tag{2.12}$$

The subset E_o of E'_o, composed of CCR-efficient DMUs, is called the *reference set* or the *peer group* to the DMU_o. It is the existence of this collection of efficient DMUs that forces the DMU_o to be inefficient. The set spanned by E_o is called the *efficient frontier* of DMU_o.

2.5 MEANING OF OPTIMAL WEIGHTS

The $(\boldsymbol{v}^*,\ \boldsymbol{u}^*)$ obtained as an optimal solution for (LP_o) results in a set of optimal weights for the DMU_o. The ratio scale is evaluated by :

$$\theta^* = \frac{\sum_{r=1}^{s} u_r^* y_{ro}}{\sum_{i=1}^{m} v_i^* x_{io}}. \tag{2.13}$$

From (2.8), the denominator is 1 and hence

$$\theta^* = \sum_{r=1}^{s} u_r^* y_{ro}. \tag{2.14}$$

As mentioned earlier, $(\boldsymbol{v}^*,\ \boldsymbol{u}^*)$ are the set of most favorable weights for the DMU_o in the sense of maximizing the ratio scale. v_i^* is the optimal weight for the input item i and its magnitude expresses how highly the item is evaluated, relatively speaking. Similarly, u_r^* does the same for the output item r. Furthermore, if we examine each item $v_i^* x_{io}$ in the virtual input

$$\sum_{i=1}^{m} v_i^* x_{io} \ (= 1), \tag{2.15}$$

then we can see the relative importance of each item by reference to the value of each $v_i^* x_{io}$. The same situation holds for $u_r^* y_{ro}$ where the u_r^* provides a measure of the relative contribution of y_{ro} to the overall value of θ^*. These values not only show which items contribute to the evaluation of DMU_o, but also to what extent they do so.

2.6 EXPLANATORY EXAMPLES

We illustrate the use of the CCR model via the following small-scale examples.

2.6.1 *Example 2.1 (1 Input and 1 Output Case)*

Table 2.1 shows 8 DMUs with 1 input and 1 output. (The first example in Chapter 1).

Table 2.1. Example 2.1

DMU	A	B	C	D	E	F	G	H
Input	2	3	3	4	5	5	6	8
Output	1	3	2	3	4	2	3	5

We can evaluate the efficiency of DMU A, by solving the LP problem below:

$$
\begin{array}{lllll}
< A > \text{ max} & \theta = u & & & \\
\text{subject to} & 2v = 1 & & & \\
 & u \le 2v & (A) & 3u \le 3v & (B) \\
 & 2u \le 3v & (C) & 3u \le 4v & (D) \\
 & 4u \le 5v & (E) & 2u \le 5v & (F) \\
 & 3u \le 6v & (G) & 5u \le 8v & (H)
\end{array}
$$

where all variables are constrained to be nonnegative.

The optimal solution, easily obtained by simple ratio calculations, is given by ($v^* = 0.5$, $u^* = 0.5$, $\theta^* = 0.5$). Thus, the CCR-efficiency of A is $\theta^* = u^* = 0.5$. The reference set for A is found to be $E_A = \{B\}$ by inserting $u^* = 0.5$ and $v^* = 0.5$, the best possible weights for DMU A, in each of the above constraints. Thus the performance of B is used to characterize A and rates it as inefficient even with the best weights that the data admit for A.

The efficiency of B can be similarly evaluated from the data in Table 2.1 by:

$$
\begin{array}{lllll}
< B > \text{ max} & \theta = 3u & & & \\
\text{subject to} & 3v = 1 & & & \\
 & u \le 2v & (A) & 3u \le 3v & (B) \\
 & 2u \le 3v & (C) & 3u \le 4v & (D) \\
 & 4u \le 5v & (E) & 2u \le 5v & (F) \\
 & 3u \le 6v & (G) & 5u \le 8v & (H)
\end{array}
$$

The optimal solution is ($v^* = 0.3333$, $u^* = 0.3333$, $\theta^* = 1$) and B is CCR-efficient. See Definition 2.1.

We can proceed in a similar way with the other DMUs to obtain the results shown in Table 2.2. Only DMU B is efficient and is in the reference set of all of the other DMUs. (See Figure 2.1.) Figure 2.1 portrays the situation geometrically. The efficient frontier represented by the solid line passes through B and no other point. The θ^* values in Table 2.2 show what is needed to bring each DMU onto the efficient frontier. For example, the value of $\theta^* = 1/2$ applied

Table 2.2. Results of Example 2.1

DMU	CCR(θ^*)	Reference Set
A	0.5000	B
B	1.0000	B
C	0.6667	B
D	0.7500	B
E	0.8000	B
F	0.4000	B
G	0.5000	B
H	0.6250	B

to A's input will bring A onto the efficient frontier by reducing its input 50% while leaving its output at its present value. Similarly $0.6667 \times 3 = 2$ will position C on the frontier. And so on.

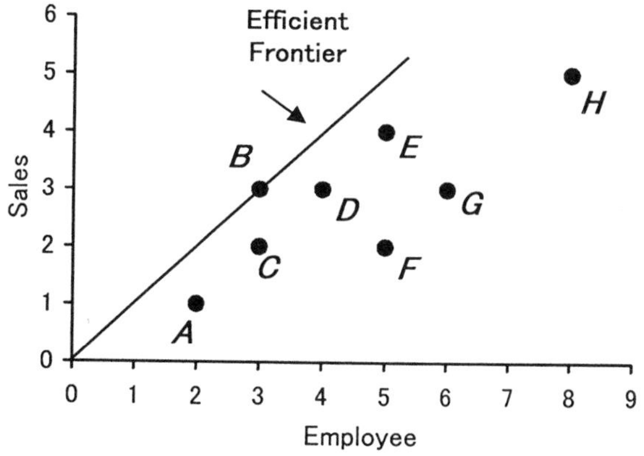

Figure 2.1. Example 2.1

2.6.2 *Example 2.2 (2 Inputs and 1 Output Case)*

Table 2.3 shows 6 DMUs with 2 inputs and 1 output where the output value is unitized to 1 for each DMU.

(1) The linear program for DMU A is:

$$
\begin{array}{lllll}
< A > \ \max & \theta = u & & & \\
\text{subject to} & 4v_1 + 3v_2 = 1 & & & \\
& u \le 4v_1 + 3v_2 & (A) & u \le 7v_1 + 3v_2 & (B) \\
& u \le 8v_1 + v_2 & (C) & u \le 4v_1 + 2v_2 & (D) \\
& u \le 2v_1 + 4v_2 & (E) & u \le 10v_1 + v_2 & (F)
\end{array}
$$

where all variables are constrained to be nonnegative.

Table 2.3. Example 2.2

	DMU	A	B	C	D	E	F
Input	x_1	4	7	8	4	2	10
	x_2	3	3	1	2	4	1
Output	y	1	1	1	1	1	1

This problem can be solved by a linear programming code. It can also be solved by simply deleting v_2 from the inequalities by inserting $v_2 = (1 - 4v_1)/3$ and observing the relationship between v_1 and u. The (unique) optimal solution is ($v_1^* = 0.1429$, $v_2^* = 0.1429$, $u^* = 0.8571$, $\theta^* = 0.8571$) and the CCR-efficiency of A is 0.8571. By applying the optimal solution to the above constraints, the reference set for A is found to be $E_A = \{D,\ E\}$.

(2) The linear program for DMU B is:

$$
\begin{array}{lllll}
<B> \ \max & \theta = u & & & \\
\text{subject to} & 7v_1 + 3v_2 = 1 & & & \\
& u \le 4v_1 + 3v_2 & (A) & u \le 7v_1 + 3v_2 & (B) \\
& u \le 8v_1 + v_2 & (C) & u \le 4v_1 + 2v_2 & (D) \\
& u \le 2v_1 + 4v_2 & (E) & u \le 10v_1 + v_2 & (F)
\end{array}
$$

The (unique) optimal solution is ($v_1^* = 0.0526$, $v_2^* = 0.2105$, $u^* = 0.6316$, $\theta^* = 0.6316$), the CCR-efficiency of B is 0.6316, and the reference set is $E_B = \{C,\ D\}$.

Now let us observe the difference between the optimal weights $v_1^* = 0.0526$ and $v_2^* = 0.2105$. The ratio $v_2^*/v_1^* = 0.2105/0.0526 = 4$ suggests that it is advantageous for B to weight *Input* x_2 four times more than *Input* x_1 in order to maximize the ratio scale measured by virtual input vs. virtual output. These values have roles as measures of the sensitivity of efficiency scores in reference to variations in input items. This topic will be dealt with in detail in Chapter 9 of this book where a systematic basis for conducting such sensitivity analyses will be provided. Here we only note that our analysis shows that a reduction in *Input* x_2 has a bigger effect on efficiency than does a reduction in *Input* x_1.

(3) An optimal solution for C is ($v_1^* = 0.0833, v_2^* = 0.3333, u^* = 1, \theta^* = 1$) and C is CCR-efficient by Definition 2.1. However, the optimal solution is not uniquely determined, as will be observed in the next section.

Likewise, D and E are CCR-efficient.

(4) The linear program for DMU F is:

$$
\begin{array}{lllll}
<F> \ \max & \theta = u & & & \\
\text{subject to} & 10v_1 + v_2 = 1 & & & \\
& u \le 4v_1 + 3v_2 & (A) & u \le 7v_1 + 3v_2 & (B) \\
& u \le 8v_1 + v_2 & (C) & u \le 4v_1 + 2v_2 & (D) \\
& u \le 2v_1 + 4v_2 & (E) & u \le 10v_1 + v_2 & (F)
\end{array}
$$

The optimal solution for F is ($v_1^* = 0$, $v_2^* = 1$, $u^* = 1$, $\theta^* = 1$) and with $\theta^* = 1$, F looks efficient. However, we notice that $v_1^* = 0$. We therefore assign a small positive value ε to v_1 and observe the change in θ^*. That is, we use the data for F and with $10\varepsilon + v_2 = 1$ to obtain $v_2 = 1 - 10\varepsilon$. By inserting this value in the above inequalities, the following constraints are obtained:

$$
\begin{array}{llll}
u \le 3 - 26\varepsilon & (A) & u \le 3 - 23\varepsilon & (B) \\
u \le 1 - 2\varepsilon & (C) & u \le 2 - 16\varepsilon & (D) \\
u \le 4 - 38\varepsilon & (E) & u \le 1 & (F)
\end{array}
$$

Noting that ε is a small positive value, the minimum of the right-hand terms is attained with

$$u = 1 - 2\varepsilon.$$

Therefore, for any $\varepsilon > 0$, it follows that $\theta^* = 1 - 2\varepsilon < 1$. Thus, v_1 must be zero in order for F to have $\theta^* = 1$. We therefore conclude that F is CCR-inefficient by Definition 2.1.

Furthermore, let us examine the inefficiency of F by comparing F with C. C has *Input* $x_1 = 8$ and *Input* $x_2 = 1$, while F has *Input* $x_1 = 10$ and *Input* $x_2 = 1$. F has 2 units of excess in *Input* x_1 compared with C. This deficiency is concealed because the optimal solution forces the weight of *Input* x_1 to zero ($v_1^* = 0$). C is in the reference set of F and hence by direct comparison we can identify the fact that F has used an excessive amount of this input.

It is not always easy to see such an excess in an input (or a shortage in output) from the optimal solution of the CCR model. In the next chapter, we will approach the CCR model from the dual side of the linear program and this will enable us to determine the excesses and shortfalls explicitly by the nonzero values with which these are identified.

A DMU such as F, with $\theta^* = 1$ and with an excess in inputs and/or a shortage in outputs, is called *ratio* efficient but *mix* inefficient.

Table 2.4 shows the CCR-efficiency (θ^*) of Example 2.2 and Figure 2.2 depicts the efficient frontier.

Table 2.4. Results of Example 2.2

DMU	x_1	x_2	y	CCR(θ^*)	Reference Set	v_1	v_2	u
A	4	3	1	0.8571	D E	.1429	.1429	.8571
B	7	3	1	0.6316	C D	.0526	.2105	.6316
C	8	1	1	1	C	.0833	.3333	1
D	4	2	1	1	D	.1667	.1667	1
E	2	4	1	1	E	.2143	.1429	1
F	10	1	1	1	C	0	1	1

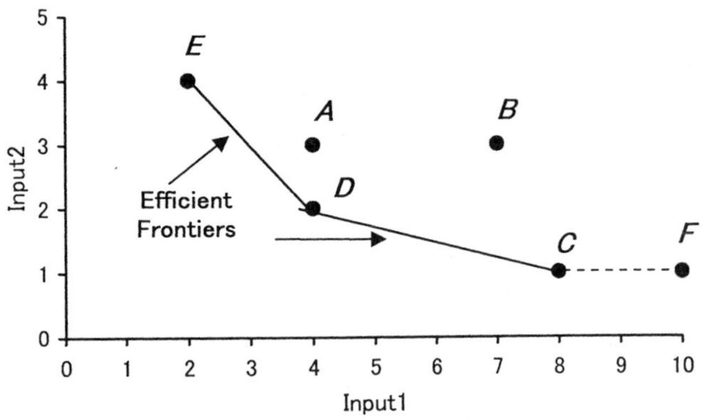

Figure 2.2. Example 2.2

2.7 ILLUSTRATION OF EXAMPLE 2.2

In order to demonstrate the role of weights $(\boldsymbol{v}, \boldsymbol{u})$ for identifying the CCR-efficiency of DMUs, we will show graphically the efficient frontier of Example 2.2 in the weight variables (=multiplier) space. Example 2.2 has 2 inputs and 1 output, whose value is unitized to 1. For this simple example we can illustrate the situations using a two dimensional graph. The linear programming constraints for each DMU have the following inequalities in common with all variables being constrained to be nonnegative.

$$
\begin{array}{llll}
u \leq 4v_1 + 3v_2 & (A) & u \leq 7v_1 + 3v_2 & (B) \\
u \leq 8v_1 + v_2 & (C) & u \leq 4v_1 + 2v_2 & (D) \\
u \leq 2v_1 + 4v_2 & (E) & u \leq 10v_1 + v_2 & (F)
\end{array}
$$

Dividing these expressions by $u > 0$, we obtain the following inequalities:

$$
\begin{array}{llll}
1 \leq 4(v_1/u) + 3(v_2/u) & (A) & 1 \leq 7(v_1/u) + 3(v_2/u) & (B) \\
1 \leq 8(v_1/u) + (v_2/u) & (C) & 1 \leq 4(v_1/u) + 2(v_2/u) & (D) \\
1 \leq 2(v_1/u) + 4(v_2/u) & (E) & 1 \leq 10(v_1/u) + (v_2/u) & (F)
\end{array}
$$

These inequalities are depicted in Figure 2.3 by taking v_1/u and v_2/u as axes. The area denoted by P then shows the feasible region for the above constraints. The boundary of P consists of three line segments and two axes. The three line segments correspond to the efficient DMUs C, D and E.

We explain this situation using D as an example and we also explain the relationship between this region and the inefficient DMUs using A as an example.

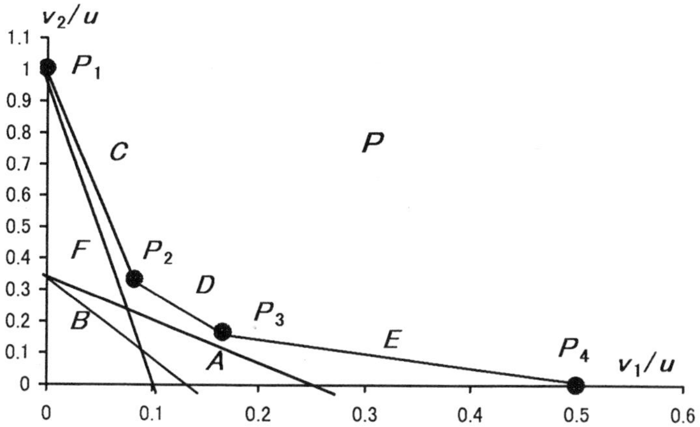

Figure 2.3. Region P

1. **Example** D
   The linear program for D consists of the preceding inequalities plus the following,

$$\begin{aligned} \max \quad & u \\ \text{subject to} \quad & 4v_1 + 2v_2 = 1. \end{aligned} \tag{2.16}$$

   Dividing (2.16) by u, we have

$$4(v_1/u) + 2(v_2/u) = 1/u. \tag{2.17}$$

   The objective function $u \to \max$ yields the same solutions as $1/u \to \min$, so the problem is to find the minimum t for which the following line touches the region P:

$$4(v_1/u) + 2(v_2/u) = t. \tag{2.18}$$

   From Figure 2.3 we see that $t = 1$ (and hence $u = 1$) represents the optimal line for D, showing that D is efficient. It is also easy to see that D is efficient for *any* weight (v_1, v_2) on the line segment (P_2, P_3). This observation leads to the conclusion that the optimal (v_1, v_2) for D is not unique. In fact, the value $(v_1 = .1667, v_2 = .1667)$ for D in Table 2.4 is an example, and actually corresponds to P_3 in Figure 2.3. See Problem 2.1 at the end of this Chapter.

   Similarly, any (v_1, v_2) on the line segment (P_1, P_2) expresses the optimal weight for C and any (v_1, v_2) on the line segment (P_3, P_4) for E.

   Thus, the optimal weights for an efficient DMU need not be unique and we should be careful to keep this in mind.

2. **Example** A
   Next, we consider the inefficient DMUs, taking A as an example. The linear

program for A consists of the following expressions added to the inequalities above:

$$\begin{aligned} \max \quad & u \\ \text{subject to} \quad & 4v_1 + 3v_2 = 1. \end{aligned} \tag{2.19}$$

As with example D, (2.19) can be transformed into

$$4(v_1/u) + 3(v_2/u) = t. \tag{2.20}$$

Then the problem is to find the minimum t within the region P. Referring to Figure 2.4, we can see that the solution is given by the point P_3, where the line parallel to line A touches the region P for the first time. P_3 is the intersection of lines D and E and this is the geometric correspond of the fact that the reference set to A consists of D and E. A simple calculation finds that $t = 1/0.8571$ and hence the efficiency of A is $u = .8571$. The value of $(v_1,\ v_2)$ at P_3 is :

$$v_1 = .1667 \times .8571 = .1429, \quad v_2 = .1667 \times .8571 = .1429, \tag{2.21}$$

which are the optimal weights for A. The optimal weights for A are unique. Usually, the optimal weights for inefficient DMUs are unique, the exception being when the line of the DMU is parallel to one of the boundaries of the region P.

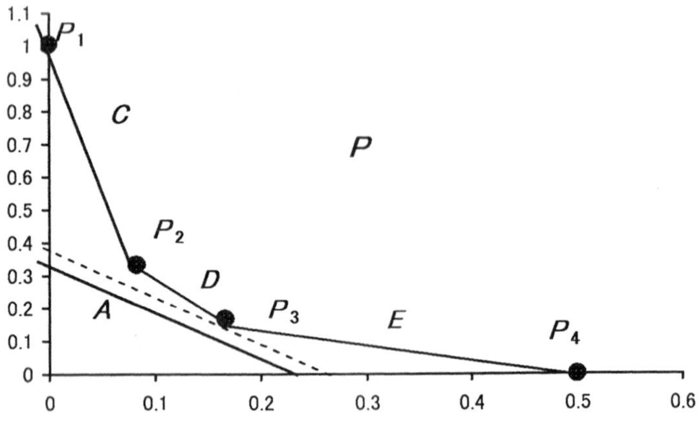

Figure 2.4. The Case of DMU A

2.8 SUMMARY OF CHAPTER 2

In this chapter, we introduced the CCR model, which is a basic DEA model.

1. For each DMU, we formed the virtual input and output by (yet unknown) weights (v_i) and (u_r):

$$\text{Virtual input} = v_1 x_{1o} + \cdots + v_m x_{mo}$$
$$\text{Virtual output} = u_1 y_{1o} + \cdots + u_s y_{so}.$$

   Then we tried to determine the weight, using linear programming so as to maximize the ratio

$$\frac{\text{virtual output}}{\text{virtual input}}.$$

   The optimal weights may (and generally will) vary from one DMU to another DMU. Thus, the "weights" in DEA are derived from the data instead of being fixed in advance. Each DMU is assigned a best set of weights with values that may vary from one DMU to another. Here, too, the DEA weights differ from customary weightings (e.g., as in index number constructions) so we will hereafter generally use the term "multiplier" to distinguish these DEA values from the other commonly used approaches.

2. CCR-efficiency was defined, along with the reference sets for inefficient DMUs.

3. Details of the linear programming solution procedure and the production function correspondence are given in Chapter 3.

2.9 SELECTED BIBLIOGRAPHY

The term 'Decision Making Unit' (DMU) was used for the first time in the CCR model proposed in Charnes, Cooper and Rhodes (1978).[4] The term DEA (Data Envelopment Analysis) was introduced in their report "A Data Envelopment Analysis Approach to Evaluation of the Program Follow Through Experiment in U.S. Public School Education," (1978),[5] Rhodes (1978)[6] and appeared in Charnes, Cooper and Rhodes' subsequent paper (1979).[7] DEA originated from efforts to evaluate results from an early 1970's project called "Program Follow Through"—a huge attempt by the U.S. Office (now Department) of Education to apply principles from the statistical design of experiments to a set of matched schools in a nationwide study. The purpose of the study was to evaluate educational programs designed to aid disadvantaged students in U.S. public schools. The data base was sufficiently large that issues of degrees of freedom, etc., were not a serious problem despite the numerous input and output variables used in the study. Nevertheless, unsatisfactory and even absurd results were secured from all of the statistical-econometric approaches that were tried. While trying to respond to this situation, Rhodes called Cooper's attention to Farrell's seminal article, "The Measurement of Productive Efficiency," in the *Journal of the Royal Statistical Society* (1957). Charnes, Cooper and Rhodes extended Farrell's work and succeeded in establishing DEA as a basis for efficiency analysis. Details of the project are described in Charnes, Cooper and Rhodes (1981).[8] A brief history of DEA can be found in Charnes and Cooper (1985).[9]

2.10 PROBLEM SUPPLEMENT FOR CHAPTER 2

Problem 2.1

In Example 2.2, determine the region of (v_1, v_2) that makes each of DMUs C, D and E efficient, by referring to Table 2.4 and Figure 2.3.

Suggested Answer : For C: line segment $\overline{P_1P_2}$. This is the line segment stretching from $(v_1/u = 0,\ v_2/u = 1)$ at P_1 to $(v_1/u = .08333,\ v_2/u = .3333)$ at P_2. For D: line segment $\overline{P_2P_3}$. This is the line segment stretching from $(v_1/u = .08333,\ v_2/u = .3333)$ at P_2 to $(v_1/u = .1667,\ v_2/u = .1667)$ at P_3. For E: line segment $\overline{P_3P_4}$. This is the line segment stretching from $(v_1/u = .1667,\ v_2/u = .1667)$ at P_3 to $(v_1/u = .5,\ v_2/u = 0)$ at P_4.

Problem 2.2

Use the data of Tables 2.3 and 2.4 to relate Figures 2.2 and 2.3.

Suggested Answer : The relation between Figures 2.2 and 2.3 is an example of what is called "the point-to-hyperplane correspondence" in mathematical programming. This means that the coordinates of the points for the representation in one of the spaces correspond to the coefficients for the hyperplanes in the dual (here=multiplier) space. The spaces in these figures are 2-dimensional, so the hyperplanes take the form of lines. For example, the coordinates for A in Figure 2.2 as obtained from Table 2.3 correspond to the coefficients in $4v_1 + 3v_2 \geq 1u$, the expression associated with DMU A as obtained from Table 2.4. In this single output case, we can use (2.19) to effect a further simplification by moving to homogenous coordinates and then try to minimize t, as given in (2.20). This minimization is undertaken subject to the similar transformation (to homogenous coordinates) for all of the other constraints obtained from Table 2.4. The result, as given in the discussion of (2.21), is $t = 1/0.8571$, so $u = 0.8571$ and $v_1 = v_2 = 0.1429$. Substitution in the expression with which we began then gives

$$1.00 \doteq 4v_1 + 3v_2 > 1u = 0.8571.$$

Hence DMU A is found to be inefficient. It should have produced more output, or used less input (or both).

This evaluation, as noted in Table 2.4, is determined relative to the corresponding expressions obtained for D and E — both of which achieve equality between both sides of their expressions, while using these same (best) weights for A — viz.

$$\begin{aligned} D: \quad 0.8574 &= 4v_1 + 2v_2 \approx 0.8571 \\ E: \quad 0.8574 &= 2v_1 + 4v_2 \approx 0.8571. \end{aligned}$$

We now note that the points for A and B in Figure 2.2 lie *above* the efficient frontier whenever the corresponding hyperplanes (here=lines) lie *below* the hyperplanes for D and E in Figure 2.3. The situation for F, which is also inefficient, differs because its hyperplane (=line) intersects the efficiency frontier at one point, P_1, but otherwise lies everywhere below it. This is the dual reflection of the hyperplane-to-point correspondence while examination of Figure 2.2 shows that F has the same coordinate as C for Input 2 but F's coordinate for Input 1 exceeds the value of the same coordinate for C.

In this chapter we examined the multiplier model in which the maximizing objective tries to move each hyperplane "up," as far as the other constraints allow, with the highest hyperplanes used to determine the boundaries of P, as exhibited in Figure 2.3. In the next chapter we shall examine the minimizing model which moves in the opposite direction in order to move the boundaries of P (the production possibility set) as far "down" as possible. Then we shall use the duality theorem of linear programming to show that the optimal values yield the same production possibility sets so either (or both) problems may be used, as desired and, even better, a suitable reading of the solution yields the solution to both problems when either one is solved.

Problem 2.3

Can you relate the CCR ratio form given by (2.3)-(2.6) in this chapter to the "ratio of ratios" form given by (1.4) in Chapter 1 for engineering efficiency?

Suggested Response : The "ratio of ratios" form given for engineering efficiency in (1.4) is for use with a single input and a single output. This can be interpreted as "virtual inputs" and "virtual outputs" for the values used in the numerators and denominators of (2.3)-(2.6) when specific values are assigned to all of the variables u_r and v_i. To put these ratios in the form of a "ratio of ratios" we will use what is called the TDT (Thompson-Dharmapala-Thrall) measure of efficiency obtained from the following problem,

$$\max_{\boldsymbol{u},\boldsymbol{v}} \quad \frac{\sum_{r=1}^{s} u_r y_{ro}}{\sum_{i=1}^{m} v_i x_{io}} \bigg/ \frac{\sum_{r=1}^{s} u_r y_{rk}}{\sum_{i=1}^{m} v_i x_{ik}} \tag{2.22}$$

$$\text{where} \quad \frac{\sum_{r=1}^{s} u_r y_{rk}}{\sum_{i=1}^{m} v_i x_{ik}} = \text{maximum}_{j=1,..,n} \left\{ \frac{\sum_{r=1}^{s} u_r y_{rj}}{\sum_{i=1}^{m} v_i x_{ij}} \right\} \tag{2.23}$$

$$u_r, v_i \geq 0 \quad \forall r, i. \tag{2.24}$$

Here "$\forall r, i$" means "for all r, i." Thus, for each choice of u_r and v_i, the maximum of the ratios in the braces on the right in the last expression is to be used as the denominator ratio in the first expression. Considering all allowable possibilities the problem is to maximize the ratio in (2.22) and "theoretical" as well as "observed" values may be used in these expressions if desired. A detailed discussion of the TDT measure and its properties may be found in W.W. Cooper, R.G. Thompson and R.M. Thrall "Extensions and New Development in DEA," *Annals of Operations Research*, 66, 1996, pp.3-45. Here we only note that no bounds are placed on the admissible values in the ratios of the last expression. The idea is to allow these to be specialized, if one wants to do so, by imposing such bounds on any or all these ratios. If we limit all the ratio values to a bound of unity, as is done in (2.4), we obtain the model given in (2.3)-(2.6). The value of the objective, (2.3), can then be interpreted as a "ratio of ratios" because a necessary condition for a solution to be optimal in (2.3)-(2.6) is that at least one of the ratios must achieve the maximal allowable value of unity. Thus, the maximization for (2.3)-(2.6) obtains an optimal value in the form of "ratio of ratios" in which the maximal term in the braces gives

$$\frac{\sum_{r=1}^{s} u_r^* y_{rk}}{\sum_{i=1}^{m} v_i^* x_{ik}} = 1.$$

Problem 2.4

Background: "Evaluating Efficiency of Turbofan Jet Engines in Multiple Input-Output Contexts: A Data Envelopment Analysis Approach," by S. Bulla, W.W. Cooper, K.S. Park and D. Wilson(1999)[10] reports results from a study of 29 jet engines produced by different manufacturers. The engineering measure of efficiency is given by $\eta = TV/\dot{Q}$ where T = Thrust of Engine, V = Cruise Velocity and $\dot{Q}$ = Heat Input from Fuel. The DEA evaluations were based on these same two outputs but the single input (in the denominator) was replaced by 3 inputs: (1) Fuel Consumption (2) Weight of Engine and (3) Drag.

Using data from commercially available sources the results obtained from these two approaches to efficiency are portrayed in Figure 2.5 on the left. Coefficients for the resulting regression were found to be statistically significant. The value of R^2, however, suggested that the explanatory power of engineering efficiency was low. More input variables are needed to bring the engineering results into closer correspondence with the DEA efficiency scores.

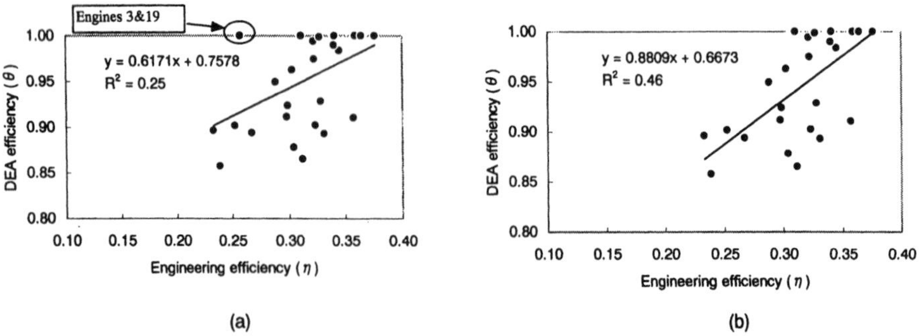

Figure 2.5. A Scatter Plot and Regression of DEA and Engineering Efficiency Ratings

As shown at the top of Figure (a), Engines 3 and 19 have the same coordinates. Their "outlier" character suggests that their removal would yield a regression with a higher R^2. This is confirmed by the new regression and the new R^2 in Figure (b) where the estimates show statistically significant improvements. Nevertheless, the same conclusion is reached: more inputs are needed for the engineering ratio to provide better "explanations" of the DEA scores, but it is not clear how this can be done.

Assignment : Is the removal of engines 3 and 19 justified or should the potential effects on the efficiency scores of other engines also be considered? Discuss.

Suggested Answer : Statistical independence for all observations is generally assumed in the removal of outliers. However, engines 3 and 19 have DEA ratings of unity and this suggests that they may be members of reference sets used to evaluate *other* engines. To check this possibility we use Figure 2.6 which is known as an "envelopment map."[11]

Eng.	1	2	3	4	5	6	7	8	9	10	11	12	13	14	15	16	17	18	19	20	21	22	23	24	25	26	27	28	29	N*
1			✓												✓														✓	3
2			✓												✓														✓	3
3			✓																											0
4			✓												✓			✓											✓	4
5											✓					✓														2
6																✓													✓	2
7															✓	✓		✓												3
8															✓	✓		✓											✓	4
9														✓		✓		✓											✓	4
10														✓		✓		✓											✓	4
11											✓																			0
12																✓													✓	2
13														✓																1
14														✓																0
15															✓															0
16																✓														0
17														✓		✓		✓											✓	4
18																		✓												0
19																			✓											0
20															✓	✓														2
21														✓															✓	2
22																✓													✓	2
23			✓												✓			✓											✓	4
24															✓	✓		✓												3
25														✓		✓		✓												3
26															✓	✓		✓												3
27														✓		✓		✓												3
28															✓	✓		✓												3
29																													✓	0
TN*	0	0	4	0	0	0	0	0	0	0	1	0	0	7	10	15	0	12	0	0	0	0	0	0	0	0	0	0	12	61

TN* ≡ Total number of times that engine *j* (= 1, ..., 29 in column) was used to evaluate *other* engines.
N* ≡ Number of times that *other* engines were used to evaluate engine *i* (= 1, ..., 29 in row).

Figure 2.6. Envelopment Map for 29 Jet Engines

The values at the bottom of each column show the number of times an engine identified at the top of this column entered into a reference set to evaluate *other* engines, and the total for each row shows the number of *other* engines used in an evaluation. Thus, engine 3 was used a total (net) of 4 times in evaluating *other* engines and the row total shows that no *other* engine was involved in the evaluation of engine 3 — a property of efficient engines when the latter are extreme points of the production possibility set. A removal of engine 3 will therefore affect the efficiency ratings of other engines — unless in each case there exists an alternate optimum in which engine 3 does *not* actively enter as a member of the basis (=reference set).

No similar interaction with other engines occurs for engine 19, however, as is clear from the fact that the row and column totals are both zero for this engine. Hence our analysis shows that 19 (but not 3) may be treated as an outlier *without affecting any other observations.* Proceeding in this manner produces Figure 2.7 which, though statistically significant, is closer to Figure 2.5 (a).

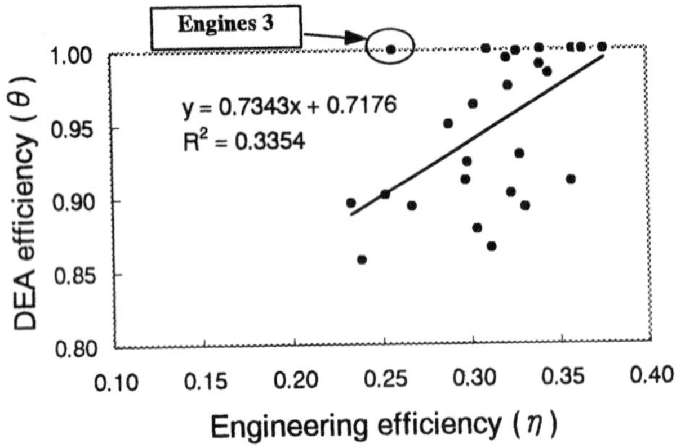

Figure 2.7. A Comparative Portrayal and Regression Analysis for 28 Engines (All Data Except for Engine 19)

Problem 2.5

Try to bring the engineering definition into play in a DEA evaluation.

Suggested Response : As noted in (1.4) the engineering "ratio of ratios" reduces to a comparison of actual to theoretically obtainable output from the amount of input actually used by any engine. This could be brought into play in a DEA evaluation by associating the theoretically obtainable output with the actual input for each engine (=DMU). This would produce n additional observations giving $2n$ DMUs from which evaluations would be made. However, there is no guarantee that a DMU will be evaluated relative to its own input. A DMU evaluation of any engine (=DMU) could utilize a reference set yielding a *lower* input and a *higher* output. This could be checked and further analyzed, however, because the members of the reference set would be identified as well as the output shortfalls and input excesses for the actual DMU being evaluated.

Notes

1. This condition will be relaxed to allow nonnegative data in Chapter 3. Furthermore, in Chapter 5 (Section 5.2), we will introduce models which can also deal with negative data.

2. *Proof of Theorem 2.2.* Let θ^*, u_r^*, v_i^* be optimal for (2.3)-(2.6). Now replace the original y_{rj} and x_{ij} by $\rho_r y_{rj}$ and $\delta_i x_{ij}$ for some choices of $\rho_r, \delta_i > 0$. But then choosing $u_r' = u_r^*/\rho_r$ and $v_i' = v_i^*/\delta_i$ we have a solution to the transformed problem with $\theta' = \theta^*$. An optimal value for the transformed problem must therefore have $\theta'^* \geq \theta^*$. Now suppose we could have $\theta'^* > \theta^*$. Then, however, $u_r = u_r'^* \rho_r$ and $v_i = v_i'^* \delta_i$ satisfy the original constraints so the assumption $\theta'^* > \theta^*$ contradicts the optimality assumed for θ^* under these constraints. The only remaining possibility is $\theta'^* = \theta^*$. This proves the invariance claimed for (2.3). Theorem 2.1 demonstrated the equivalence of (LP_o) to (FP_o) and thus the same result must hold and the theorem is therefore proved. □

3. We use the notations $\boldsymbol{v}$ and $\boldsymbol{u}$ instead of $\boldsymbol{\nu}$ and $\boldsymbol{\mu}$ in (LP_o).

4. A. Charnes, W.W. Cooper and E. Rhodes (1978), "Measuring the Efficiency of Decision Making Units," *European Journal of Operational Research* 2, pp.429-444.

5. A. Charnes, W.W. Cooper and E. Rhodes (1978), "A Data Envelopment Analysis Approach to Evaluation of the Program Follow Through Experiments in U.S. Public School Education," Management Science Research Report No. 432, Carnegie-Mellon University, School of Urban and Public Affairs, Pittsburgh, PA.

6. E.L. Rhodes (1978), "Data Envelopment Analysis and Related Approaches for Measuring the Efficiency of Decision-Making Units with an Application to Program Follow Through in U.S. Education," unpublished Ph.D. thesis, Carnegie-Mellon University, School of Urban and Public Affairs, Pittsburgh, PA.

7. A. Charnes, W.W. Cooper and E. Rhodes (1979), "Short Communication: Measuring the Efficiency of Decision Making Units," *European Journal of Operational Research* 3, p.339.

8. A. Charnes, W.W. Cooper and E. Rhodes (1981), "Evaluating Program and Managerial Efficiency: An Application of Data Envelopment Analysis to Program Follow Through," *Management Science* 27, pp.668-697.

9. A. Charnes and W.W. Cooper (1985), "Preface to Topics in Data Envelopment Analysis," *Annals of Operations Research* 2, pp.59-94

10. S. Bulla, W.W. Cooper, K.S. Park and D. Wilson (2000), *Journal of Propulsion and Power* 16, pp.431-439.

11. See Research Report CCS 532(Austin, Texas: University of Texas at Austin, Graduate School of Business, 1986) "Data Envelopment Analysis Approaches to Policy Evaluation and Management of Army Recruiting Activities I: Tradeoff between Joint Service and Army Advertising" by A. Charnes, W.W. Cooper, B. Golany, Major R. Halek, G. Klopp, E. Schmitz and Captain D. Thomas.

3 THE CCR MODEL AND PRODUCTION CORRESPONDENCE

3.1 INTRODUCTION

In this chapter, we relax the *positive* data set assumption. We shall instead assume the data are semipositive. That is, we assume that some (but not all) inputs and outputs are positive. This will allow us to deal with applications which involve zero data in inputs and/or outputs. The *production possibility set* composed of these input and output data (X, Y) will also be introduced. The *dual* problem of the CCR model will then be constructed and it will be shown that the dual problem evaluates efficiency based on a linear programming problem applied to the data set (X, Y). The CCR-efficiency will be redefined, taking into account all input excesses and output shortfalls.

The observations that form the production possibility set are very fundamental in that they make it possible to assess the CCR model from a broader point of view and to extend this model to other models that will be introduced in succeeding chapters.

One version of a CCR model aims to minimize inputs while satisfying at least the given output levels. This is called the *input-oriented* model. There is another type of model called the *output-oriented* model that attempts to maximize outputs without requiring more of any of the observed input values. In the last section, the latter will be introduced along with a combination of the two models.

Computational aspects of the CCR model are also covered in this chapter. The computer code DEA-Solver accompanying this book (as mentioned in Problem 1.1 of Chapter 1) will be utilized on some of the problems we provide.

3.2 PRODUCTION POSSIBILITY SET

We have been dealing with the pairs of positive input and output vectors $(\boldsymbol{x}_j, \boldsymbol{y}_j)$ $(j = 1, \ldots, n)$ of n DMUs. In this chapter, the positive data assumption is relaxed. All data are assumed to be nonnegative but at least one component of every input and output vector is positive. We refer to this as *semipositive* with a mathematical characterization given by $\boldsymbol{x}_j \geq \mathbf{0}, \boldsymbol{x}_j \neq \mathbf{0}$ and $\boldsymbol{y}_j \geq \mathbf{0}, \boldsymbol{y}_j \neq \mathbf{0}$ for some $j = 1, \ldots, n$. Therefore, each DMU is supposed to have at least one positive value in both input and output. We will call a pair of such semipositive input $\boldsymbol{x} \in R^m$ and output $\boldsymbol{y} \in R^s$ *an activity* and express them by the notation $(\boldsymbol{x}, \boldsymbol{y})$. The components of each such vector pair can be regarded as a semipositive orthant point in $(m + s)$ dimensional linear vector space in which the superscript m and s specify the number of dimensions required to express inputs and outputs, respectively. The set of feasible activities is called *the production possibility set* and is denoted by P. We postulate the following

Properties of P (the Production Possibility Set)

(A1) The observed activities $(\boldsymbol{x}_j, \boldsymbol{y}_j)$ $(j = 1, \ldots, n)$ belong to P.

(A2) If an activity $(\boldsymbol{x}, \boldsymbol{y})$ belongs to P, then the activity $(t\boldsymbol{x}, t\boldsymbol{y})$ belongs to P for any positive scalar t. We call this property the *constant returns-to-scale* assumption.

(A3) For an activity $(\boldsymbol{x}, \boldsymbol{y})$ in P, any semipositive activity $(\bar{\boldsymbol{x}}, \bar{\boldsymbol{y}})$ with $\bar{\boldsymbol{x}} \geq \boldsymbol{x}$ and $\bar{\boldsymbol{y}} \leq \boldsymbol{y}$ is included in P. That is, any activity with input no less than $\boldsymbol{x}$ in any component and with output no greater than $\boldsymbol{y}$ in any component is feasible.

(A4) Any semipositive linear combination of activities in P belongs to P.[1]

Arranging the data sets in matrices $X = (\boldsymbol{x}_j)$ and $Y = (\boldsymbol{y}_j)$, we can define the production possibility set P satisfying (A1) through (A4) by

$$P = \{(\boldsymbol{x}, \boldsymbol{y}) \mid \boldsymbol{x} \geq X\boldsymbol{\lambda},\ \boldsymbol{y} \leq Y\boldsymbol{\lambda},\ \boldsymbol{\lambda} \geq \mathbf{0}\}, \tag{3.1}$$

where $\boldsymbol{\lambda}$ is a semipositive vector in R^n.

Figure 3.1 shows a typical production possibility set in two dimensions for the CCR model in the single input and single output case, so that $m = 1$ and $s = 1$, respectively. In this example the possibility set is determined by B and the ray from the origin through B is the efficient frontier.

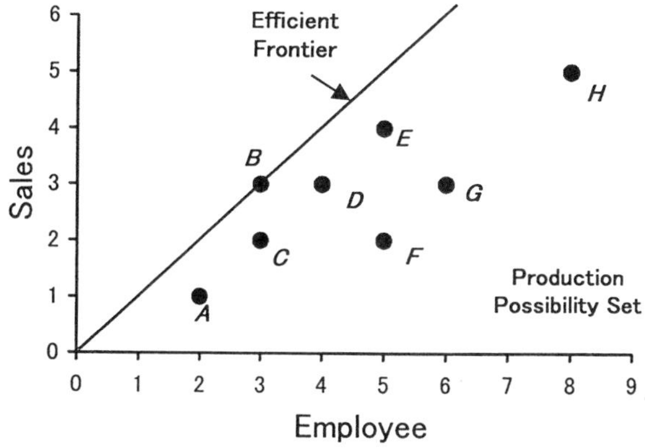

Figure 3.1. Production Possibility Set

3.3 THE CCR MODEL AND DUAL PROBLEM

Based on the matrix (X, Y), the CCR model was formulated in the preceding chapter as an LP problem with row vector $\boldsymbol{v}$ for input multipliers and row vector $\boldsymbol{u}$ as output multipliers. These multipliers are treated as variables in the following LP problem ([Multiplier form]):

$$(LP_o) \quad \max_{\boldsymbol{v},\boldsymbol{u}} \quad \boldsymbol{u}\boldsymbol{y}_o \tag{3.2}$$

$$\text{subject to} \quad \boldsymbol{v}\boldsymbol{x}_o = 1 \tag{3.3}$$

$$-\boldsymbol{v}X + \boldsymbol{u}Y \leq \mathbf{0} \tag{3.4}$$

$$\boldsymbol{v} \geq \mathbf{0}, \quad \boldsymbol{u} \geq \mathbf{0}. \tag{3.5}$$

This is the same as (2.7)-(2.11), (LP_o) in the preceding chapter, which is now expressed in vector-matrix notation.

The dual problem[2] of (LP_o) is expressed with a real variable θ and the transpose, T, of a nonnegative vector $\boldsymbol{\lambda} = (\lambda_1, \ldots, \lambda_n)^T$ of variables as follows ([Envelopment form]):

$$(DLP_o) \quad \min_{\theta,\boldsymbol{\lambda}} \quad \theta \tag{3.6}$$

$$\text{subject to} \quad \theta\boldsymbol{x}_o - X\boldsymbol{\lambda} \geq \mathbf{0} \tag{3.7}$$

$$Y\boldsymbol{\lambda} \geq \boldsymbol{y}_o \tag{3.8}$$

$$\boldsymbol{\lambda} \geq \mathbf{0}. \tag{3.9}$$

Correspondences between the primal (LP_o) and the dual (DLP_o) constraints and variables are displayed in Table 3.1.

(DLP_o) has a feasible solution $\theta = 1$, $\lambda_o = 1$, $\lambda_j = 0$ $(j \neq o)$. Hence the optimal θ, denoted by θ^*, is not greater than 1. On the other hand, due to

Table 3.1. Primal and Dual Correspondences

Multiplier form Constraint (LP_o)	Envelopment form Variable (DLP_o)	Envelopment form Constraint (DLP_o)	Multiplier form Variable (LP_o)
$\boldsymbol{v}\boldsymbol{x}_o = 1$	θ	$\theta\boldsymbol{x}_o - X\boldsymbol{\lambda} \geq \boldsymbol{0}$	$\boldsymbol{v} \geq \boldsymbol{0}$
$-\boldsymbol{v}X + \boldsymbol{u}Y \leq \boldsymbol{0}$	$\boldsymbol{\lambda} \geq \boldsymbol{0}$	$Y\boldsymbol{\lambda} \geq \boldsymbol{y}_o$	$\boldsymbol{u} \geq \boldsymbol{0}$

the nonzero (i.e., semipositive) assumption for the data, the constraint (3.8) forces $\boldsymbol{\lambda}$ to be nonzero because $\boldsymbol{y}_o \geq \boldsymbol{0}$ and $\boldsymbol{y}_o \neq \boldsymbol{0}$. Hence, from (3.7), θ must be greater than zero. Putting this all together, we have $0 < \theta^* \leq 1$. Now we observe the relation between the production possibility set P and (DLP_o). The constraints of (DLP_o) require the activity $(\theta\boldsymbol{x}_o,\ \boldsymbol{y}_o)$ to belong to P, while the objective seeks the minimum θ that reduces the input vector $\boldsymbol{x}_o$ radially to $\theta\boldsymbol{x}_o$ while remaining in P. In (DLP_o), we are looking for an activity in P that guarantees at least the output level $\boldsymbol{y}_o$ of DMU$_o$ in all components while reducing the input vector $\boldsymbol{x}_o$ proportionally (radially) to a value as small as possible. Under the assumptions of the preceding section, it can be said that $(X\boldsymbol{\lambda},\ Y\boldsymbol{\lambda})$ outperforms $(\theta\boldsymbol{x}_o,\ \boldsymbol{y}_o)$ when $\theta^* < 1$. With regard to this property, we define the input *excesses* $\boldsymbol{s}^- \in R^m$ and the output *shortfalls* $\boldsymbol{s}^+ \in R^s$ and identify them as "slack" vectors by:

$$\boldsymbol{s}^- = \theta\boldsymbol{x}_o - X\boldsymbol{\lambda}, \quad \boldsymbol{s}^+ = Y\boldsymbol{\lambda} - \boldsymbol{y}_o, \tag{3.10}$$

with $\boldsymbol{s}^- \geq \boldsymbol{0},\ \boldsymbol{s}^+ \geq \boldsymbol{0}$ for any feasible solution $(\theta,\ \boldsymbol{\lambda})$ of (DLP_o).

To discover the possible input excesses and output shortfalls, we solve the following two-phase LP problem.

Phase I

We solve (DLP_o). Let the optimal objective value be θ^*. By the duality theorem of linear programming,[3] θ^* is equal to the optimal objective value of (LP_o) and is the CCR-efficiency value, also called "Farrell Efficiency," after M.J. Farrell (1957). See below. This value of θ^* is incorporated in the following Phase II extension of (DLP_o).

Phase II

Using our knowledge of θ^*, we solve the following LP using $(\boldsymbol{\lambda},\ \boldsymbol{s}^-,\ \boldsymbol{s}^+)$ as variables:

$$\max_{\boldsymbol{\lambda},\boldsymbol{s}^-,\boldsymbol{s}^+} \quad \omega = \boldsymbol{e}\boldsymbol{s}^- + \boldsymbol{e}\boldsymbol{s}^+ \tag{3.11}$$

$$\text{subject to} \quad \boldsymbol{s}^- = \theta^*\boldsymbol{x}_o - X\boldsymbol{\lambda} \tag{3.12}$$

$$\boldsymbol{s}^+ = Y\boldsymbol{\lambda} - \boldsymbol{y}_o \tag{3.13}$$

$$\lambda \geq 0, \quad s^- \geq 0, \quad s^+ \geq 0,$$

where $e = (1, \ldots, 1)$ (a vector of ones) so that $es^- = \sum_{i=1}^m s_i^-$ and $es^+ = \sum_{r=1}^s s_r^+$.

The objective of Phase II is to find a solution that maximizes the sum of input excesses and output shortfalls while keeping $\theta = \theta^*$.

We should note that we could replace the objective term in (3.11) with any weighted sum of input excesses and output shortfalls such as:

$$\omega = w_x s^- + w_y s^+, \tag{3.14}$$

where the weights w_x and w_y are positive row vectors. The modified objective function may result in a different optimal solution for Phase II. However, we can have the optimal $\omega^* > 0$ in (3.11) if and only if a nonzero value is also obtained when the objective in (3.11) is replaced with (3.14). Thus the objective in (3.11) will identify some nonzero slacks with inefficiency if and only if some nonzero (possibly different) slacks are identified with inefficiency in (3.14).

Definition 3.1 (Max-slack Solution, Zero-slack Activity)
An optimal solution $(\lambda^*, s^{-*}, s^{+*})$ *of* Phase II *is called the* max-slack solution. *If the max-slack solution satisfies* $s^{-*} = 0$ *and* $s^{+*} = 0$, *then it is called* zero-slack.

Definition 3.2 (CCR-Efficiency, Radial Efficiency, Technical Efficiency)
If an optimal solution $(\theta^*, \lambda^*, s^{-*}, s^{+*})$ *of the two LPs above satisfies* $\theta^* = 1$ *and is zero-slack* ($s^{-*} = 0$, $s^{+*} = 0$), *then the* DMU_o *is called* CCR- efficient. *Otherwise, the* DMU_o *is called* CCR-inefficient, *because*

(*i*) $\theta^* = 1$
(*ii*) *All slacks are zero*

must both be satisfied if full efficiency is to be attained.

The first of these two conditions is referred to as "radial efficiency." It is also referred to as "technical efficiency" because a value of $\theta^* < 1$ means that all inputs can be simultaneously reduced without altering the mix (=proportions) in which they are utilized. Because $(1-\theta^*)$ is the maximal proportionate reduction allowed by the production possibility set, any further reductions associated with nonzero slacks will necessarily change the input proportions. Hence the inefficiencies associated with any nonzero slack identified in the above two-phase procedure are referred to as "mix inefficiencies." Other names are also used to characterize these two sources of inefficiency. For instance, the term "weak efficiency" is sometime used when attention is restricted to (*i*) in Definition 3.2. The conditions (*i*) and (*ii*) taken together describe what is also called "Pareto-Koopmans" or "strong" efficiency, which can be verbalized as follows,

Definition 3.3 (Pareto-Koopmans Efficiency)
A DMU is fully efficient if and only if it is not possible to improve any input or output without worsening some other input or output.

This last definition recognizes the contributions of the economists Vilfredo Pareto and Tjalling Koopmans. Implementable form was subsequently given to it by M.J. Farrell, another economist, who showed how to apply these concepts to observed data. However, Farrell was only able to carry his developments to a point which satisfied condition (*i*) but not condition (*ii*) in Definition 3.2. Hence he did not fully satisfy the conditions for Pareto-Koopmans efficiency but stopped short, instead, with what we just referred to as "weak efficiency" (also called "Farrell efficiency") because nonzero slack, when present in any input or output, can be used to effect additional improvements without worsening any other input or output. Farrell, we might note, was aware of this shortcoming in his approach which he tried to treat by introducing new (unobserved) "points at infinity" but was unable to give his concept implementable form.[4] In any case, this was all accomplished by Charnes, Cooper and Rhodes in a mathematical formulation that led to the two-phase procedure described above. Hence we also refer to this as "CCR-efficiency" when this procedure is utilized on empirical data to fulfill both (*i*) and (*ii*) in Definition 3.2.

We have already given a definition of CCR-efficiency in Chapter 2. For the data set (X, Y) under the assumption of semipositivity, we can also define CCR- efficiency by Definition 2.1 in that chapter. We now prove that the CCR-efficiency above, gives the same efficiency characterization as is obtained from Definition 2.1 in Chapter 2. This is formalized by:

Theorem 3.1 *The CCR-efficiency given in Definition 3.2 is equivalent to that given by Definition 2.1.*

Proof. First, notice that the vectors $\boldsymbol{v}$ and $\boldsymbol{u}$ of (LP_o) are dual multipliers corresponding to the constraints (3.7) and (3.8) of (DLP_o), respectively. See Table 3.1. Now the following "complementary conditions"[5] hold between any optimal solutions $(\boldsymbol{v}^*,\ \boldsymbol{u}^*)$ of (LP_o) and $(\boldsymbol{\lambda}^*,\ \boldsymbol{s}^{-*},\ \boldsymbol{s}^{+*})$ of (DLP_o).

$$\boldsymbol{v}^*\boldsymbol{s}^{-*} = 0 \ \text{ and } \ \boldsymbol{u}^*\boldsymbol{s}^{+*} = 0. \tag{3.15}$$

Known as the "complementary slackness" condition, this means that if any component of $\boldsymbol{v}^*$ or $\boldsymbol{u}^*$ is positive then the corresponding component of $\boldsymbol{s}^{-*}$ or $\boldsymbol{s}^{+*}$ must be zero, and conversely, with the possibility also allowed in which both components may be zero simultaneously.

Now we demonstrate that Definition 3.2 implies Definition 2.1

(*i*) If $\theta^* < 1$, then DMU_o is CCR-inefficient by Definition 2.1, since (LP_o) and (DLP_o) have the same optimal objective value θ^*.

(*ii*) If $\theta^* = 1$ and is not zero-slack ($\boldsymbol{s}^{-*} \neq \boldsymbol{0}$ and/or $\boldsymbol{s}^{+*} \neq \boldsymbol{0}$), then, by the complementary conditions above, the elements of $\boldsymbol{v}^*$ or $\boldsymbol{u}^*$ corresponding to the positive slacks must be zero. Thus, DMU_o is CCR- inefficient by Definition 2.1.

(*iii*) Lastly if $\theta^* = 1$ and zero-slack, then, by the "strong theorem of complementarity,"[6] (LP_o) is assured of a positive optimal solution $(\boldsymbol{v}^*, \boldsymbol{u}^*)$ and hence DMU_o is CCR-efficient by Definition 2.1.

The reverse is also true by the complementary relation and the strong complementarity theorem between $(\boldsymbol{v}^*, \boldsymbol{u}^*)$ and $(\boldsymbol{s}^{-*}, \boldsymbol{s}^{+*})$. □

3.4 THE REFERENCE SET AND IMPROVEMENT IN EFFICIENCY

Definition 3.4 (Reference Set)
For an inefficient DMU_o*, we define its* reference set E_o*, based on the max-slack solution as obtained in phases one and two —see (3.11)— by*

$$E_o = \{j \mid \lambda_j^* > 0\} \ (j \in \{1, \ldots, n\}). \tag{3.16}$$

An optimal solution can be expressed as

$$\begin{aligned} \theta^* \boldsymbol{x}_o &= \sum_{j \in E_o} \boldsymbol{x}_j \lambda_j^* + \boldsymbol{s}^{-*} \\ \boldsymbol{y}_o &= \sum_{j \in E_o} \boldsymbol{y}_j \lambda_j^* - \boldsymbol{s}^{+*}, \end{aligned} \tag{3.17}$$

where $j \in E_o$ means the index j is included in the set E_o. This can be interpreted as follows,

$$\boldsymbol{x}_o \geq \theta^* \boldsymbol{x}_o - \boldsymbol{s}^{-*} = \sum_{j \in E_o} \boldsymbol{x}_j \lambda_j^*$$

which means

$$\begin{aligned} &\boldsymbol{x}_o \geq \text{technical} - \text{mix inefficiency} \\ &= \text{a positive combination of observed input values.} \end{aligned} \tag{3.18}$$

Also

$$\boldsymbol{y}_o \leq \boldsymbol{y}_o + \boldsymbol{s}^{+*} = \sum_{j \in E_o} \boldsymbol{y}_j \lambda_j^*$$

means

$$\begin{aligned} &\boldsymbol{y}_o \leq \text{observed outputs + shortfalls} \\ &= \text{a positive combination of observed output values.} \end{aligned} \tag{3.19}$$

These relations suggest that the efficiency of $(\boldsymbol{x}_o, \boldsymbol{y}_o)$ for DMU_o can be improved if the input values are reduced radially by the ratio θ^* and the input excesses recorded in $\boldsymbol{s}^{-*}$ are eliminated. Similarly efficiency can be attained if the output values are augmented by the output shortfalls in $\boldsymbol{s}^{+*}$. Thus, we have a method for improving an inefficient DMU that accords with Definition 3.2. The gross input improvement $\Delta \boldsymbol{x}_o$ and output improvement $\Delta \boldsymbol{y}_o$ can be calculated from:

$$\Delta \boldsymbol{x}_o = \boldsymbol{x}_o - (\theta^* \boldsymbol{x}_o - \boldsymbol{s}^{-*}) = (1 - \theta^*) \boldsymbol{x}_o + \boldsymbol{s}^{-*} \tag{3.20}$$
$$\Delta \boldsymbol{y}_o = \boldsymbol{s}^{+*}. \tag{3.21}$$

Hence, we have a formula for improvement, which is called *the CCR projection:*[7]

$$\widehat{\boldsymbol{x}}_o = \boldsymbol{x}_o - \Delta \boldsymbol{x}_o = \theta^* \boldsymbol{x}_o - \boldsymbol{s}^{-*} \leq \boldsymbol{x}_o \tag{3.22}$$
$$\widehat{\boldsymbol{y}}_o = \boldsymbol{y}_o + \Delta \boldsymbol{y}_o = \boldsymbol{y}_o + \boldsymbol{s}^{+*} \geq \boldsymbol{y}_o. \tag{3.23}$$

However, note that there are other formulae for improvement as will be described later.

In Theorems 3.2, 3.3 and 3.4 in the next section, we will show that the improved activity $(\widehat{\boldsymbol{x}}_o, \widehat{\boldsymbol{y}}_o)$ projects DMU$_o$ into the reference set E_o and any nonnegative combination of DMUs in E_o is efficient.

3.5 THEOREMS ON CCR-EFFICIENCY

Theorem 3.2 *The improved activity $(\widehat{\boldsymbol{x}}_o,\ \widehat{\boldsymbol{y}}_o)$ defined by (3.22) and (3.23) is CCR-efficient.*

Proof. The efficiency of $(\widehat{\boldsymbol{x}}_o,\ \widehat{\boldsymbol{y}}_o)$ is evaluated by solving the LP problem below:

$$(DLP_e) \qquad \min_{\theta,\boldsymbol{\lambda},\boldsymbol{s}^-,\boldsymbol{s}^+} \quad \theta \tag{3.24}$$

$$\text{subject to} \quad \theta\widehat{\boldsymbol{x}}_o - X\boldsymbol{\lambda} - \boldsymbol{s}^- = \boldsymbol{0} \tag{3.25}$$

$$Y\boldsymbol{\lambda} - \boldsymbol{s}^+ = \widehat{\boldsymbol{y}}_o \tag{3.26}$$

$$\boldsymbol{\lambda} \geq \boldsymbol{0}, \quad \boldsymbol{s}^- \geq \boldsymbol{0}, \quad \boldsymbol{s}^+ \geq \boldsymbol{0}. \tag{3.27}$$

Let an optimal (max-slack) solution for (DLP_e) be $(\widehat{\theta}, \widehat{\boldsymbol{\lambda}}, \widehat{\boldsymbol{s}}^-, \widehat{\boldsymbol{s}}^+)$. By inserting the formulae (3.22) and (3.23) into the constraints, we have

$$\widehat{\theta}\theta^*\boldsymbol{x}_o = X\widehat{\boldsymbol{\lambda}} + \widehat{\boldsymbol{s}}^- + \widehat{\theta}\boldsymbol{s}^{-*}$$
$$\boldsymbol{y}_o = Y\widehat{\boldsymbol{\lambda}} - \widehat{\boldsymbol{s}}^+ - \boldsymbol{s}^{+*}.$$

Now we can also write this solution as

$$\widetilde{\theta}\boldsymbol{x}_o = X\widehat{\boldsymbol{\lambda}} + \widetilde{\boldsymbol{s}}^-$$
$$\boldsymbol{y}_o = Y\widehat{\boldsymbol{\lambda}} - \widetilde{\boldsymbol{s}}^+$$

where $\widetilde{\theta} = \widehat{\theta}\theta^*$ and $\widetilde{\boldsymbol{s}}^- = \widehat{\boldsymbol{s}}^- + \widehat{\theta}\boldsymbol{s}^{-*} \geq \boldsymbol{0}$, $\widetilde{\boldsymbol{s}}^+ = \widehat{\boldsymbol{s}}^+ + \boldsymbol{s}^{+*} \geq \boldsymbol{0}$. However, θ^* is part of an optimal solution so we must have $\widetilde{\theta} = \widehat{\theta}\theta^* = \theta^*$ so $\widehat{\theta} = 1$. Furthermore, with $\widehat{\theta} = 1$ we have

$$\boldsymbol{e}\widetilde{\boldsymbol{s}}^- + \boldsymbol{e}\widetilde{\boldsymbol{s}}^+ = (\boldsymbol{e}\widehat{\boldsymbol{s}}^- + \boldsymbol{e}\boldsymbol{s}^{-*}) + (\boldsymbol{e}\widehat{\boldsymbol{s}}^+ + \boldsymbol{e}\boldsymbol{s}^{+*}) \leq \boldsymbol{e}\boldsymbol{s}^{-*} + \boldsymbol{e}\boldsymbol{s}^{+*}$$

since $\boldsymbol{e}\boldsymbol{s}^{-*} + \boldsymbol{e}\boldsymbol{s}^{+*}$ is maximal. It follows that we must have $\boldsymbol{e}\widehat{\boldsymbol{s}}^- + \boldsymbol{e}\widehat{\boldsymbol{s}}^+ = \boldsymbol{0}$ which implies that all components of $\widehat{\boldsymbol{s}}^-$ and $\widehat{\boldsymbol{s}}^+$ are zero. Hence conditions (i) and (ii) of Definition 3.2 are both satisfied and CCR-efficiency is achieved as claimed. □

Corollary 3.1 (Corollary to Theorem 3.2)
The point with coordinates $\widehat{\boldsymbol{x}}_o$, $\widehat{\boldsymbol{y}}_o$ defined by (3.22) and (3.23), viz.,

$$\widehat{\boldsymbol{x}}_o = \theta^*\boldsymbol{x}_o - \boldsymbol{s}^{-*} = \sum_{j\in E_o} \boldsymbol{x}_j\lambda_j^* \tag{3.28}$$

$$\widehat{\boldsymbol{y}}_o = \boldsymbol{y}_o + \boldsymbol{s}^{+*} = \sum_{j\in E_o} \boldsymbol{y}_j\lambda_j^* \tag{3.29}$$

is the point on the efficient frontier used to evaluate the performance of DMU_o.

In short the CCR projections identify the point either as a positive combination of other DMUs with $\boldsymbol{x}_o \geq \widehat{\boldsymbol{x}}_o$ and $\widehat{\boldsymbol{y}}_o \geq \boldsymbol{y}_o$ unless $\theta^* = 1$ and all slacks are zero in which case $\boldsymbol{x}_o = \widehat{\boldsymbol{x}}_o$ and $\widehat{\boldsymbol{y}}_o = \boldsymbol{y}_o$ so the operation in (3.28) performed on the observation for DMU_o identifies a new DMU positioned on the efficient frontier. Conversely, the point associated with the thus generated DMU evaluates the performance of DMU_o as exhibiting input excesses $\boldsymbol{x}_o - \widehat{\boldsymbol{x}}_o$ and output shortfalls $\widehat{\boldsymbol{y}}_o - \boldsymbol{y}_o$.

We notice that the improvement by the formulae (3.22) and (3.23) should be achieved by using the *max-slack* solution. If we do it based on another (not max-slack) solution, the improved activity $(\widehat{\boldsymbol{x}}_o,\ \widehat{\boldsymbol{y}}_o)$ is not necessarily CCR-efficient. This is shown by the following example with 4 DMUs A, B, C and D, each with 3 inputs x_1, x_2 and x_3 and all producing 1 output in amount $y = 1$.

	A	B	C	D
x_1	2	2	2	1
x_2	1	1.5	2	1
x_3	1	1	1	2
y	1	1	1	1

Here the situation for DMU C is obvious. We can observe, for instance, that DMU C has two possible slacks in x_2, 1 against A and 0.5 against B. If we choose B as the reference set of C, then the improved activity coincides with B, which still has a slack of 0.5 in x_2 against A. Therefore, the improved activity is not CCR-efficient. However, if we improve C by using the max-slack solution, then we move to the CCR-efficient DMU A directly.

Lemma 3.1 *For the improved activity* $(\widehat{\boldsymbol{x}}_o,\ \widehat{\boldsymbol{y}}_o)$, *there exists an optimal solution* $(\widehat{\boldsymbol{v}}_o, \widehat{\boldsymbol{u}}_o)$ *for the problem* (LP_e), *which is dual to* (DLP_e), *such that*

$$\widehat{\boldsymbol{v}}_o > \mathbf{0} \text{ and } \widehat{\boldsymbol{u}}_o > \mathbf{0}$$

$$\widehat{\boldsymbol{v}}_o \boldsymbol{x}_j = \widehat{\boldsymbol{u}}_o \boldsymbol{y}_j \quad (j \in E_o) \tag{3.30}$$

$$\widehat{\boldsymbol{v}}_o X \geq \widehat{\boldsymbol{u}}_o Y \tag{3.31}$$

Proof. Since $(\widehat{\boldsymbol{x}}_o, \widehat{\boldsymbol{y}}_o)$ is zero-slack, the strong theorem of complementarity means that there exists a positive optimal solution $(\widehat{\boldsymbol{v}}_o, \widehat{\boldsymbol{u}}_o)$ for (LP_e). The equality (3.30) is the complementarity condition between primal-dual optimal solutions. The inequality (3.31) is a part of the constraints of (LP_e). □

Theorem 3.3 *The DMUs in* E_o *as defined in (3.16) are CCR-efficient.*

Proof. As described in Lemma 3.1, there exists a positive multiplier $(\widehat{\boldsymbol{v}}_o, \widehat{\boldsymbol{u}}_o)$. These vectors also satisfy

$$\widehat{\boldsymbol{v}}_o \boldsymbol{x}_j = \widehat{\boldsymbol{u}}_o \boldsymbol{y}_j \quad (j \in E_o) \tag{3.32}$$

$$\widehat{\boldsymbol{v}}_o X \geq \widehat{\boldsymbol{u}}_o Y \tag{3.33}$$

For each $j \in E_o$, we can adjust $(\hat{\boldsymbol{v}}_o, \hat{\boldsymbol{u}}_o)$ by using a scalar multiplier so that the relation $\hat{\boldsymbol{v}}_o \boldsymbol{x}_j = \hat{\boldsymbol{u}}_o \boldsymbol{y}_j = 1$ holds, while keeping (3.33) satisfied. Thus, activity $(\boldsymbol{x}_j, \boldsymbol{y}_j)$ is CCR-efficient by Definition 2.1. (A proof via the envelopment model may be found in Problem 7.5 of Chapter 7, and this proof comprehends the BCC as well as the CCR model). □

Theorem 3.4 *Any semipositive combination of DMUs in E_o is CCR-efficient.*

Proof. Let the combined activity be

$$\boldsymbol{x}_c = \sum_{j \in E_o} c_j \boldsymbol{x}_j \text{ and } \boldsymbol{y}_c = \sum_{j \in E_o} c_j \boldsymbol{y}_j \text{ with } c_j \geq 0 \quad (j \in E_o). \tag{3.34}$$

The multiplier $(\hat{\boldsymbol{v}}_o, \hat{\boldsymbol{u}}_o)$ in Lemma 3.1 satisfies

$$\hat{\boldsymbol{v}}_o \boldsymbol{x}_c = \hat{\boldsymbol{u}}_o \boldsymbol{y}_c \tag{3.35}$$

$$\hat{\boldsymbol{v}}_o X \geq \hat{\boldsymbol{u}}_o Y \tag{3.36}$$

$$\hat{\boldsymbol{v}}_o > \boldsymbol{0}, \ \hat{\boldsymbol{u}}_o > \boldsymbol{0} \tag{3.37}$$

Thus, $(\boldsymbol{x}_c, \boldsymbol{y}_c)$ is CCR-efficient by Definition 2.1. □

3.6 COMPUTATIONAL ASPECTS OF THE CCR MODEL

In this section, we discuss computational procedures for solving linear programs for the CCR model in detail. Readers who are not familiar with the terminology of linear programs and are not interested in the computational aspects of DEA can skip this section.

3.6.1 Computational Procedure for the CCR Model

As described in Section 3.3, the computational scheme of the CCR model for DMU_o results in the following two stage LP problem.

(DLP_o)

$$\begin{array}{llrcl} \text{Phase I objective} & \min & & & \theta & (3.38)\\ \text{Phase II objective} & \min & & & -\boldsymbol{e}\boldsymbol{s}^- - \boldsymbol{e}\boldsymbol{s}^+ & (3.39)\\ & \text{subject to} & \theta \boldsymbol{x}_o & = & X\boldsymbol{\lambda} + \boldsymbol{s}^- & (3.40)\\ & & \boldsymbol{y}_o & = & Y\boldsymbol{\lambda} - \boldsymbol{s}^+ & (3.41)\\ & \theta \geq 0, \ \boldsymbol{\lambda} \geq \boldsymbol{0}, & \boldsymbol{s}^- & \geq & \boldsymbol{0}, \ \boldsymbol{s}^+ \geq \boldsymbol{0}, & (3.42) \end{array}$$

where Phase II replaces the variable θ with a fixed value of min $\theta = \theta^*$.

Using the usual LP notation, we now rewrite (DLP_o) as follows,

(DLP'_o)

$$\begin{array}{llrcl} \text{Phase I objective} & \min & z_1 & = & cx & (3.43)\\ \text{Phase II objective} & \min & z_2 & = & dx & (3.44)\\ & \text{subject to} & Ax & = & b & (3.45)\\ & & x & \geq & 0, & (3.46) \end{array}$$

where $\boldsymbol{c}$ and $\boldsymbol{d}$ are row vectors.

The correspondences between (DLP_o) and (DLP'_o) are:

$$\boldsymbol{x} = (\theta,\ \boldsymbol{\lambda}^T,\ \boldsymbol{s}^{-T},\ \boldsymbol{s}^{+T})^T \tag{3.47}$$

$$\boldsymbol{c} = (1,\ \boldsymbol{0},\ \boldsymbol{0},\ \boldsymbol{0}) \tag{3.48}$$

$$\boldsymbol{d} = (0,\ \boldsymbol{0},\ -\boldsymbol{e},\ -\boldsymbol{e}) \tag{3.49}$$

$$A = \begin{pmatrix} \boldsymbol{x}_o & -X & -I & O \\ \boldsymbol{0} & Y & O & -I \end{pmatrix} \tag{3.50}$$

$$\boldsymbol{b} = \begin{pmatrix} \boldsymbol{0} \\ \boldsymbol{y}_o \end{pmatrix}, \tag{3.51}$$

where $\boldsymbol{e}$ is the vector with all elements unity. See the discussion immediately following (3.13).

Phase I

First, we solve the LP problem with the Phase I objective. Letting an optimal basis be B we use this matrix B as follows. First we compute several values, as noted below, where R is the nonbasic part of the matrix A and the superscript B (or R) shows the columns of A corresponding to B (or R).

$$\text{basic solution} \quad \boldsymbol{x}^B = \bar{\boldsymbol{b}} = B^{-1}\boldsymbol{b} \tag{3.52}$$

$$\text{simplex multiplier} \quad \boldsymbol{\pi} = \boldsymbol{c}^B B^{-1} \tag{3.53}$$

$$\text{Phase I simplex criterion} \quad \boldsymbol{p}^R = \boldsymbol{\pi}R - \boldsymbol{c}^R \tag{3.54}$$

$$\text{Phase II simplex criterion} \quad \boldsymbol{q}^R = \boldsymbol{\pi}R - \boldsymbol{d}^R, \tag{3.55}$$

where B^{-1} is the inverse of B, $\boldsymbol{\pi}$ is a vector of "multipliers" derived (from the data) as in (3.53), and $\boldsymbol{p}^R$ and $\boldsymbol{q}^R$ are referred to as "reduced costs."

From optimality of the basis B, we have:

$$\bar{\boldsymbol{b}} \geq \boldsymbol{0}, \tag{3.56}$$

because the conditions for non-negativity are satisfied by $\boldsymbol{x}^B$ in (3.52) and

$$\boldsymbol{p}^R \leq \boldsymbol{0}, \tag{3.57}$$

as required for optimality in (3.54) and, for the columns of the basis B, we have: $\boldsymbol{p}^B = \boldsymbol{\pi}B - \boldsymbol{c}^B = \boldsymbol{0}$.

Phase II

We exclude the columns with $p_j < 0$ in the optimal simplex tableau at the end of Phase I from the tableaux for further consideration. The remainder is called the *restricted problem* or *restricted tableau* and dealt with in the next computational step. At Phase II, we solve the LP problem with the second objective. If the simplex criterion of Phase II objective satisfies $\boldsymbol{q}^R \leq \boldsymbol{0}$, we then halt the iterations. The basic solution thus obtained is a max-slack solution. This is the procedure for finding θ^* and a max-slack solution $(\boldsymbol{s}^{-*},\ \boldsymbol{s}^{+*})$ for DMU$_o$. As has been already pointed out, (LP_o) in (3.2)-(3.5) is dual to (DLP_o) in (3.6)-(3.9), and we will discuss its solution procedure in the next section.

3.6.2 Data Envelopment Analysis and the Data

The characterizations given by (3.52)-(3.55) represent borrowings from the terminology of linear programming in which the components of π are referred to as "simplex multipliers" because they are associated with a use of the simplex method which generates such multipliers in the course of solving LP problems such as (DLP_o'). "Simplex" and "dual simplex" methods are also used in DEA. We therefore refer to the problem (LP_o) as being in "multiplier form." See (3.2)-(3.5). Problem (DLP_o) is then said to be in "envelopment form." This is the source of the name "Data Envelopment Analysis."

Reference to Figure 3.2 in the next section will help to justify this usage by noting, first, that all data are inside the frontier stretching from F through R. Also, at least one observation is touching the frontier. Hence, in the manner of an envelope all data are said to be "enveloped" by the frontier which achieves its importance because it is used to analyze and evaluate the performances of DMUs associated with observations like those portrayed in Figure 3.2. More compactly, we say that DEA is used to evaluate the performances of each observation relative to the frontier that envelops all of the observations.

3.6.3 Determination of Weights (=Multipliers)

The simplex multipliers in π at the end of Phase I are associated with an optimal solution of its dual problem (LP_o)[8] as given in (3.2)-(3.5). In fact, π is an $(m+s)$ vector in which the first m components assign optimal weights $\boldsymbol{v}^*$ to the inputs and the remaining s components assign optimal weights $\boldsymbol{u}^*$ to the outputs. By observing the dual problem of (DLP_o') in (3.38)-(3.42), it can be shown that $\boldsymbol{v}^*$ and $\boldsymbol{u}^*$ satisfy (3.3) through (3.5). The symbol I, for the identity matrix in A of (3.50), relates to the (input and output) slack. Let the simplex criteria to be applied to these columns be represented by "pricing vectors" $\boldsymbol{p}^{s^-}$ and $\boldsymbol{p}^{s^+}$. These vectors relate to $\boldsymbol{v}^*$ and $\boldsymbol{u}^*$ in the following way:

$$\boldsymbol{v}^* = -\boldsymbol{p}^{s^-} \quad (\geq \boldsymbol{0}) \tag{3.58}$$

$$\boldsymbol{u}^* = -\boldsymbol{p}^{s^+} \quad (\geq \boldsymbol{0}). \tag{3.59}$$

3.6.4 Reasons for Solving the CCR Model Using the Envelopment Form

It is not advisable to solve (LP_o) directly. The reasons are:

(1) The computational effort of LP is apt to grow in proportion to powers of the number of constraints. Usually in DEA, n, the number of DMUs is considerably larger than $(m+s)$, the number of inputs and outputs and hence it takes more time to solve (LP_o) which has n constraints than to solve (DLP_o) which has $(m+s)$ constraints. In addition, since the memory size needed for keeping the basis (or its inverse) is the square of the number of constraints, (DLP_o) is better fitted for memory saving purposes.

(2) We cannot find the pertinent max-slack solution by solving (LP_o).

(3) The interpretations of (DLP_o) are more straightforward because the solutions are characterized as inputs and outputs that correspond to the original

data whereas the multipliers provided by solutions to (LP_o) represent evaluations of these observed values. These values are also important, of course, but they are generally best reserved for supplementary analyses after a solution to (DLP_o) is achieved.

3.7 EXAMPLE

We will apply (DLP_o) to a sample problem and comment on the results. For this purpose we use Example 3.1 as shown in Table 3.2, which is Example 2.2 with an added activity G. All computations were done using the DEA-Solver which comes with this book and the results are stored in the Excel 97 Workbook "Sample-CCR-I.xls" in the sample file. So, readers interested in using DEA-Solver should try to trace the results by opening this file or running DEA-Solver for this problem.

Table 3.2. Example 3.1

	DMU	A	B	C	D	E	F	G
Input	x_1	4	7	8	4	2	10	3
	x_2	3	3	1	2	4	1	7
Output	y	1	1	1	1	1	1	1

(1) (DLP_A) for A is:

$$\begin{aligned}
&\text{Phase I} \quad \min\ \theta \\
&\text{Phase II} \quad \min\ -s_1^- - s_2^- - s^+ \\
&\text{subject to} \\
&4\theta - 4\lambda_A - 7\lambda_B - 8\lambda_C - 4\lambda_D - 2\lambda_E - 10\lambda_F - 3\lambda_G - s_1^- = 0 \\
&3\theta - 3\lambda_A - 3\lambda_B - \lambda_C - 2\lambda_D - 4\lambda_E - \lambda_F - 7\lambda_G - s_2^- = 0 \\
&\lambda_A + \lambda_B + \lambda_C + \lambda_D + \lambda_E + \lambda_F + \lambda_G - s^+ = 1
\end{aligned}$$

where the variables are restricted to nonnegative values in the vectors $\boldsymbol{\lambda}$, $\boldsymbol{s}^-$ and $\boldsymbol{s}^+$.
The optimal solution for (DLP_A) is:

$\theta^* = 0.8571$ (in Worksheet "Sample-CCR-I.Score")
$\lambda_D^* = 0.7143$, $\lambda_E^* = 0.2857$, other $\lambda_j^* = 0$ (in Worksheet "Sample-CCR-I.Score")
$s_1^{-*} = s_2^{-*} = s^{+*} = 0$ (in Worksheet "Sample-CCR-I.Slack").

Since $\lambda_D^* > 0$ and $\lambda_E^* > 0$, the reference set for A is

$$E_A = \{D,\ E\}.$$

And $\lambda_D^* = 0.7143$, $\lambda_E^* = 0.2857$ show the proportions contributed by D and E to the point used to evaluate A. Hence A is technically inefficient. No mix inefficiencies are present because all slacks are zero. Thus removal of all inefficiencies is achieved by reducing all inputs by 0.1429 or, approximately, 15% of their observed values. In fact, based on this reference set and $\boldsymbol{\lambda}^*$, we can express the input and output values needed to bring A into efficient status as

$$0.8571 \times (\text{Input of } A) = 0.7143 \times (\text{Input of } D) + 0.2857 \times (\text{Input of } E)$$
$$(\text{Output of } A) = 0.7143 \times (\text{Output of } D) + 0.2857 \times (\text{Output of } E).$$

From the magnitude of coefficients on the right hand side, A has more similarity to D than E. A can be made efficient either by using these coefficients, $\lambda_D^* = 0.7143$, $\lambda_E^* = 0.2857$ or by reducing both of its inputs—viz., by reducing the input value radially in the ratio 0.8571. It is the latter (radial contraction) that is used in DEA-Solver. Thus, as seen in Worksheet "Sample-CCR-I.Projection," the CCR-projection of (3.32) and (3.33) is achieved by,

$$\widehat{x}_1 \leftarrow \theta^* x_1 = 0.8571 \times 4 = 3.4286 \quad (14.29\% \text{ reduction})$$
$$\widehat{x}_2 \leftarrow \theta^* x_2 = 0.8571 \times 3 = 2.5714 \quad (14.29\% \text{ reduction})$$
$$\widehat{y} \leftarrow y = 1 \quad (\text{no change}).$$

The optimal solution for the multiplier problem (LP_A) can be found in Worksheet "Sample-CCR-I.Weight" as follows,

$$v_1^* = 0.1429, \quad v_2^* = 0.1429, \quad u^* = 0.8571.$$

This solution satisfies constraints (3.3)-(3.5) and maximizes the objective in (3.2), i.e., $u^*y = 0.8571 \times 1 = 0.8571 = \theta^*$ in the optimal objective value of (DLP_A). (See Problem 3.4 for managerial roles of these optimal multipliers (weights) for improving A.)

The Worksheet "Sample-CCR-I.WeightedData" includes optimal weighted inputs and output, i.e.,

$$v_1^* x_1 = 0.1429 \times 4 = 0.5714$$
$$v_2^* x_2 = 0.1429 \times 3 = 0.4286$$
$$u^* y = 0.8571 \times 1 = 0.8571.$$

The sum of the first two terms is 1 which corresponds to the constraint (3.3). The last term is the optimal objective value in this single output case.

(2) Moving from DMU A to DMU B, the optimal solution for B is:

$$\theta^* = 0.6316$$
$$\lambda_A^* = \lambda_B^* = 0, \ \lambda_C^* = 0.1053, \ \lambda_D^* = 0.8947, \ \lambda_E^* = \lambda_F^* = \lambda_G^* = 0$$
$$s_1^{-*} = s_2^{-*} = s^{+*} = 0$$
$$v_1^* = 0.0526, \ v_2^* = 0.2105, \ u^* = 0.6316.$$

Since $\lambda_C^* > 0,\ \lambda_D^* > 0$, the reference set for B is:

$$E_B = \{C,\ D\}.$$

B can be expressed as:

$$\begin{aligned}0.6316 \times (\text{Input of } B) &= 0.1053 \times (\text{Input of } C) + 0.8947 \times (\text{Input of } D)\\ (\text{Output of } B) &= 0.1053 \times (\text{Output of } C) + 0.8947 \times (\text{Output of } D).\end{aligned}$$

That is, it can be expressed in ratio form, as shown on the left or as a nonnegative combination of $\lambda_j^* > 0$ values as shown on the right. Using the expression on the left, the CCR-projection for B is

$$\begin{aligned}&\widehat{x}_1 \leftarrow \theta^* x_1 = 0.6316 \times 7 = 4.4211 \quad (36.84\%\ \text{reduction})\\ &\widehat{x}_2 \leftarrow \theta^* x_2 = 0.6316 \times 3 = 1.8974 \quad (36.84\%\ \text{reduction})\\ &\widehat{y} \leftarrow y = 1 \quad (\text{no change}).\end{aligned}$$

(3) $C,\ D$ and E

These 3 DMUs are found to be efficient. (See Worksheet "Sample-CCR-I.Score.")

(4) The optimal solution of the LP problem for F is:

$$\begin{aligned}&\theta^* = 1\\ &\lambda_A^* = \lambda_B^* = 0,\ \lambda_C^* = 1,\ \lambda_D^* = \lambda_E^* = \lambda_F^* = \lambda_G^* = 0\\ &s_1^{-*} = 2,\ s_2^{-*} = s^{+*} = 0\\ &v_1^* = 0,\ v_2^* = 1,\ u^* = 1\end{aligned}$$

where, again, the last set of values refer to the multiplier problem. For the envelopment model we have $\lambda_C^* > 0$ as the only positive value of λ. Hence the reference set for F is:

$$E_F = \{C\}.$$

Considering the excess in Input 1 ($s_1^{-*} = 2$), F can be expressed as:

$$\begin{aligned}(\text{Input 1 of } F) &= (\text{Input 1 of } C) + 2\\ (\text{Input 2 of } F) &= (\text{Input 2 of } C)\\ (\text{Output of } F) &= (\text{Output of } C).\end{aligned}$$

Although F is "radial-efficient," it is nevertheless "CCR-inefficient" due to this excess (mix inefficiency) associated with $s_1^{-*} = 2$. Thus the performance of F can be improved by subtracting 2 units from Input 1. This can be accomplished by subtracting 2 units from input 1 on the left and setting $s_1^{-*} = 0$ on the right without worsening any other input and output. Hence condition (*ii*) in Definition 3.2 is not satisfied until this is done, so F did *not* achieve Pareto-Koopmans efficiency in its performance. See Definition 3.3.

(5) The optimal solution of the LP problem for G is:

$$\begin{aligned}
&\theta^* = 0.6667\\
&\lambda_A^* = \lambda_B^* = \lambda_C^* = \lambda_D^* = 0,\ \lambda_E^* = 1,\ \lambda_F^* = \lambda_G^* = 0\\
&s_1^{-*} = 0,\ s_2^{-*} = 0.6667,\ s^{+*} = 0\\
&v_1^* = 0.3333,\ v_2^* = 0,\ u^* = 0.6667
\end{aligned}$$

Since $\lambda_E^* > 0$, the reference set for G is:

$$E_G = \{E\}.$$

Considering the excess in Input 2 ($s_2^{-*} = 0.6667$), G can be expressed by:

$$\begin{aligned}
0.6667 \times (\text{Input 1 of } G) &= (\text{Input 1 of } E)\\
0.6667 \times (\text{Input 2 of } G) &= (\text{Input 2 of } E) + 0.6667\\
(\text{Output of } G) &= (\text{Output of } E).
\end{aligned}$$

One plan for the improvement of G is to reduce all input values by multiplying them by 0.6667 and further subtracting 0.6667 from Input 2. When this is done the thus altered values coincide with the coordinates of E. Geometrically then, the CCR-projection for G is

$$\begin{aligned}
&\widehat{x}_1 \leftarrow \theta^* x_1 - s_1^{-*} = 0.6667 \times 3 - 0 = 2\ \ (33.33\%\ \text{reduction})\\
&\widehat{x}_2 \leftarrow \theta^* x_2 - s_2^{-*} = 0.6667 \times 7 - 0.6667 = 4\ \ (42.86\%\ \text{reduction})\\
&\widehat{y} \leftarrow y = 1\ \ (\text{no change}),
\end{aligned}$$

where $\widehat{x}_1 = 2,\ \widehat{x}_2 = 4,\ \widehat{y} = 1$ which values are the same as for E in Table 3.2.

The above observations are illustrated by Figure 3.2, which depicts Input 1 and Input 2 values of all DMUs. Since the output value is 1 for all the DMUs, we can compare their efficiencies via the input values.

The efficient frontier consists of the bold line CDE and the production possibility set is the region enclosed by this efficient frontier line plus the vertical line going up from E and the horizontal line extending to the right from C. Let the intersection point of OA and DE be Q. The activity Q has input proportional to that of A (4, 3) and is the least input value point on OA in the production possibility set and

$$\frac{OQ}{OA} = 0.8571$$

corresponds to radial (or ratio) efficiency of A. [9] Also, we have:

$$\begin{aligned}
\text{Input 1 of } Q &= 0.8571 \times 4\ (\text{Input 1 of } A) = 3.428 = \widehat{x}_1\\
\text{Input 2 of } Q &= 0.8571 \times 3\ (\text{Input 2 of } A) = 2.571 = \widehat{x}_2.
\end{aligned}$$

However, Q is the point that divides D and E in the ratio 0.7143 to 0.2857.

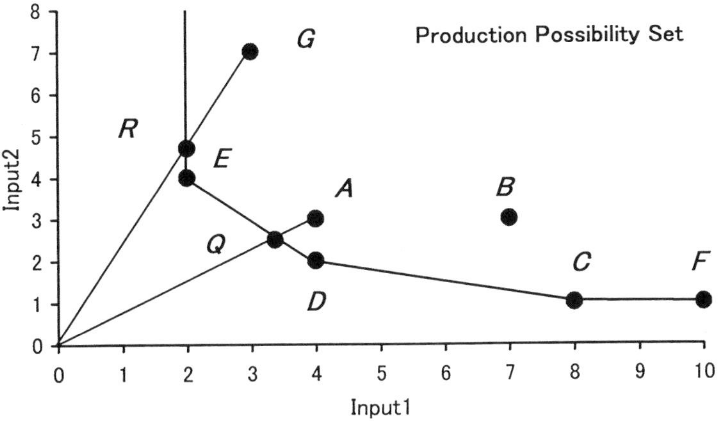

Figure 3.2. Example 3.1

Hence its values are calculated again as:

$$\begin{aligned}\text{Input 1 of } Q &= 0.7143 \times 4 \text{ (Input 1 of } D) + 0.2857 \times 2 \text{ (Input 1 of } E) = 3.428\\ \text{Input 2 of } Q &= 0.7143 \times 2 \text{ (Input 2 of } D) + 0.2857 \times 4 \text{ (Input 2 of } E) = 2.571,\end{aligned}$$

where $\lambda_D^* = 0.7143$ and $\lambda_E^* = 0.2857$. Comparing these results we see that the coordinates of Q, the DMU used to evaluate A, can be derived in either of these two ways. There are also still more ways of effecting such improvements, as we will later see when changes in mix proportions are permitted.

This brings us to the role of nonzero slacks which we can illustrate with F and G. From Figure 3.2, it is evident that if we reduce Input 1 of F by the nonzero slack value of 2, then F coincides with C and is efficient. The presence of this nonzero slack means that F is not Pareto-Koopmans efficient even though its radial value is $\theta^* = 1$.

As a further illustration of the two conditions for efficiency in Definition 3.2 we turn to the evaluation of G. The CCR-efficiency of G is calculated by $OR/OG = 0.6667$. Thus G is not radially (or weakly) efficient as evaluated at R. However R is also not efficient. We can make it efficient by further reducing Input 2 by 0.6667 and shifting to E as we have seen in the CCR-projection of G. Hence G fails both of the conditions specified in Definition 3.2.

Table 3.3 summarizes the results obtained by applying DEA-Solver to all of the data in Table 3.2. Only C, D and E are fully efficient. A and B fail because $\theta^* < 1$. Their intersection with the frontier gives their radial (weak) inefficiency score with zero slacks because they intersect an efficient portion of the frontier radially. F has a value of $\theta^* = 1$ because it is on the frontier. This portion of frontier is not efficient, however, as evidenced by the nonzero slack for F under s_1^- in Table 3.3. Finally, G fails to be efficient both because $\theta^* < 1$ *and* nonzero slacks are involved in its evaluation by E, a point on the efficient portion of the frontier.

Table 3.3. Results of Example 3.1

DMU	CCR-Eff θ^*	Ref Set	Excess s_1^-	s_2^-	Shortfall s^+
A	0.8571	D E	0	0	0
B	0.6316	C D	0	0	0
C	1.0000	C	0	0	0
D	1.0000	D	0	0	0
E	1.0000	E	0	0	0
F	1.0000	C	2	0	0
G	0.6667	E	0	.6667	0

3.8 THE OUTPUT-ORIENTED MODEL

Up to this point, we have been dealing mainly with a model whose objective is to minimize inputs while producing at least the given output levels. This type of model is called *input-oriented.* There is another type of model that attempts to maximize outputs while using no more than the observed amount of any input. This is referred to as the *output-oriented* model, formulated as:

$$(DLPO_o) \quad \max_{\eta, \boldsymbol{\mu}} \quad \eta \tag{3.60}$$

$$\text{subject to} \quad \boldsymbol{x}_o - X\boldsymbol{\mu} \geq \mathbf{0} \tag{3.61}$$

$$\eta \boldsymbol{y}_o - Y\boldsymbol{\mu} \leq \mathbf{0} \tag{3.62}$$

$$\boldsymbol{\mu} \geq \mathbf{0}. \tag{3.63}$$

An optimal solution of $(DLPO_o)$ can be derived directly from an optimal solution of the input-oriented CCR model given in (3.6)-(3.9) as follows. Let us define

$$\boldsymbol{\lambda} = \boldsymbol{\mu}/\eta, \quad \theta = 1/\eta. \tag{3.64}$$

Then $(DLPO_o)$ becomes

$$(DLP_o) \quad \min_{\theta, \boldsymbol{\lambda}} \quad \theta$$

$$\text{subject to} \quad \theta \boldsymbol{x}_o - X\boldsymbol{\lambda} \geq \mathbf{0}$$

$$\boldsymbol{y}_o - Y\boldsymbol{\lambda} \leq \mathbf{0}$$

$$\boldsymbol{\lambda} \geq \mathbf{0},$$

which is the input-oriented CCR model. Thus, an optimal solution of the output-oriented model relates to that of the input-oriented model via:

$$\eta^* = 1/\theta^*, \quad \boldsymbol{\mu}^* = \boldsymbol{\lambda}^*/\theta^*. \tag{3.65}$$

The slack $(\boldsymbol{t}^-,\ \boldsymbol{t}^+)$ of the output-oriented model is defined by:

$$X\boldsymbol{\mu} + \boldsymbol{t}^- = \boldsymbol{x}_o$$
$$Y\boldsymbol{\mu} - \boldsymbol{t}^+ = \eta \boldsymbol{y}_o.$$

These values are also related to the input-oriented model via

$$\boldsymbol{t}^{-*} = \boldsymbol{s}^{-*}/\theta^*, \quad \boldsymbol{t}^{+*} = \boldsymbol{s}^{+*}/\theta^*. \tag{3.66}$$

Now, $\theta^* \leq 1$, so returning to (3.64), η^* satisfies

$$\eta^* \geq 1. \tag{3.67}$$

The higher the value of η^*, the less efficient the DMU. θ^* expresses the input reduction rate, while η^* describes the output enlargement rate. From the above relations, we can conclude that an input-oriented CCR model will be efficient for any DMU if and only if it is also efficient when the output-oriented CCR model is used to evaluate its performance.

The dual problem of $(DLPO_o)$ is expressed in the following model, with components of the vectors $\boldsymbol{p}$ and $\boldsymbol{q}$ serving as variables.

$$(LPO_o) \quad \min_{\boldsymbol{p},\boldsymbol{q}} \quad \boldsymbol{p}\boldsymbol{x}_o \tag{3.68}$$
$$\text{subject to} \quad \boldsymbol{q}\boldsymbol{y}_o = 1 \tag{3.69}$$
$$-\boldsymbol{p}X + \boldsymbol{q}Y \leq \mathbf{0} \tag{3.70}$$
$$\boldsymbol{p} \geq \mathbf{0}, \quad \boldsymbol{q} \geq \mathbf{0}. \tag{3.71}$$

On the multiplier side we have:

Theorem 3.5 *Let an optimal solution of (LP_o) be $(\boldsymbol{v}^*,\ \boldsymbol{u}^*)$, then an optimal solution of the output-oriented model (LPO_o) is obtained from*

$$\boldsymbol{p}^* = \boldsymbol{v}^*/\theta^*, \qquad \boldsymbol{q}^* = \boldsymbol{u}^*/\theta^*. \tag{3.72}$$

Proof. It is clear that $(\boldsymbol{p}^*,\ \boldsymbol{q}^*)$ is feasible for (LPO_o). Its optimality comes from the equation below.

$$\boldsymbol{p}^*\boldsymbol{x}_o = \boldsymbol{v}^*\boldsymbol{x}_o/\theta^* = \eta^*. \tag{3.73}$$

□

Thus, the solution of the output-oriented CCR model may be obtained from that of the input oriented CCR model. The improvement using this model is expressed by:

$$\widehat{\boldsymbol{x}}_o \Leftarrow \boldsymbol{x}_o - \boldsymbol{t}^{-*} \tag{3.74}$$
$$\widehat{\boldsymbol{y}}_o \Leftarrow \eta^*\boldsymbol{y}_o + \boldsymbol{t}^{+*}. \tag{3.75}$$

Carrying this a stage further we note that (LPO_o) is equivalent to the following fractional programming problem:

$$\min_{\boldsymbol{\pi},\boldsymbol{\rho}} \quad \frac{\boldsymbol{\pi}\boldsymbol{x}_o}{\boldsymbol{\rho}\boldsymbol{y}_o} \tag{3.76}$$

$$\text{subject to} \quad \frac{\boldsymbol{\pi}\boldsymbol{x}_j}{\boldsymbol{\rho}\boldsymbol{y}_j} \geq 1 \quad (j = 1, \ldots, n) \tag{3.77}$$

$$\boldsymbol{\pi} \geq \mathbf{0}, \quad \boldsymbol{\rho} \geq \mathbf{0}. \tag{3.78}$$

That is, we exchanged the numerator and the denominator of (2.3) and (2.4) as given in Chapter 2 and minimized the objective function. It is therefore quite natural that the solutions are found to be linked by a simple rule. This mathematical transformation does not imply that there is no managerial significance to be assigned to the choice of models since, *inter alia*, different corrections may be associated with output maximization and input minimization. The difference can be substantial so this choice always deserves consideration. Furthermore, later in this book, we shall also introduce other models, where outputs are maximized and inputs are simultaneously minimized so that still further choices may need to be considered.

3.9 DISCRETIONARY AND NON-DISCRETIONARY INPUTS

Up to this point we have assumed that all inputs and outputs can be varied at the discretion of management or other users. These may be called "discretionary variables." "Non-discretionary variables," not subject to management control, may also need to be considered. In evaluating performances of different bases for the Fighter Command of U.S. Air Forces, for instance, it was necessary to consider weather as an input since the number of "sorties" (successfully completed missions) and "aborts" (non-completed mission)[10] treated as outputs, could be affected by the weather (measured in degree days and numbers of "flyable" days) at different bases.

Even though Non-Discretionary, it is important to take account of such inputs in a manner that is reflected in the measures of efficiency used. We follow the route provided by Banker and Morey (1986)[11] who refer to such variables as "exogenously fixed," in forms like "age of store," in their use of DEA to evaluate the performances of 60 DMUs in a network of fast-food restaurants. Reverting to algebraic notation, we can represent their formulation by the following modification of the CCR model.

$$\min \quad \theta - \varepsilon \left(\sum_{i \in D} s_i^- + \sum_{r=1}^{s} s_r^+ \right) \tag{3.79}$$

$$\text{subject to} \quad \theta x_{io} = \sum_{j=1}^{n} x_{ij}\lambda_j + s_i^-, \quad i \in D \tag{3.80}$$

$$x_{io} = \sum_{j=1}^{n} x_{ij}\lambda_j + s_i^-, \quad i \in ND \tag{3.81}$$

$$y_{ro} = \sum_{j=1}^{n} y_{rj}\lambda_j - s_r^+, \quad r = 1, \ldots, s. \tag{3.82}$$

where all variables (except θ) are constrained to be nonnegative.

Here the symbols $i \in D$ and $i \in ND$ refer to the sets of "Discretionary" and "Non-Discretionary" inputs, respectively. To be noted in the constraints is the fact that the variable θ is not applied to the latter inputs because these values are exogenously fixed and it is therefore not possible to vary them at the discretion of management. This is recognized by entering all x_{io}, $i \in ND$ at their fixed (observed) value. Turning to the objective (3.79) we utilize the symbol[12] $\varepsilon > 0$ to mean that the slack variables (shown in the parenthses) are to be handled at a second stage where, as previously described, they are to be maximized in a manner that does not disturb the previously determined first-stage minimization of θ to achieve $\theta = \theta^*$. Finally, we note that the slacks $s_i^-, i \in ND$ are omitted from the objective. Hence these Non-Discretionary inputs do not enter directly into the efficiency measures being optimized in (3.79). They can, nevertheless, affect the efficiency evaluations by virtue of their presence in the constraints.

We can further clarify the way these Non-Discretionary variables affect the efficiency scores by writing the dual of (3.79)-(3.82) in the form of the following (modified) multiplier model,

$$\max \quad \sum_{r=1}^{s} u_r y_{ro} - \sum_{i \in ND} v_i x_{io} \tag{3.83}$$

$$\text{subject to} \quad \sum_{r=1}^{s} u_r y_{rj} - \sum_{i \in ND} v_i x_{ij} - \sum_{i \in D} v_i x_{ij} \leq 0, \ j = 1, \ldots, n \tag{3.84}$$

$$\sum_{i \in D} v_i x_{io} = 1 \tag{3.85}$$

$$v_i \geq \varepsilon, \ i \in D \tag{3.86}$$

$$v_i \geq 0, \ i \in ND \tag{3.87}$$

$$u_r \geq \varepsilon, \ r = 1, \ldots, s. \tag{3.88}$$

As can be seen, the Non-Discretionary but not the Discretionary inputs, enter into the objective (3.83). The multiplier values associated with these Non-Discretionary inputs may be zero, as in (3.87), but the other variables must always be positive, as in (3.86) and (3.88). The interpretations we now provide flow from the "complementary slackness principle" of linear programming. If (3.81) is satisfied strictly at an optimum then $v_i^* = 0$ is associated with this constraint and this x_{io} does not affect the evaluation recorded in (3.79). On the other hand, if $v_i^* > 0$ for any $i \in ND$ then the efficiency score recorded in (3.79) is reduced by the multiplier of x_{io} for this DMU_o.

This follows from the dual theorem of linear programming—viz.,

$$\theta^* - \varepsilon\left(\sum_{i\in D} s_i^{-*} + \sum_{r=1}^{s} s_r^{+*}\right) = \sum_{r=1}^{s} u_r^* y_{ro} - \sum_{i\in ND} v_i^* x_{io}. \tag{3.89}$$

Via this same relation we see that a *decrease* in this same x_{io} will *increase* the efficiency score recorded in the expression or the left of the equality (3.89).

For managerial use, the sense of this mathematical characterization may be interpreted in the following manner. The output achieved, as recorded in the y_{ro}, deserve a higher efficiency rating when they have been achieved under a relatively tighter constraint and a lower efficiency score when this constraint is loosened by increasing this x_{io}.

This treatment of Non-Discretionary variables must be qualified, at least to some extent, since, *inter alia,* allowance must be made for ranges over which the values of v_i^* and u_r^* remain unaltered. Modifications to allow for $v_i^* \geq \varepsilon \geq 0$ must also be introduced, as described in Problem 3.2 at the end of this chapter and effects like those treated in Problem 3.3 need to recognized when achievement of the efficient frontier is a consideration.

After additional backgrounds have been supplied, Chapter 7, later in this book, introduces a variety of extensions and alternate ways of treating Non-Discretionary outputs as well as inputs. Here we only need to note that modifications of a relatively straightforward variety may also be made as in the following example.

A study of 638 public secondary schools in Texas was undertaken by a consortium of 3 universities in collaboration with the Texas Education Agency (TEA). The study was intended to try to develop improved methods for accountability and evaluation of school performances. In addition to discretionary inputs like teacher salaries, special instruction, etc., the following Non-Discretionary inputs had to be taken into account,

1. **Minority** : Number of minority students, expressed as a percentage of total student enrollment.

2. **Disadvantage** : Number of economically disadvantaged students, expressed as a percentage of total student enrollment.

3. **LEP** : Number of Limited English Proficiency students, expressed as a percentage of total student enrollment.

These inputs differ from the ones that Banker and Morey had in mind. For instance, a regression calculation made at an earlier stage of the consortium study yielded *negative* coefficients that tested statistically significant for every one of these 3 inputs in terms of their effects on academic test scores. In the subsequent DEA study it was therefore deemed desirable to reverse the sign associated with the x_{io} in the expression on the right of (3.89)

This was accomplished by reversing (3.81)

$$\text{from } \sum_{j=1}^{n} x_{ij}\lambda_j \leq x_{io} \text{ to } \sum_{j=1}^{n} x_{ij}\lambda_j \geq x_{io} \tag{3.90}$$

for each of these $i \in ND$. In this manner an ability to process more of these inputs, when critical, was recognized in the form of higher (rather than lower) efficiency scores. Full details are supplied in I.R. Bardhan (1995)[13] and summarized in Arnold *et al.* (1997).[14] Here we note that the results were sufficiently satisfactory to lead to a recommendation to accord recognition to "efficiency" as well as "effectiveness" in state budgetary allocations to different school districts. Although not accepted, the recommendation merits consideration at least to the extent of identifying shortcomings in the usual statements (and reward structures) for academic performance. The following scheme can help to clarify what is being said,[15]

Effectiveness implies

- Ability to state desired goals
- Ability to achieve desired goals

Efficiency relates to

- Benefits realized
- Resources used

Consider, for instance, the State-mandated Excellence Standards for Texas recorded in Table 3.4. These represent statements of desired goals and schools are rewarded (or not rewarded) on the basis of their achievements. Nothing is

Table 3.4. State-mandated Excellence Standards on Student Outcomes

	Outcome Indicator	State-mandated Excellence Standard
1.	Texas Assessment of Academic Skills (TAAS) Test	90% of students passing on all standardized tests
2.	Attendance	97% of total enrollment in the school
3.	Dropout Rate	Less than or equal to 1% of total enrollment
4.	Graduation Rate	99% of graduating class
5.	College Admission Tests	• 35% of graduates scoring above the criterion score which is equal to 25 on the ACT[a] and 1000 on the SAT[b] • 70% of graduates taking either the ACT or the SAT

[a]ACT = American Collegiate Tests
[b]SAT = Scholastic Aptitude Tests

said about the amounts (or varieties) of resources used. Hence it should be no surprise that only 1 of the excellent-rated schools included in this study was found to be efficient. The other schools rated as excellent by the State of Texas had all expended excessive resources in achieving these goals. On the other hand many schools that failed to achieve excellence were nevertheless found to be efficient in producing desired outputs under very difficult conditions. Therefore some way of recognizing this kind of achievements is needed if "efficiency" as well as "effectiveness" is to be rewarded.[16]

Table 3.5 as taken from Bardhan (1994), contains portions of the printout from a DEA study involving one of these "excellent" schools. The row labelled Disadv (= Economically Disadvantaged) shows a slack value of 15 students. Judging from the performances of its peer group of efficient schools, this school

Table 3.5. Non-Discretionary Inputs

	Current Level	*Slack*	*Value if Efficient*
Minority	47.0	—	47.0
Disadv	14.0	15.0	29.0
LEP	3.5	—	3.5

should therefore have been able to more than double the number of such students that it processed *without affecting its efficiency score* because (a) the slack for $i \in ND$ is not in the objective of (3.79) and (b) the presence of nonzero slack for any $i \in ND$ means the associated $v_i^* = 0$. In addition, the other ND inputs are associated with positive multiplier values so this school would have been able to *increase* its efficiency score by *increasing* the Minority and LEP (Low English Proficiency = Mainly Hispanic) students it processed. As shown in Arnold *et al.* (1997) this can be done by introducing constraints to insure that no worsening of any already achieved excellence is avoided.

This is not the end of the line for what can be done. See Section 7.3 in Chapter 7, below, for further developments. Additional extensions could proceed to a two-stage operation in which school "outputs" are transformed into "outcomes" where the latter includes things like success in securing employment *after* school has been completed.[17] For instance, see C.A.K. Lovell *et al.* (1994)[18] who employ cohort data in such a two-stage analysis where "outputs" are converted to "outcomes" to find that the record for American Public Schools is substantially better for the latter than the former.

3.10 SUMMARY OF CHAPTER 3

In this chapter we described the CCR model in some detail in both its input-oriented and output-oriented versions.

1. We also relaxed assumptions of a positive data set to semipositivity.

2. We defined the production possibility set based on the constant returns-to-scale assumption and developed the CCR model under this assumption.

3. The dual problem of the original CCR model was introduced as (DLP_o) in (3.6)-(3.9) and the existence of input excesses and output shortfalls clarified by solving this model. (To avoid confusion and to align our terminology with the DEA literature, we referred to the dual as the "envelopment model" and the primal (LP_o) introduced in Chapter 2 as the "multiplier model.")

4. In Definition 3.2 we identified a DMU as *CCR-efficient* if and only if it is (i) radial-efficient and (ii) has zero-slack in the sense of Definition 3.1. Hence a DMU is CCR-efficient if and only if it has no input excesses and no output shortfalls.

5. Improvement of inefficient DMUs was discussed and formulae were given for effecting the improvements needed to achieve full CCR efficiency in the form of the CCR-projections given in (3.22) and (3.23).

6. Detailed computational procedures for the CCR model were presented in Section 3.6 and an optimal multiplier values $(\boldsymbol{v}, \boldsymbol{u})$ were obtained as the simplex multipliers for an optimal tableau obtained from the simplex method of linear programming.

3.11 NOTES AND SELECTED BIBLIOGRAPHY

As noted in Section 3.3 the term "Pareto-Koopmans" efficiency refers to Vilfredo Pareto and Tjalling Koopmans. The former, i.e., Pareto, was concerned with welfare economics which he visualized in terms of a vector-valued function with its components representing the utilities of all consumers. He then formulated the so-called Pareto condition of welfare maximization by noting that such a function could not be at a maximum if it was possible to increase one of its components without worsening other components of such a vector valued function. He therefore suggested this as a criterion for judging any proposed social policy: "the policy should be adopted if it made some individuals better off without decreasing the welfare (measured by their utilities) of other individuals." This, of course, is a necessary but not a sufficient condition for maximizing such a function. Proceeding further, however, would involve comparing utilities to determine whether a decrease in the utilities of some persons would be more than compensated by increases in the utilities of other individuals. Pareto, however, wanted to proceed as far as possible without requiring such interpersonal comparisons of utilities. See Vilfredo Pareto, *Manuel d'economie politique,* deuxieme edition, Appendix, pp. 617 ff., Alfred Bonnet, ed.(Paris: Marcel Giard, 1927).

Tjalling Koopmans adapted these concepts to production. In an approach that he referred to as "activity analysis," Koopmans altered the test of a vector optimum by reference to whether it was possible to increase any output without worsening some other output under conditions allowed by available resources such as labor, capital, raw materials, etc. See pp. 33-97 in T.C. Koopmans, ed., *Activity Analysis of Production and Allocation,* (New York: John Wiley & Sons, Inc., 1951).

The approaches by Pareto and Koopmans were entirely conceptual. No empirical applications were reported before the appearance of the 1957 article by M.J. Farrell in the *Journal of the Royal Statistical Society* under the title "The Measurement of Productive Efficiency." This article showed how these methods could be applied to data in order to arrive at relative efficiency evaluations. This, too, was in contrast with Koopmans and Pareto who conceptualized matters in terms of theoretically known efficient responses without much (if any) attention to how inefficiency, or more precisely, technical inefficiency, could be identified. Koopmans, for instance, assumed that producers would respond optimally to prices which he referred to as "efficiency prices." Pareto assumed that all consumers would (and could) maximize their utilities under the social policies being considered. The latter topic, i.e., the identification of inefficiencies, seems to have been, first, brought into view in an article published as "The Coefficient of Resource Utilization," by G. Debreu in *Econometrica* (1951) pp.273-292. Even though their models took the form of linear programming problems both Debreu and Farrell formulated their models in the tradition of "activity analysis." Little, if any, attention had been paid to computational implementation in the activity analysis literature. Farrell, therefore, undertook a massive and onerous series of matrix inversions in his first efforts. The alternative of linear programming algorithms was called to Farrell's attention by A.J. Hoffman, who served as a commentator in this same issue of the *Journal of the Royal Statistical Society*. Indeed, the activity analysis approach had already been identified with linear programming and reformulated and extended in the 1957 article A. Charnes and W.W. Cooper published in "On the Theory and Computation of Delegation-Type Models: K-Efficiency, Functional Efficiency and Goals," *Proceedings of the Sixth International Meeting of The Institute of Management Science* (London: Pergamon Press). See also Chapter IX in A. Charnes and W.W. Cooper, *Management Models and Industrial Applications of Linear Programming* (New York: John Wiley & Sons, Inc., 1961).

The modern version of DEA originated in two articles by A. Charnes, W.W. Cooper and E. Rhodes: (1) "Measuring the Efficiency of Decision Making Units," *European Journal of Operational Research* 2, 1978, pp.429-444 and (2) "Evaluating Program and Managerial Efficiency: An Application of Data Envelopment Analysis to Program Follow Through," *Management Science* 27, 1981, pp.668-697. The latter article not only introduced the name Data Envelopment Analysis for the concepts introduced in the former article, it also exploited the duality relations as well as the computational power that the former had made available. In addition, extensions were made that included the CCR projection operations associated with (3.32)-(3.33) and used these projections to evaluate programs (such as the U.S. Office of Education's "Program Follow Through") that made it possible to identify a "program's efficiency" separately from the way the programs had been managed. (Hence the distinction between the "program" and "managerial" efficiencies had been confounded in the observations generated from this education study.) Equally important, the latter article started a tradition in which applications were used to guide subsequent research in developing new concepts and methods of analyses. This, in turn, led to new applications, and so on, with many hundreds now reported in the literature.

The first of the above two articles also introduced the ratio form of DEA that is represented as (FP_o) in Section 2.3 of Chapter 2. This ratio form with its enhanced interpretive power and its contacts with other definitions of efficiency in engineering and science was made possible by prior research in which Charnes and Cooper opened the field of fractional programming. See Problem 3.1, below. It also led to other new models and measures of efficiency which differ from (DLP_o) in Section 3.3, above, which had been the only model form that was previously used.

3.12 RELATED DEA-SOLVER MODELS FOR CHAPTER 3

CCR-I (Input-oriented Charnes-Cooper-Rhodes model).

This code solves the CCR model expressed by (3.6)-(3.9) or by (3.38)-(3.42). The data set should be prepared in an Excel Workbook by using an appropriate Workbook name prior to execution of this code. See the sample format displayed in Figure B.1 in Section B.5 of Appendix B and refer to explanations above the figure. This style is the most basic one and is adopted in other models as the standard main body of data. The main results will be obtained in the following Worksheets as displayed in Table 3.6. The worksheet "Summary" includes statistics on data — average, standard deviation of each input and output, and correlation coefficients between observed items. It also reports DMUs with inappropriate data for evaluation and summarizes the results.

CCR-O (Output-oriented CCR model).

This code solves the output-oriented CCR model expressed by (3.60)-(3.63). In this model the optimal efficiency score η^* describes the output enlargement rate and satisfies $\eta^* \geq 1$. However, we display this value by its inverse as $\theta^* = 1/\eta^* (\leq 1)$ and call it the "CCR-O efficiency score." This will facilitate comparisons of scores between the input-oriented and the output-oriented models. In the CCR model, both models have related efficiency values as shown by (3.64). The other results are exhibited in the Worksheets in Table 3.6. In Worksheet "Weight," $\boldsymbol{v}$ and $\boldsymbol{u}$ correspond to $\boldsymbol{p}$ and $\boldsymbol{q}$ in (3.68)-(3.71), respectively. "Projection" is based on the formulas in (3.74) and (3.75).

Table 3.6. Worksheets Containing Main Results

Worksheet name	Contents
Summary	Summary on data and results.
Score	The efficiency score θ^*, the reference set($\boldsymbol{\lambda}^*$), ranking, etc.
Rank	The descending order ranking of efficiency scores.
Weight	The optimal (dual) multipliers $\boldsymbol{v}^*$, $\boldsymbol{u}^*$ in (3.2)-(3.5).
WeightedData	The weighted data $\{x_{ij}v_i^*\}$ and $\{y_{rj}u_r^*\}$.
Slack	The input excesses $\boldsymbol{s}^-$ and the output shortfalls $\boldsymbol{s}^+$ in (3.10).
Projection	Projection onto the efficient frontiers by (3.22)-(3.23).
Graph1	The bar chart of the CCR scores.
Graph2	The bar chart of scores in ascending order.

3.13 PROBLEM SUPPLEMENT FOR CHAPTER 3

Problem 3.1 (Ratio Forms and Strong and Weak Disposal)

The suggested response to Problem 2.3 in Chapter 2 showed how the "ratio of ratios" definition of efficiency in engineering could be subsumed under the CCR ratio form of DEA given for (FP_o) in Section 2.3 of that chapter. Can you now extend this to show how the linear programming formulation given in (LP_o) relates to the ratio form given in (FP_o)?

Suggested Response : A proof of equivalence is provided with Theorem 2.1 in Section 2.4 of Chapter 2. The following proof is adopted from A. Charnes and W.W. Cooper, "Programming with Linear Fractional Functionals" *Naval Research Logistics Quarterly* 9, 1962, pp.181-185, which initiated (and named) the field of fractional programming. We use this as an alternative because it provides contact with the (now) extensive literature on fractional programming. See S. Schaible (1994) "Fractional Programming" in S. Gass and C.M. Harris, eds., *Encyclopedia of Operations Research and Management Science* (Norwell, Mass. Kluwer Academic Publishers) who notes 900 articles that have appeared since Charnes-Cooper (1962). To start we reproduce (FP_o) from section 2.3 of Chapter 2, as follows

$$\begin{aligned} \max \quad & \theta = \frac{\sum_{r=1}^{s} u_r y_{ro}}{\sum_{i=1}^{m} v_i x_{io}} \\ \text{subject to} \quad & \frac{\sum_{r=1}^{s} u_r y_{rj}}{\sum_{i=1}^{m} v_i x_{ij}} \leq 1, \quad j = 1, \ldots, n \\ & u_r, v_i \geq 0. \ \forall r, i \end{aligned} \tag{3.91}$$

Now we can choose a new variable t in such a way that

$$t \sum_{i=1}^{m} v_i x_{io} = 1, \tag{3.92}$$

which implies $t > 0$. Multiplying all numerators and denominators by this t does not change the value of any ratio. Hence setting

$$\begin{aligned} \mu_r &= tu_r, \quad r = 1, \ldots, s \\ \nu_i &= tv_i, \quad i = 1, \ldots, m \end{aligned} \tag{3.93}$$

we have replaced the above problem by the following equivalent,

$$\begin{aligned} \max \quad & \theta = \sum_{r=1}^{s} \mu_r y_{ro} \\ \text{subject to} \quad & \sum_{i=1}^{m} \nu_i x_{io} = 1 \\ & \sum_{r=1}^{s} \mu_r y_{rj} - \sum_{i=1}^{m} \nu_i x_{ij} \leq 0 \;\; j = 1, ..., n \\ & \mu_r, \; \nu_i \geq 0. \;\; \forall r, \; i \end{aligned} \tag{3.94}$$

This is the same as (LP_o) in Section 2.4 which we have transformed using what is referred to as the "Charnes-Cooper transformation" in fractional programming. This reduction of (3.91)to the linear programming equivalent in (3.94) also makes available (DLP_o) the dual problem which we reproduce here as

$$\begin{aligned} \min \quad & \theta \\ \text{subject to} \quad & \theta x_{io} = \sum_{j=1}^{n} x_{ij}\lambda_j + s_i^-, \quad i = 1, \ldots, m \\ & y_{ro} = \sum_{j=1}^{n} y_{rj}\lambda_j - s_r^+. \;\; r = 1, \ldots, s \end{aligned} \tag{3.95}$$

This is the form used by Farrell (1978) which we employed for Phase I, as described in Section 3.3, followed by a Phase II in which the slacks are maximized in the following problem

$$\begin{aligned} \max \quad & \sum_{i=1}^{m} s_i^- + \sum_{r=1}^{s} s_r^+ \\ \text{subject to} \quad & \theta^* x_{io} = \sum_{j=1}^{n} x_{ij}\lambda_j + s_i^-, \quad i = 1, \ldots, m \\ & y_{ro} = \sum_{j=1}^{n} y_{rj}\lambda_j - s_r^+, \quad r = 1, \ldots, s \\ & 0 \leq \lambda_j, s_i^-, s_r^+, \quad \forall j, i, r \end{aligned} \tag{3.96}$$

where θ^* is the value obtained by solving (3.95) in Phase I. As noted in the text the solution to (3.95) is referred to as "Farrell efficiency." It is also referred to as "weak efficiency," as measured by θ^*, since this measure does not comprehend the non-zero slacks that may be present. In the economics literature (3.95) is said to assume "strong disposal." If we omit the s_i^- in the first m constraints in (3.95) we then have what is called "weak disposal" which we can write in the following form

$$\begin{aligned} \min \quad & \theta \\ \text{subject to} \quad & \theta x_{io} = \sum_{j=1}^{n} x_{ij}\lambda_j, \quad i = 1, \ldots, m \\ & y_{ro} = \sum_{j=1}^{n} y_{rj}\lambda_j - s_r^+, \quad r = 1, \ldots, s \end{aligned} \tag{3.97}$$

which means that the input inequalities are replaced with equalities so there is no possibility of positive input slacks that may have to be disposed of. Sometimes this is referred to as the assumption of "weak" and "strong" *input* disposal in order to distinguish it from corresponding formulations in *output* oriented models. In either case, these weak and strong disposal assumptions represent refinements of the "free disposal" assumption introduced by T.C. Koopmans (1951) for use in activity analysis. This assumption means that there is no cost associated with disposing of excess slacks in inputs or outputs. That is, slacks in the objective are all to be assigned a zero coefficient. Hence, all nonzero slacks are to be ignored, whether they occur in inputs or outputs. For a fuller treatment of "weak" and "strong" disposal, see R. Färe, S. Grosskopf and C.A.K. Lovell, *The Measurement of Efficiency of Production* (Boston: Kluwer Academic Publishers Group, 1985).

Problem 3.2

Can you provide a mathematical formulation that will serve to unify the Phase I and Phase II procedures in a single model?

Suggested Response : One way to do this is to join (3.6) and (3.11) together in a single objective as follows:

$$\begin{aligned} \min \quad & \theta - \varepsilon \left(\sum_{i=1}^{m} s_i^- + \sum_{r=1}^{s} s_r^+ \right) \\ \text{subject to} \quad & \theta x_{io} = \sum_{j=1}^{n} x_{ij}\lambda_j + s_i^-, \quad i = 1, \ldots, m \\ & y_{ro} = \sum_{j=1}^{n} y_{rj}\lambda_j - s_r^+. \quad r = 1, \ldots, s \\ & 0 \le \lambda_j, s_i^-, s_r^+. \quad \forall j, i, r \end{aligned} \tag{3.98}$$

It is tempting to represent $\varepsilon > 0$ by a small (real) number such as $\varepsilon = 10^{-6}$. However, this is not advisable. It can lead to erroneous results and the situation may be *worsened* by replacing $\varepsilon = 10^{-6}$ by even smaller values. See I. Ali and L. Seiford, "The Mathematical Programming Approach to Efficiency Analysis" in H.O. Fried, C.A.K. Lovell and S.S. Schmidt, ed., *The Measurement of Productive Efficiency* (New York: Oxford University Press, 1993) or Ali and Seiford (1993) "Computational Accuracy and Infinitesimals in Data Envelopment Analysis," *INFOR* 31, pp.290-297.

As formulated in the Charnes, Cooper and Rhodes article in the *European Journal of Operational Research*, cited in the Notes and Selected Bibliography, above, $\varepsilon > 0$ is formulated as a "non-Archimedean infinitesimal." That is, $\varepsilon > 0$ is smaller than *any* positive real number and, in fact, the product of ε by *any* real number, so that, however large the multiplier, $k > 0$, the value of $k\varepsilon > 0$ remains smaller than *any* positive real number. This means that $\varepsilon > 0$ is not a real number because the latter all have the Archimedean property — viz., given any real number $n > 0$ there exists another real number $n/2$ such that $n > n/2 > 0$. Thus, to deal with non-Archimedean elements it is necessary to embed the field of real numbers in a still larger field. However, it is not necessary to go into the further treatments of this kind of (non-standard) mathematics. It is not even necessary to specify a value of $\varepsilon > 0$ explicitly. The two-phase procedure described in this chapter accomplishes all that is required. Phase I accords priority to $\min\theta = \theta^*$ with $\theta^* > 0$ when the data are semipositive. Fixing $\theta = \theta^*$ as is done in (3.12) for Phase II in Section 3.3 of this chapter, we must then have, by definition,

$$0 < \theta^* - \varepsilon\left(\sum_{i=1}^{m} s_i^{-*} + \sum_{r=1}^{s} s_r^{+*}\right).$$

A more detailed treatment of these non-Archimedean elements and their relations to mathematical programming may be found in V. Arnold, I. Bardhan, W.W. Cooper and A. Gallegos, "Primal and Dual Optimality in Computer Codes Using Two-Stage Solution Procedures in DEA" in J.E. Aronson and S. Zionts, ed., *Operations Research: Methods, Models and Applications* (Westport, Conn.: Quorum Books, 1998). Here we only note that the further implications of this non-Archimedean element, which appears in the objective of (3.98) can be brought forth by writing its dual as

$$\begin{aligned} \max \quad & \theta = \sum_{r=1}^{s} \mu_r y_{ro} && (3.99)\\ \text{subject to} \quad & \sum_{r=1}^{s} \mu_r y_{rj} - \sum_{i=1}^{m} \nu_i x_{ij} \leq 0 \;\; j = 1, \ldots, n \\ & \sum_{i=1}^{m} \nu_i x_{io} = 1 \\ & \mu_r,\ \nu_i \geq \varepsilon > 0.\ \forall r,\ i \end{aligned}$$

This means that all variables are constrained to positive values. Hence, they are to be accorded "some" worth, even though it is not specified explicitly.

Finally, as shown in Arnold *et al.*, the principle of complementary slackness used in Section 3.3 of this chapter is modified to the following

$$0 \leq s_i^{-*}\nu_i^* \leq s_i^{-*}\varepsilon \tag{3.100}$$

$$0 \leq s_r^{+*}\mu_r^* \leq s_r^{+*}\varepsilon. \tag{3.101}$$

Hence, unlike what is done with free disposal, one cannot justify ignoring nonzero slacks by assuming that corresponding multiplier values will be zero. Indeed, the stage two optimization maximizes the slacks, as is done in the Phase II procedure associated with (3.96), in order to try to wring out the maximum possible inefficiency values associated with nonzero slacks.

Problem 3.3

We might note that (3.99), above, differs from (3.94) by its inclusion of the non-Archimedean conditions $\mu_r, \nu_i \geq \varepsilon > 0$. Can you show how these conditions should be reflected in a similarly altered version of the ratio model in (3.91)?

Suggested Answer : Multiply and divide the objective in (3.99) by $t > 0$. Then multiply all constraints by the same t to obtain

$$\max \frac{\sum_{r=1}^{s}(t\mu_r)y_{ro}}{t} \tag{3.102}$$

$$\begin{aligned} \text{subject to} \quad & \sum_{r=1}^{s}(t\mu_r)y_{rj} - \sum_{i=1}^{m}(t\nu_i)x_{ij} \leq 0 \quad j = 1, \ldots, n \\ & \sum_{i=1}^{m}(t\nu_i x_{io}) = t \\ & (t\mu_r),\ (t\nu_i) \geq \varepsilon > 0. \quad \forall r,\ i \end{aligned}$$

Set $u_r = t\mu_r$, $v_i = t\nu_i$ for each r and i. Then substitute in these last expressions to obtain

$$\max \frac{\sum_{r=1}^{s} u_r y_{ro}}{\sum_{i=1}^{m} v_i x_{io}} \tag{3.103}$$

$$\begin{aligned} \text{subject to} \quad & \frac{\sum_{r=1}^{s} u_r y_{rj}}{\sum_{i=1}^{m} v_i x_{ij}} \leq 1, \quad j = 1, \ldots, n \\ & u_r \Big/ \sum_{i=1}^{m} v_i x_{io} \geq \varepsilon \quad \forall r \\ & v_i \Big/ \sum_{i=1}^{m} v_i x_{io} \geq \varepsilon. \quad \forall i \end{aligned}$$

Note that we have reflected the condition $\sum_{i=1}^{m} v_i x_{io} = t$ in the above objective but not in the constraints because it can *always* be satisfied. We can also now write

$$\frac{\sum_{r=1}^{s} u_r^* y_{ro}}{\sum_{i=1}^{m} v_i^* x_{io}} = \sum_{r=1}^{s} u_r^* y_{ro} = \theta^* - \varepsilon\left(\sum_{i=1}^{m} s_i^{-*} + \sum_{r=1}^{s} s_r^{+*}\right). \tag{3.104}$$

The equality on the right follows from the dual theorem of linear programming. We have added the equality on the left by virtue of our derivation and, of course, the values of these expressions are bounded by zero and one.

Problem 3.4

The optimal solution for (LP_o) , the multiplier problem, for DMU A in Example 3.1 is

$$v_1^* = 0.1429, v_2^* = 0.1429, u^* = 0.8571.$$

Show how these multiplier values could be used as a management guide to improve efficiency in A's performance.

Suggested Response : Applying the above solution to the data in Table 3.2 we utilize the multiplier model to write the constraints for DMU A as follows:

	DMU	Output Value	Output Variable	Input Variable	Input Value
	A	0.8571	$= 1u^*$	$\le 4v_1^* + 3v_2^*$	=1.000
	B	0.8571	$= 1u^*$	$\le 7v_1^* + 3v_2^*$	=1.429
	C	0.8571	$= 1u^*$	$\le 8v_1^* + v_2^*$	=1.284
$\longrightarrow$	D	0.8571	$= 1u^*$	$\le 4v_1^* + 2v_2^*$	=0.8571
$\longrightarrow$	E	0.8571	$= 1u^*$	$\le 2v_1^* + 4v_2^*$	=0.8571
	F	0.8571	$= 1u^*$	$\le 10v_1^* + v_2^*$	=1.5719
	G	0.8571	$= 1u^*$	$\le 3v_1^* + 7v_2^*$	=1.429

$$\text{Add constraint}: \quad 1 = 4v_1^* + 3v_2^*.$$

Note, first, that $1u^* = \theta^* = 0.8571$ so the value of the optimal solution of the multiplier model is equal to the optimal value of the envelopment model in accordance with the duality theory of linear programming and, similarly, the reference set is $\{D, E\}$, as indicated by the arrows. Here we have $v_1^* = v_2^* = 0.1429$ so a unit reduction in the $4+3 = 7$ units of input used by A would bring it into $4 + 2 = 6$ units used by D and E. Thus, $v^* = v_1^* = v_2^* = 0.1429 = 1/7$ is the reduction required to bring A to full efficiency. We can provide added perspective by reducing $x_1 = 4$ to $\widehat{x}_1 = 3$ which positions A half way between D and E on the efficient frontier in Figure 3.2. It also produces equality of the relations for A as in the following expression

$$u^* = 3v_1^* + 3v_2^* \doteq 0.8575.$$

Division then produces

$$\frac{u^*}{3v_1^* + 3v_2^*} = 1,$$

which is the condition for efficiency prescribed for (FP_o) as given in (2.3)-(2.6) in Chapter 2. Here we have $v_1^* = v_2^* = 0.1429$ so a use of either x_1 or x_2 produces equal results per unit change en route to achieving full efficiency. More generally we will have $v_1^* \neq v_2^*$, etc., so these values may be used to guide managerial priorities for the adjustments to the model.

There are two exceptions that require attention. One exception is provided by Definition 2.1 in Chapter 2, which notes that all components of the vectors $\boldsymbol{v}^*$

and $\boldsymbol{u}^*$ must be positive in at least one optimal solution. We use the following solution for G as an illustration,

$$v_1^* = 0.3333, \quad v_2^* = 0, \quad u^* = 0.6666.$$

From the data of Example 3.1, we then have

$$0.6666 = 1u^* < 3v_1^* + 7v_2^* = 1.00.$$

Replacing $x_1 = 3$ by $\widehat{x}_1 = 2$ produces equality, but efficiency cannot be claimed because $v_2^* = 0$ is present.

Reference to Figure 3.2 shows that the inefficiency associated with the nonzero slack in going from R to E is not attended to and, indeed, the amount of slack associated with x_2 is worsened by the horizontal movement associated with the reduction in going from $x_1 = 3$ to $\widehat{x}_1 = 2$. Such a worsening would not occur for F in Figure 3.2, but in either case full efficiency associated with achieving a ratio of unity does not suffice unless all multipliers are positive in the associated adjustments.

The second exception noted above involves evaluations of the DMUs that are fully efficient and hence are used to evaluate *other* DMUs. This complication is reflected in the fact that the data in the basis matrix B is changed by any such adjustment and so the inverse B^{-1} used to evaluate other DMUs, as given in (3.53), is also changed. This topic involves complexities which cannot be treated here, but is attended to later in this book after the requisite background has been supplied. It is an important topic that enters into further uses of DEA and hence has been the subject of much research which we summarize as follows.

The problem of sensitivity to data variations was first addressed in the DEA literature by A. Charnes, W.W. Cooper, A.Y. Lewin, R.C. Morey and J. Rousseau "Sensitivity and Stability Analysis in DEA," *Annals of Operations Research* 2, 1985, pp.139-156. This paper was restricted to sensitivity analyses involving changes in a single input or output for an efficient DMU. Using the CCR ratio model, as in (3.91), above, this was subsequently generalized to simultaneous changes in all inputs and outputs for any DMU in A. Charnes and L. Neralic, "Sensitivity Analysis of the Proportionate Change of Inputs (or Outputs) in Data Envelopment Analysis," *Glasnik Matematicki* 27, 1992, pp.393-405. Also using the ratio form of the multiplier model, R.G. Thompson and R.M. Thrall provided sensitivity analyses in which *all* data for *all* DMUs are varied simultaneously. The basic idea is as follows. All inputs and all outputs are worsened for the efficient DMUs until at least one DMU changes its status. The multipliers and ratio values are then recalculated and the process is reiterated until all DMUs become inefficient. See R.G. Thompson, P.S. Dharmapala, J. Diaz, M. Gonzalez-Lima and R.M. Thrall, "DEA Multiplier Analytic Center Sensitivity Analysis with an Illustrative Application to Independent Oil Companies," *Annals of Operations Research* 66, 1996, pp.163-167. An alternate and more exact approach is given in L. Seiford and J. Zhu, "Sensitivity Analysis of DEA Models for Simultaneous Changes in All the Data," *Journal of the Operational Research Society* 49, 1998, pp.1060-1071. Finally, L. Seiford and

J. Zhu in "Stability Regions for Maintaining Efficiency in Data Envelopment Analysis," *European Journal of Operational Research* 108, 1998, pp. 127-139 develop a procedure for determining exact stability regions within which the efficiency of a DMU remains unchanged. More detail on the topic of sensitivity is covered in Chapter 9. See also W.W. Cooper, L.M. Seiford and J. Zhu (2004) Chapter 3, "Sensitivity Analysis in DEA," in W.W. Cooper, L.M. Seiford and J. Zhu, eds., *Handbook on Data Envelopment Analysis* (Norwell, Mass., Kluwer Academic Publishers).

Problem 3.5 (Complementary Slackness and Sensitivity Analysis)

Table 3.7 displays a data set of 5 stores (A, B, C, D and E) with two inputs (the number of employees and the floor area) and two outputs (the volume of sales and profits).

Assignment: (i) Using the code CCR-I in the DEA-Solver, obtain the efficiency score (θ^*), reference set ($\boldsymbol{\lambda}^*$), the optimal weights ($\boldsymbol{v}^*, \boldsymbol{u}^*$) and slacks ($\boldsymbol{s}^{-*}, \boldsymbol{s}^{+*}$) for each store. (ii) Interpret the complementary slackness conditions (3.15) between the optimal weights and the optimal slacks. (iii) Discuss the meaning of u_1^* of C, and use sensitivity analysis to verify your discussion. (iv) Check the output shortfall s_2^{+*} of B and identify the influence of its increase on the value of the corresponding multiplier (weight) u_2 using the CCR-I code. (v) Regarding the store B, discuss the meaning of v_1^*.

Suggested Response : (i) The efficiency score (θ^*) and reference set ($\boldsymbol{\lambda}^*$) are listed in Table 3.7 under the heading "Score." Store E is the only efficient

Table 3.7. Data and Scores of 5 Stores

	Data				Score		
Store	Employee	Area	Sales	Profits	θ^*	Rank	Reference
A	10	20	70	6	0.933333	2	E (0.77778)
B	15	15	100	3	0.888889	3	E (1.11111)
C	20	30	80	5	0.533333	5	E (0.88889)
D	25	15	100	2	0.666667	4	E (1.11111)
E	12	9	90	8	1	1	E (1)

DMU and is referent to all other stores. The optimal weights and slacks are displayed in Table 3.8.

(ii) The complementary slackness conditions in (3.15) assert that, for an optimal solution we have

$$v_i^* s_i^{-*} = 0 \text{ (for } i = 1, 2) \text{ and } u_r^* s_r^{+*} = 0 \text{ (for } r = 1, 2) .$$

Table 3.8. Optimal Weights and Slacks

	Weights				Slacks			
Store	Emply.	Area	Sales	Profits	Emply.	Area	Sales	Profits
	v_1^*	v_2^*	u_1^*	u_2^*	s_1^{-*}	s_2^{-*}	s_1^{+*}	s_s^{+*}
A	0.1	0	0.0133	0	0	11.6667	0	0.2222
B	0.0667	0	0.00889	0	0	3.3333	0	5.8889
C	0.05	0	0.00667	0	0	8	0	2.1111
D	0	0.0667	0.00667	0	3.3333	0	0	6.8889
E	0.0702	0.0175	0.00526	0.0658	0	0	0	0

This means that if $v_i^* > 0$ then $s_i^{-*} = 0$ and if $s_i^{-*} > 0$ then $v_i^* = 0$, and the same relations hold between $\boldsymbol{u}^*$ and $\boldsymbol{s}^{+*}$. This can be interpreted thus: if a DMU has an excess (s^{-*}) in an input against the referent DMU, then the input item has no value in use to the DMU_o being evaluated, so the optimal (multiplier) solution assigns a value of zero to the corresponding weight (v^*). Similarly we can see that for a positive weight to be assigned to a multiplier, the corresponding optimal slack must be zero.
(iii) C has $u_1^* = 0.00667$ and this value can be interpreted in two ways. First, with recourse to the fractional program described by (2.3)-(2.6) in Chapter 2, we can interpret u_1^* (the optimal weight to "sales") as the degree of influence which one unit of sales has on the optimal efficiency score. Thus, if C increases its sales by one unit, then it is expected that θ^* will increase by 0.00667 ($=u_1^*$) to 0.533333+0.00667=0.54, since the denominator of (2.3) does not change and retains the value 1. On the other hand, $u_2^* = 0$ implies that a one unit increase in "profits" has no effect on the efficiency score of C. The second interpretation is that in (3.59) we introduced $\boldsymbol{u}^*$ as "pricing vectors" and in the envelopment form of LP, u_r^* is the reduced cost induced by one unit change in output r. Notice that $\boldsymbol{y}_o$ appears as the constant in (3.41). Hence $u_1^* = 0.00667$ shows the degree of contribution that one unit change of sales can make to use the efficiency score. The above observations contribute to identifying which output has the largest influence on the efficiency score. However, such analysis of influence is valid only in a limited range of observations. To check this range you can change the efficiency score of C by adding to the data set a virtual store C', with sales of 101 and other data equal to C, and then applying the CCR-I code again.
(iv) B has the slack $s_2^{+*} = 5.889$ (shortfall in "profits") and, by the complementary slackness theorem, $u_2^{+*} = 0$, showing that this level of profits has no effect on the efficiency evaluation. Adding this value to current profits will raise profits to 8.889. We augmented the data set by adding B', which has profits of 8.889, and other observations equal to B and tried CCR-I for this data set. As a result, we have, for B', $s_2^{+*} = 0$ and $u_2^{+*} = 0.1$. This means that, at this level

of profits, this output has a positive influence on the efficiency score, which in this case is a 0.1 per unit increase.
(v) B has the optimal weight $v_1^* = 0.0667$. It can therefore be said that changes in the number of employees affects the efficiency score. In this case, reduced costs analysis using LP is not as straightforward as in u^*. Hence, dealing with the fractional program in (2.3)-(2.6) is appropriate for the sensitivity analysis of data on employees. If B reduces its employees by one to 14, then the denominator of (2.3) decreases by $v_1^* = 0.0667$. Since the other terms are unchanged, the objective function (2.3) takes the value

$$(0.00889 \times 100 + 0 \times 3)/(0.0667 \times 14 + 0 \times 15) = 0.952.$$

Since the value is still less than one and all constraints in (2.4) are satisfied by this set of $(\boldsymbol{v}^*, \boldsymbol{u}^*)$, the above objective value gives a lower bound of θ^* for this adjusted problem. Application of CCR-I code for this new problem showed that 0.952 is the optimal objective value. Thus, it may be expected that an input item with a large v_i^* value greatly affects the efficiency value.

Problem 3.6

Solve Example 3.1 in Table 3.2 (Section 3.7) using the output-oriented model (CCR-O model in DEA-Solver) and compare the corresponding CCR-projection with the input-oriented case.

Suggested Response : For ease of comparison with the input-oriented case, the optimal objective value η^* is demonstrated by its inverse θ^* $(= 1/\eta^*)$.

The optimal solution to A reads

$$\begin{aligned} &\theta^* = 0.8571 \quad (\eta^* = 1.1667) \\ &\lambda_D^* = 0.8333, \ \lambda_E^* = 0.3333, \ \text{other } \lambda_j^* = 0 \\ &s_1^{-*} = s_2^{-*} = s^{+*} = 0 \end{aligned}$$

The output-oriented CCR-projection was performed following (3.74)-(3.75), which resulted in

$$\begin{aligned} &\widehat{x}_1 \leftarrow x_1 = 4 \quad \text{(no change)} \\ &\widehat{x}_2 \leftarrow x_2 = 3 \quad \text{(no change)} \\ &\widehat{y} \leftarrow \eta^* y = 1.167 \times 1 = 1.167 \quad \text{(16.7\% increase)}. \end{aligned}$$

These results differ from the input-oriented case, reflecting the difference in model orientation. Table 3.9 exhibits the CCR-projection in both input and output orientations.

Problem 3.7

Suppose the activity $(\boldsymbol{x}_o, \ \boldsymbol{y}_o)$ is inefficient. Let the improved activity obtained by the input-oriented model (3.22) and (3.23) be $(\widehat{\boldsymbol{x}}_o, \ \widehat{\boldsymbol{y}}_o)$,

$$\widehat{\boldsymbol{x}}_o = \theta^* \boldsymbol{x}_o - \boldsymbol{s}^{-*} = X\boldsymbol{\lambda}^* \tag{3.105}$$

Table 3.9. CCR-projection in Input and Output Orientations

DMU	Data			Score	Input orientation			Output orientation		
	x_1	x_2	y	θ^*	x_1	x_2	y	x_1	x_2	y
A	4	3	1	0.8571	3.43	2.57	1	4	3	1.17
B	7	3	1	0.6316	4.42	1.89	1	7	3	1.58
C	8	1	1	1.0000	8	1	1	8	2	1
D	4	2	1	1.0000	4	2	1	4	2	1
E	2	4	1	1.0000	2	4	1	2	4	1
F	10	1	1	1.0000	8	1	1	8	1	1
G	3	7	1	0.6667	2	4	1	3	6	1.5

$$\widehat{\boldsymbol{y}}_o = \boldsymbol{y}_o + \boldsymbol{s}^{+*} = Y\boldsymbol{\lambda}^*, \tag{3.106}$$

while that of the output-oriented model (3.74) and (3.75) be $(\breve{\boldsymbol{x}}_o,\ \breve{\boldsymbol{y}}_o)$.

$$\breve{\boldsymbol{x}}_o = \boldsymbol{x}_o - \boldsymbol{t}^{-*} = X\boldsymbol{\mu}^* \tag{3.107}$$

$$\breve{\boldsymbol{y}}_o = \eta^*\boldsymbol{y}_o + \boldsymbol{t}^{+*} = Y\boldsymbol{\mu}^*. \tag{3.108}$$

An activity on the line segment connecting these two points can be expressed by

$$(\boldsymbol{x}'_o,\ \boldsymbol{y}'_o) = \alpha_1\ (\widehat{\boldsymbol{x}}_o,\ \widehat{\boldsymbol{y}}_o)\ +\ \alpha_2\ (\breve{\boldsymbol{x}}_o,\ \breve{\boldsymbol{y}}_o) \tag{3.109}$$

$$\alpha_1 + \alpha_2 = 1,\quad \alpha_1 \geq 0,\quad \alpha_2 \geq 0.$$

Prove the following

Proposition 3.1 *The activity* $(\boldsymbol{x}'_o,\ \boldsymbol{y}'_o)$ *is CCR-efficient.*

Suggested Answer :
Proof. Since by (3.65) $\boldsymbol{\mu}^* = \boldsymbol{\lambda}^*/\theta^*$, we have

$$\begin{aligned}(\boldsymbol{x}'_o,\ \boldsymbol{y}'_o) &= \alpha_1\ (X,\ Y)\boldsymbol{\lambda}^*\ +\ \alpha_2\ (X,\ Y)\boldsymbol{\lambda}^*/\theta^* \\ &= (\alpha_1 + \alpha_2/\theta^*)\ (X\boldsymbol{\lambda}^*,\ Y\boldsymbol{\lambda}^*)\end{aligned}$$

By applying Theorem 3.4, we can see that $(\boldsymbol{x}'_o,\ \boldsymbol{y}'_o)$ is CCR-efficient. □
This proposition is valid for *any* nonnegative combination of two points ($\alpha_1 \geq 0$ and $\alpha_2 \geq 0$). That is, it holds even when relaxing the convex-combination condition $\alpha_1 + \alpha_2 = 1$. Also, the case $\alpha_1 = 1,\ \alpha_2 = 0$ corresponds to the input-oriented improvement and the case $\alpha_1 = 0,\ \alpha_2 = 1$ to the output-oriented one. The case $\alpha_1 = 1/2,\ \alpha_2 = 1/2$ is a compromise between the two points.

Problem 3.8

Solve the hospital example in Chapter 1 (see Table 1.5) by input-oriented and output-oriented CCR models. Compare the CCR-projections of hospital C in both cases. Based on the previous Proposition 3.1, determine the combination of both projections using (3.109) with $\alpha_1 = \alpha_2 = 1/2$.

Suggested Answer : The data sheet for this problem is recorded in Figure B.1 in Appendix B. Prepare two Excel Workbooks, e.g., "Hospital-CCR-I.xls" and "Hospital-CCR-O.xls," each containing Figure B.1 as the data sheet. Then run CCR-I (the input-oriented CCR model) and CCR-O (the output-oriented CCR model). After the computations, the CCR-projections are stored in "Projection" sheets as follows:
For input orientation:

Doctor	25	→	20.9	(16% reduction)
Nurse	160	→	141	(12% reduction)
Outpatient	160	→	160	(no change)
Inpatient	55	→	55	(no change)

For output orientation:

Doctor	25	→	23.6	(5% reduction)
Nurse	160	→	160	(no change)
Outpatient	160	→	181	(13% increase)
Inpatient	55	→	62.3	(13% increase)

We have a compromise of the two models as the averages of the projected values as follows (this is equivalent to the case $\alpha_1 = \alpha_2 = 1/2$ in Proposition 3.1):

Doctor	25	→	22	(12% reduction)
Nurse	160	→	150	(6% reduction)
Outpatient	160	→	171	(7% increase)
Inpatient	55	→	59	(7% increase)

These improvements will put C on the efficient frontier.

Problem 3.9

Assume that the data set (X, Y) is semipositive (see the definition in Section 3.2) and let an optimal solution of (DLP_o) in Section 3.3 be $(\theta^*, \boldsymbol{\lambda}^*, \boldsymbol{s}^{-*}, \boldsymbol{s}^{+*})$. Prove that the reference set defined by $E_o = \{j | \lambda^* > 0\}$ $(j \in \{1, \ldots, n\})$ is not empty.
Suggested Answer :
We have the equation:

$$\theta^* \boldsymbol{x}_o = X\boldsymbol{\lambda}^* + \boldsymbol{s}^{-*}.$$

Let an optimal solution of (LP_o) in Section 3.3 be $(\boldsymbol{v}^*, \boldsymbol{u}^*)$. By multiplying $\boldsymbol{v}^*$ from left to the above equation, we have

$$\theta^* \boldsymbol{v}^* \boldsymbol{x}_o = \boldsymbol{v}^* X\boldsymbol{\lambda}^* + \boldsymbol{v}^* \boldsymbol{s}^{-*}.$$

Since $\boldsymbol{v}^* \boldsymbol{x}_o = 1$ is a constraint in (LP_o) and $\boldsymbol{v}^* \boldsymbol{s}^{-*} = 0$ by the complementarity condition, it holds

$$\theta^* = v^* X \boldsymbol{\lambda}^*.$$

Here, θ^* is positive as claimed in Section 3.3. Thus, $\boldsymbol{\lambda}^*$ must be semipositive and E_o is not empty. □

Problem 3.10

Part 1. Using model (3.95) show that the output inequality will always be satisfied as an equation in the single output case. That is, the output slack will be zero at an optimum.

Proof :

It is easy to see what is happening if we start with the case of no outputs. In this case the solution will be $\theta_o^* = 0$ and all $\lambda_j^* = 0$ since only the output constraints keep this from happening. It follows that for the case of one output a solution with $y_o < \sum_{j=1}^n y_j \lambda_j$ cannot be optimal since this choice of λs would prevent θ from achieving its minimum value. To see that this is so note that a choice of the minimizing value of θ is determined by

$$\theta = \max_i \left\{ \frac{\sum_{j=1}^n x_{ij}\lambda_j}{x_{io}} | i = 1, \ldots, m \right\} = \frac{\sum_{j=1}^n x_{kj}\lambda_j}{x_{ko}}$$

and this maximum value can be lowered until $y_o = \sum_{j=1}^n y_j \lambda_j$. (Here for simplicity we are assuming that all data are positive.) Hence optimality requires $y_o = \sum_{j=1}^n y_j \lambda_j^*$ so the output slack is zero in an optimum solution. □

Part 2. Extend the above to show that at least one output inequality must be satisfied as an equation in the case of multiple outputs $r = 1, \ldots, s$, like the ones represented in (3.95).

Proof :

Replace (3.95) by the following model

$$\min_{\boldsymbol{\lambda},\theta} \max_r \left\{ \theta, \sum_{j=1}^n y_{rj}\lambda_j \geq y_{ro} | r = 1, \ldots, s \right\} = \left(\theta^*, \sum_{j=1}^n y_{kj}\lambda_j^* = y_{ko} \right)$$

subject to (3.110)

$$\theta x_{io} \geq \sum_{j=1}^n x_{ij}\lambda_j, \; i = 1, \ldots, m$$

$$0 \leq \theta \leq 1, \; \lambda_j \geq 0, \; j = 1, \ldots, n.$$

This is a multiple criteria programming problem which seeks to minimize the maximum of the $r = 1, \ldots, s$ inequalities representing the output values set by the y_{ro} as lower limits for each of the s outputs. Note that the maximum output values are limited by the input constraints. The condition $0 \leq \theta \leq 1$ which eliminates the possibility of infinite solutions is not needed in (3.95) since its minimizing objective guarantees its fulfillment.

Now the maximal values may be decreased in (3.110) until equality is achieved in at least one of the output inequalities and the minimization of θ eliminates the possibility of being misled by the choice of a set of $\boldsymbol{\lambda}$ values that does not minimize this maximum because of the presence of alternate optima. The optimizing values $\theta^*, \boldsymbol{\lambda}^*$ represented on the right of (3.110) therefore minimize θ with

$$\theta^* = \max_i \left\{ \frac{\sum_{j=1}^{n} x_{ij}\lambda_j^*}{x_{io}},\ i = 1, \ldots, m \right\}$$

and all constraints are satisfied in (3.95) as well as in (3.110). □

Corollary 3.2 *At least one input as well as one output constraint will be satisfied as an equality at an optimum so these constraints have zero slack in (3.95) and (3.110).*

We also have following

Theorem 3.6 *Model (3.110) is equivalent to model (3.95).*

Similar results hold for the "output-oriented" version of the CCR model with a "Farrell" measure of efficiency that is represented by the value of θ^* in (3.95). Such equivalences may also be established for other DEA models, like the BCC model that will be presented in the next chapter. Finally, we also note that we earlier established the relation of (3.95) to a fractional programming model so our linear programming DEA formulation provides a link between this multiple objective nonlinear programming problem and a nonlinear nonconvex fractional programming problem. See also Cooper (2005)[19] for a discussion of relations between multiple criteria programming and goal programming formulations which are also equivalent to a nonlinear problem directed to minimizing a sum of absolute values.

Notes

1. It might be noted that Postulate (A2) is included in (A4) but is separated out for special attention.
2. See Appendix A.4.
3. See Appendix A.4.
4. Farrell also restricted his treatments to the single output case. See M.J. Farrell (1957) "The Measurement of Production Efficiency, " *Journal of the Royal Statistical Society* A, 120, pp.253-281.
5. In the linear programming literature this is called the "complementary slackness" condition. This terminology is due to A.W. Tucker who is responsible for formulating and proving this. See E.D. Nering and A.W. Tucker *Linear Programs and Related Problems,* (Harcourt Brace, 1993). See our Appendix A.6 for a detailed development.
6. See Appendix A.8.
7. Although empirical studies show the uniqueness of the reference set for most DMUs, there may be multiple reference sets and improvement plans (projections) in the presence of multiple optimal solutions.
8. See Appendix A.4.

9. We notice that OQ and OA are measured by some "distance measure." If we employ the "Euclidian measure" – also called the "l_2 metric" – we have

$$\frac{d(OQ)}{d(OA)} = \frac{\sqrt{3.428^2 + 2.571^2}}{\sqrt{4^2 + 3^2}} = \frac{\sqrt{18.36}}{\sqrt{25}} = 8.57.$$

However, the measure is not restricted to Euclidian measure. Any l_k measure gives the same result. See Appendix A in Charnes and Cooper, *Management Models and Industrial Applications of Linear Programming* (New York, John Wiley, Inc., 1961). See also W.W. Cooper, L.M. Seiford, K. Tone and J. Zhu "DEA: Past Accomplishments and Future Prospects," *Journal of Productivity Analysis* (submitted, 2005).

10. "Aborts" were treated as reciprocals so that an increase in their output would reduce the value of the numerator in the (FP_o) objective represented in (2.3). They could also have been subtracted from a dominatingly large positive constant. See A. Charnes, T. Clark, W.W. Cooper and B. Golany "A Development Study of Data Envelopment Analysis in Measuring the Efficiency of Maintenance Units in the U.S. Air Forces," *Annals of Operational Research* 2, 1985, pp.59-94.

11. R.D. Banker and R.C. Morey (1986), "Efficiency Analysis for Exogenously Fixed Inputs and Outputs," *Operations Research* 34, 1986, pp.513-521. See also Chapter 10 in R. Färe, S. Grosskopf and C.A. Knox Lovell *Production Frontiers* (Cambridge University Press, 1994) where this is referred to as "sub-vector optimizations."

12. A fuller treatment of these $\varepsilon > 0$ values is provided in Problem 3.2 at the end of this chapter.

13. I.R. Bardhan (1995), "DEA and Stochastic Frontier Regression Approaches Applied to Evaluating Performances of Public Secondary Schools in Texas," Ph.D. Thesis. Austin Texas: Graduate School of Business, the University of Texas at Austin. Also available from University Microfilms, Inc. in Ann Arbor, Michigan.

14. V.L. Arnold, I.R. Bardhan and W.W. Cooper "A Two-Stage DEA Approach for Identifying and Rewarding Efficiency in Texas Secondary Schools" in W.W. Cooper, S. Thore, D. Gibson and F. Phillips, eds., IMPACT: *How IC^2 Research Affects Public Policy and Business Practices* (Westport, Conn.: Quorum Books, 1997)

15. An additional category may also be identified as *Property* which applies to (a) Choice of goals or objectives, and (b) Choice of means to achieve goals or objectives.

16. See Arnold *et al.* (1997) for further suggestions and discussions.

17. See Governmental Accounting Standards Board (GASB) Research Report: "Service Efforts and Accomplishments Reporting: Its Time Has Come," H.P. Hatry, J.M. Sullivan, J.M. Fountain and L. Kremer, eds. (Norwell, Conn., 1990).

18. C.A.K. Lovell, L.C. Walters and Lisa Wood, "Stratified Models of Education Production Using Modified DEA and Regression Analysis," in A. Charnes, W.W. Cooper, A.Y. Lewin and L.M. Seiford, eds., *Data Envelopment Analysis: Theory, Methodology and Applications* (Norwell, Mass.: Kluwer Academic Publishers, 1994).

19. W.W. Cooper (2005), "Origin, Uses of and Relations Between Goal Programming and DEA (Data Envelopment Analysis)" *Journal of Multiple Criteria Decision Analysis* (to appear).

4 ALTERNATIVE DEA MODELS

4.1 INTRODUCTION

In the preceding chapters, we discussed the CCR model, which is built on the assumption of *constant* returns to scale of activities as depicted for the production frontier in the single input-single output case shown in Figure 4.1. More generally, it is assumed that the production possibility set has the following property: If $(\boldsymbol{x}, \boldsymbol{y})$ is a feasible point, then $(t\boldsymbol{x}, t\boldsymbol{y})$ for any positive t is also feasible. This assumption can be modified to allow production possibility sets with different postulates. In fact, since the very beginning of DEA studies, various extensions of the CCR model have been proposed, among which the BCC (Banker-Charnes-Cooper) model[1] is representative. The BCC model has its production frontiers spanned by the convex hull of the existing DMUs. The frontiers have piecewise linear and concave characteristics which, as shown in Figure 4.2, leads to *variable* returns-to-scale characterizations with (a) increasing returns-to-scale occurring in the first solid line segment followed by (b) decreasing returns-to-scale in the second segment and (c) constant returns-to-scale occurring at the point where the transition from the first to the second segment is made.

In this chapter, we first introduce the BCC model in Section 2. Then, in Section 3, the "Additive Model" will be described. This model has the same production possibility set as the BCC and CCR models and their variants but treats the slacks (the input excesses and output shortfalls) directly in the objective function.

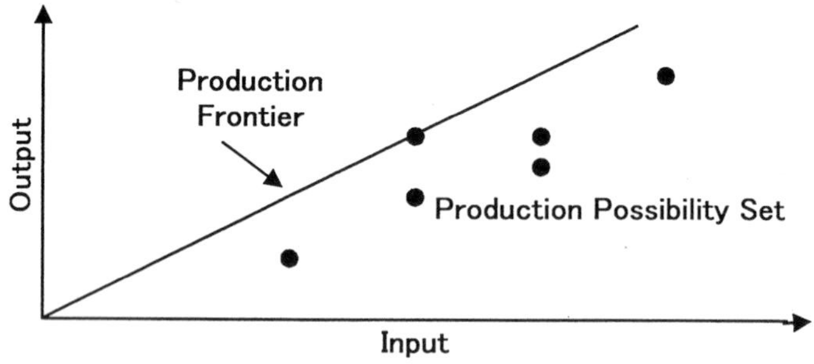

Figure 4.1. Production Frontier of the CCR Model

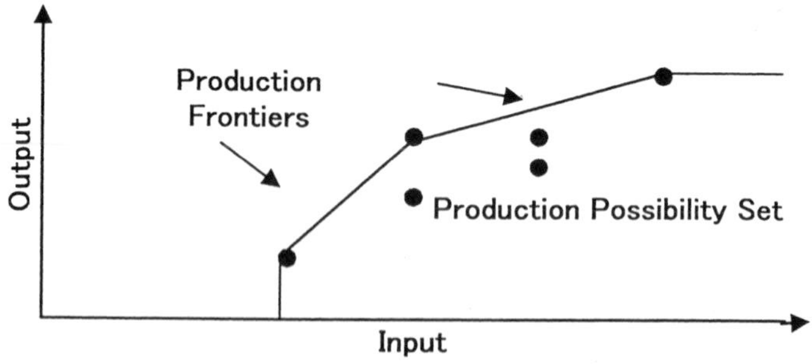

Figure 4.2. Production Frontiers of the BCC Model

CCR-type models, under weak efficiency, evaluate the radial (proportional) efficiency θ^* but do not take account of the input excesses and output shortfalls that are represented by non-zero slacks. This is a drawback because θ^* does not include the nonzero slacks. Although the Additive model deals with the input excesses and output shortfalls directly and can discriminate efficient and inefficient DMUs, it has no means to gauge the depth of inefficiency by a scalar measure similar to the θ^* in the CCR-type models. To eliminate this deficiency, we will introduce a slacks-based measure of efficiency (SBM), which was proposed by Tone (1997) [2] and is also related to the "Enhanced Russell Measure," in Section 4 in the form of a scalar with a value not greater than the corresponding CCR-type measure θ^*. This measure reflects nonzero slacks in inputs and outputs when they are present.

The BCC and CCR models differ only in that the former, but not the latter, includes the convexity condition $\sum_{j=1}^{n} \lambda_j = 1,\ \lambda_j \geq 0, \forall j$ in its constraints. Thus, as might be expected, they share properties in common and exhibit differences. They also share properties with the corresponding Additive models. Thus, the Additive model without the convexity constraint will characterize a DMU as efficient if and only if the CCR model characterizes it as efficient. Similarly, the BCC model will characterize a DMU as efficient if and only if the corresponding Additive model also characterizes it as efficient.

The concept of "translation invariance" that will be introduced in this chapter deals with lateral shifts of the constraints so that negative data, for instance, may be converted to positive values that admit of treatment by our solution methods, which assume that all data are non-negative. As we will see, the Additive models which include the convexity constraint, are translation invariant but this is not true when the convexity constraint is omitted. CCR models are also not translation invariant while BCC models are translation invariant to changes in the data for only some of their constraints.

These (and other) topics treated in this chapter will provide an overview of model selection possibilities and this will be followed up in further detail in chapters that follow.

The CCR, BCC, Additive and SBM models do not exhaust the available DEA models. Hence in the Appendix to this chapter we introduce a variant, called the Free Disposal Hull (FDH) model which assumes a nonconvex (staircase) production possibility set. Finally, in the problems we ask readers to use their knowledge of the Additive model to develop yet another, known as the "multiplicative model."

4.2 THE BCC MODELS

Let us begin this section with a simple example. Figure 4.3 exhibits 4 DMUs, A, B, C and D, each with one input and one output.

The efficient frontier of the CCR model is the dotted line that passes through B from the origin. The frontiers of the BCC model consists of the bold lines connecting A, B and C. The production possibility set is the area consisting of the frontier together with observed or possible activities with an excess of input and/or a shortfall in output compared with the frontiers. A, B and C are on the frontiers and BCC-efficient. The same is true for all points on the solid lines connecting A and B, and B and C. However, only B is CCR-efficient.

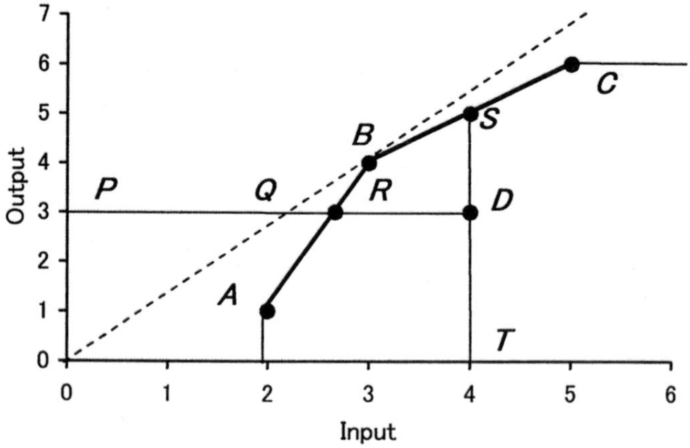

Figure 4.3. The BCC Model

Reading values from the this graph, the BCC-efficiency of D is evaluated by

$$\frac{PR}{PD} = \frac{2.6667}{4} = 0.6667,$$

while its CCR-efficiency is smaller with value

$$\frac{PQ}{PD} = \frac{2.25}{4} = 0.5625.$$

Generally, the CCR-efficiency does not exceed BCC-efficiency.

In the output-oriented BCC model, we read from the vertical axis of Figure 4.3 to find D evaluated by

$$\frac{ST}{DT} = \frac{5}{3} = 1.6667.$$

This means that achievement of efficiency would require augmenting D's output from its observed value to $1.6667 \times 3 = 5$ units. The comparable augmentation for the CCR model is obtained from the reciprocal of its input inefficiency — viz., $1/0.5625 = 1.7778$ so, as the diagram makes clear, a still greater augmentation is needed to achieve efficiency. (Note: this simple "reciprocal relation" between input and output efficiencies is not available for the BCC model.)

Banker, Charnes and Cooper (1984) published the BCC model whose production possibility set P_B is defined by:

$$P_B = \left\{(\boldsymbol{x}, \boldsymbol{y}) | \boldsymbol{x} \geq X\boldsymbol{\lambda}, \boldsymbol{y} \leq Y\boldsymbol{\lambda}, \boldsymbol{e}\boldsymbol{\lambda} = 1, \boldsymbol{\lambda} \geq \boldsymbol{0}\right\}, \tag{4.1}$$

where $X = (\boldsymbol{x}_j) \in R^{m \times n}$ and $Y = (\boldsymbol{y}_j) \in R^{s \times n}$ are a given data set, $\boldsymbol{\lambda} \in R^n$ and $\boldsymbol{e}$ is a row vector with all elements equal to 1. The BCC model differs from the CCR model only in the adjunction of the condition $\sum_{j=1}^{n} \lambda_j = 1$ which we also write $\boldsymbol{e\lambda} = 1$ where $\boldsymbol{e}$ is a row vector with all elements unity and $\boldsymbol{\lambda}$ is a column vector with all elements non-negative. Together with the condition $\lambda_j \geq 0$, for all j, this imposes a convexity condition on allowable ways in which the observations for the n DMUs may be combined.

4.2.1 The BCC Model

The input-oriented BCC model evaluates the efficiency of DMU$_o$ ($o = 1, \ldots, n$) by solving the following (envelopment form) linear program:

$$(BCC_o) \qquad \min_{\theta_B, \boldsymbol{\lambda}} \ \theta_B \tag{4.2}$$

$$\text{subject to} \quad \theta_B \boldsymbol{x}_o - X\boldsymbol{\lambda} \geq \mathbf{0} \tag{4.3}$$

$$Y\boldsymbol{\lambda} \geq \boldsymbol{y}_o \tag{4.4}$$

$$\boldsymbol{e\lambda} = 1 \tag{4.5}$$

$$\boldsymbol{\lambda} \geq \mathbf{0}, \tag{4.6}$$

where θ_B is a scalar.

The dual multiplier form of this linear program (BCC_o) is expressed as:

$$\max_{\boldsymbol{v},\boldsymbol{u},u_0} \ z = \boldsymbol{u}\boldsymbol{y}_o - u_0 \tag{4.7}$$

$$\text{subject to} \quad \boldsymbol{v}\boldsymbol{x}_o = 1 \tag{4.8}$$

$$-\boldsymbol{v}X + \boldsymbol{u}Y - u_0\boldsymbol{e} \leq \mathbf{0} \tag{4.9}$$

$$\boldsymbol{v} \geq \mathbf{0}, \ \boldsymbol{u} \geq \mathbf{0}, \ u_0 \ \text{free in sign}, \tag{4.10}$$

where $\boldsymbol{v}$ and $\boldsymbol{u}$ are vectors and z and u_o are scalars and the latter, being "free in sign," may be positive or negative (or zero). The equivalent BCC fractional program is obtained from the dual program as:

$$\max \ \frac{\boldsymbol{u}\boldsymbol{y}_o - u_0}{\boldsymbol{v}\boldsymbol{x}_o} \tag{4.11}$$

$$\text{subject to} \quad \frac{\boldsymbol{u}\boldsymbol{y}_j - u_0}{\boldsymbol{v}\boldsymbol{x}_j} \leq 1 \ \ (j = 1, \ldots, n) \tag{4.12}$$

$$\boldsymbol{v} \geq \mathbf{0}, \ \boldsymbol{u} \geq \mathbf{0}, \ u_0 \ \text{free}. \tag{4.13}$$

Correspondences between the primal-dual constraints and variables can be represented as in Table 4.1.

It is clear that a difference between the CCR and BCC models is present in the free variable u_0, which is the dual variable associated with the constraint $\boldsymbol{e\lambda} = 1$ in the envelopment model that also does not appear in the CCR model.

The primal problem (BCC_o) is solved using a two-phase procedure similar to the CCR case. In the first phase, we minimize θ_B and, in the second phase, we maximize the sum of the input excesses and output shortfalls, keeping $\theta_B = \theta_B^*$ (the optimal objective value obtained in Phase one). The evaluations secured

Table 4.1. Primal and Dual Correspondences in BCC Model

Linear Programming Form

Envelopment form constraints	Multiplier form variables	Multiplier form constraints	Envelopment form variables
$\theta_B \boldsymbol{x}_o - X\boldsymbol{\lambda} \geq \boldsymbol{0}$	$\boldsymbol{v} \geq \boldsymbol{0}$	$\boldsymbol{v}\boldsymbol{x}_o = 1$	θ
$Y\boldsymbol{\lambda} \geq \boldsymbol{y}_o$	$\boldsymbol{u} \geq \boldsymbol{0}$	$-\boldsymbol{v}X + \boldsymbol{u}Y - u_0\boldsymbol{e} \leq \boldsymbol{0}$	$\boldsymbol{\lambda} \geq \boldsymbol{0}$
$\boldsymbol{e}\boldsymbol{\lambda} = 1$	u_0		

from the CCR and BCC models are also related to each other as follows. An optimal solution for (BCC_o) is represented by $(\theta_B^*, \boldsymbol{\lambda}^*, \boldsymbol{s}^{-*}, \boldsymbol{s}^{+*})$, where $\boldsymbol{s}^{-*}$ and $\boldsymbol{s}^{+*}$ represent the maximal input excesses and output shortfalls, respectively. Notice that θ_B^* is not less than the optimal objective value θ^* of the CCR model, since (BCC_o) imposes one additional constraint, $\boldsymbol{e}\boldsymbol{\lambda} = 1$, so its feasible region is a subset of feasible region for the CCR model.

Definition 4.1 (BCC-Efficiency)
If an optimal solution $(\theta_B^*, \boldsymbol{\lambda}^*, \boldsymbol{s}^{-*}, \boldsymbol{s}^{+*})$ obtained in this two-phase process for (BCC_o) satisfies $\theta_B^* = 1$ and has no slack ($\boldsymbol{s}^{-*} = \boldsymbol{0}$, $\boldsymbol{s}^{+*} = \boldsymbol{0}$), then the DMU_o is called BCC-efficient, *otherwise it is* BCC-inefficient.

Definition 4.2 (Reference Set)
For a BCC-inefficient DMU_o, we define its reference set, *E_o, based on an optimal solution $\boldsymbol{\lambda}^*$ by*

$$E_o = \{\, j|\ \lambda_j^* > 0\}\ (j \in \{1, \ldots, n\}). \tag{4.14}$$

If there are multiple optimal solutions, we can choose any one to find that

$$\theta_B^* \boldsymbol{x}_o = \sum_{j \in E_o} \lambda_j^* \boldsymbol{x}_j + \boldsymbol{s}^{-*} \tag{4.15}$$

$$\boldsymbol{y}_o = \sum_{j \in E_o} \lambda_j^* \boldsymbol{y}_j - \boldsymbol{s}^{+*}. \tag{4.16}$$

Thus, we have a formula for improvement via the *BCC-projection*,

$$\widehat{\boldsymbol{x}}_o \Leftarrow \theta_B^* \boldsymbol{x}_o - \boldsymbol{s}^{-*} \tag{4.17}$$

$$\widehat{\boldsymbol{y}}_o \Leftarrow \boldsymbol{y}_o + \boldsymbol{s}^{+*}. \tag{4.18}$$

The following two theorems and the lemma can be proved in a way similar to proofs used for the CCR model in Chapter 3. Thus, analogous to Theorem 3.2 for the CCR model, we have

Theorem 4.1 *The improved activity* $(\widehat{\boldsymbol{x}}_o, \widehat{\boldsymbol{y}}_o)$ *is BCC-efficient.*

Similarly, the following adapts Lemma 3.1 to the BCC model,

Lemma 4.1 *For the improved activity* $(\widehat{\boldsymbol{x}}_o, \widehat{\boldsymbol{y}}_o)$*, there exists an optimal solution* $(\widehat{\boldsymbol{v}}_o, \widehat{\boldsymbol{u}}_o, \widehat{u}_0)$ *for its dual problem such that*

$$\widehat{\boldsymbol{v}}_o > \mathbf{0} \text{ and } \widehat{\boldsymbol{u}}_o > \mathbf{0} \tag{4.19}$$

$$\widehat{\boldsymbol{v}}_o \boldsymbol{x}_j = \widehat{\boldsymbol{u}}_o \boldsymbol{y}_j - \widehat{u}_0 \quad (j \in E_o) \tag{4.20}$$

$$\widehat{\boldsymbol{v}}_o X \geq \widehat{\boldsymbol{u}}_o Y - \widehat{u}_0 \boldsymbol{e}. \tag{4.21}$$

Theorem 4.2 *Every DMU in* E_o *associated with a* $\lambda_j^* > 0$ *as defined in* (4.14), *is BCC-efficient.*

This is an extension of Theorem 3.3. Finally, however, the next theorem exposes a property of BCC-efficiency for the input-oriented version of the model. This property is not secured by the CCR model, so the two may be used to check whether this property is present.

Theorem 4.3 *A DMU that has a minimum input value for any input item, or a maximum output value for any output item, is BCC-efficient.*

Proof. Suppose that DMU$_o$ has a minimum input value for input 1, i.e., $x_{1o} < x_{1j}$ $(\forall j \neq o)$. Then, from (4.3) and (4.5), DMU$_o$ has the unique solution $(\theta_B^* = 1, \lambda_o^* = 1, \lambda_j^* = 0\ (\forall j \neq o))$. Hence, DMU$_o$ has $\theta_B^* = 1$ with no slacks and is BCC-efficient. The maximum output case can be proved analogously. □

4.2.2 The Output-oriented BCC Model

Turning to the output-oriented BCC model we write

$$(BCC - O_o) \qquad \max_{\eta_B, \boldsymbol{\lambda}} \ \eta_B \tag{4.22}$$

$$\text{subject to} \quad X\boldsymbol{\lambda} \leq \boldsymbol{x}_o \tag{4.23}$$

$$\eta_B \boldsymbol{y}_o - Y\boldsymbol{\lambda} \leq \mathbf{0} \tag{4.24}$$

$$\boldsymbol{e}\boldsymbol{\lambda} = 1 \tag{4.25}$$

$$\boldsymbol{\lambda} \geq \mathbf{0}. \tag{4.26}$$

This is the envelopment form of the output-oriented BCC model. The dual (multiplier) form associated with the above linear program $(BCC - O_o)$ is expressed as:

$$\min_{\boldsymbol{v}, \boldsymbol{u}, v_0} \ z = \boldsymbol{v}\boldsymbol{x}_o - v_0 \tag{4.27}$$

$$\text{subject to} \quad \boldsymbol{u}\boldsymbol{y}_o = 1 \tag{4.28}$$

$$\boldsymbol{v}X - \boldsymbol{u}Y - v_0\boldsymbol{e} \geq \mathbf{0} \tag{4.29}$$

$$\boldsymbol{v} \geq \mathbf{0}, \ \boldsymbol{u} \geq \mathbf{0}, \ \ v_0 \text{ free in sign,} \tag{4.30}$$

where v_0 is the scalar associated with $e\lambda = 1$ in the envelopment model. Finally, we have the equivalent (BCC) fractional programming formulation for the latter (multiplier) model:

$$\min \frac{vx_o - v_0}{uy_o} \tag{4.31}$$

$$\text{subject to} \quad \frac{vx_j - v_0}{uy_j} \geq 1 \ (j = 1, \ldots, n) \tag{4.32}$$

$$v \geq 0, \quad u \geq 0, \quad v_0 \text{ free in sign.} \tag{4.33}$$

4.3 THE ADDITIVE MODEL

The preceding models required us to distinguish between input-oriented and output-oriented models. Now, however, we combine both orientations in a single model, called the *Additive model.*

4.3.1 The Basic Additive Model

There are several types of Additive models, from which we select following:

$$(ADD_o) \quad \max_{\lambda, s^-, s^+} \quad z = es^- + es^+ \tag{4.34}$$

$$\text{subject to} \quad X\lambda + s^- = x_o \tag{4.35}$$

$$Y\lambda - s^+ = y_o \tag{4.36}$$

$$e\lambda = 1 \tag{4.37}$$

$$\lambda \geq 0, \quad s^- \geq 0, \ s^+ \geq 0. \tag{4.38}$$

A variant, which we do not explore here, is an Additive model[3] which omits the condition $e\lambda = 1$.

The dual problem to the above can be expressed as follows:

$$\min_{v, u, u_0} \quad w = vx_o - uy_o + u_0 \tag{4.39}$$

$$\text{subject to} \quad vX - uY + u_0 e \geq 0 \tag{4.40}$$

$$v \geq e \tag{4.41}$$

$$u \geq e \tag{4.42}$$

$$u_0 \ \text{ free.} \tag{4.43}$$

To explain this model we use Figure 4.4, where four DMUs A, B, C and D, each with one input and one output, are depicted. Since, by (4.35)– (4.38), the model (ADD_o) has the same production possibility set as the BCC model, the efficient frontier, which is continuous, consists of the line segments $\overline{AB}$ and $\overline{BC}$. Now consider how DMU D might be evaluated. A feasible replacement of D with s^- and s^+ is denoted by the arrows s^- and s^+ in the figure. As shown by the dotted line in the figure, the maximal value of $s^- + s^+$ is attained at B. It is clear that this model considers the input excess and the output shortfall simultaneously in arriving at a point on the efficient frontier which is most distant from D.[4]

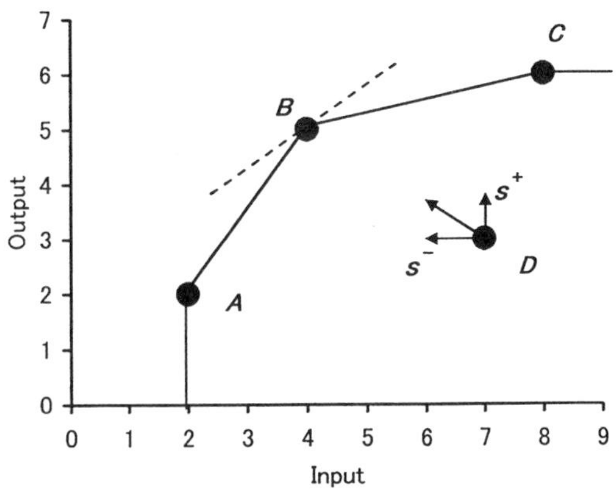

Figure 4.4. The Additive Model

Taking these considerations into account we can obtain a definition of efficiency as follows for the Additive model.

Let the optimal solutions be $(\boldsymbol{\lambda}^*,\ \boldsymbol{s}^{-*},\ \boldsymbol{s}^{+*})$. The definition of efficiency for an efficient DMU in the Additive model is then given by:

Definition 4.3 (ADD-efficient DMU)
DMU_o is ADD-efficient *if and only if* $\boldsymbol{s}^{-*} = \mathbf{0}$ *and* $\boldsymbol{s}^{+*} = \mathbf{0}$.

Theorem 4.4 *DMU_o is ADD-efficient if and only if it is BCC-efficient.*

A proof of this theorem may be found in Ahn *et al.*[5] Here, however, it suffices to note that the efficiency score θ^* is not measured explicitly but is implicitly present in the slacks $\boldsymbol{s}^{-*}$ and $\boldsymbol{s}^{+*}$. Moreover, whereas θ^* reflects only Farrell (=weak) efficiency, the objective in (ADD_o) reflects all inefficiencies that the model can identify in *both* inputs and outputs.

Theorem 4.5 *Let us define* $\widehat{\boldsymbol{x}}_o = \boldsymbol{x}_o - \boldsymbol{s}^{-*}$ *and* $\widehat{\boldsymbol{y}}_o = \boldsymbol{y}_o + \boldsymbol{s}^{+*}$. *Then,* $(\widehat{\boldsymbol{x}}_o,\ \widehat{\boldsymbol{y}}_o)$ *is ADD-efficient.*

By this theorem, improvement to an efficient activity is attained by the following formulae (Projection for the Additive model):

$$\widehat{\boldsymbol{x}}_o \Leftarrow \boldsymbol{x}_o - \boldsymbol{s}^{-*} \tag{4.44}$$
$$\widehat{\boldsymbol{y}}_o \Leftarrow \boldsymbol{y}_o + \boldsymbol{s}^{+*}, \tag{4.45}$$

with $(\widehat{\boldsymbol{x}}_o,\ \widehat{\boldsymbol{y}}_o)$ serving as the coordinates for the point on the efficient frontier used to evaluate DMU_o.

Example 4.1

We clarify the above with Table 4.2 which shows 8 DMUs with one input and one output. The solution of the Additive model and that of the BCC model are both exhibited. The reference set is determined as the set of DMUs which are in the optimal basic set of the LP problem. It is thus of interest to note that B, C and E are all fully efficient under both the BCC and (ADD_o) model, as asserted in Theorem 4.4. Furthermore, the nonzero slack under s^{+*} for A shows that it is not fully efficient for the BCC model (as well as (ADD_o)) even though it is weakly efficient with a value of $\theta^* = 1$. This means that A is on a part of the frontier that is not efficient, of course, while all of the other values with $\theta^* < 1$ mean that they fail even to achieve the frontier and, as can be observed, they also have nonzero slacks in their optimum (2nd-stage) solutions. For the latter observations, therefore, a projection to $\hat{\boldsymbol{x}}_o = \theta^* \boldsymbol{x}_o$ need not achieve efficiency. It could merely bring the thus adjusted performance onto a portion of the frontier which is not efficient — as was the case for A.

Table 4.2. Data and Results of Example 4.1

	Input	Output	BCC	Additive Model		
DMU	x	y	θ^*	s^{-*}	s^{+*}	Ref.
A	2	1	1	0	1	C
B	3	3	1	0	0	B
C	2	2	1	0	0	C
D	4	3	0.75	1	0	B
E	6	5	1	0	0	E
F	5	2	0.40	3	0	C
G	6	3	0.50	3	0	B
H	8	5	0.75	2	0	E

The following definition brings up another important distinction between the Additive model and BCC (or CCR) models;

Definition 4.4 (Mix) *We define "Mix" as proportions in which inputs are used or in which outputs are produced.*

Returning to the BCC (or CCR) models, it can be observed that $(1 - \theta^*)$ represents reductions which can be achieved without altering the input mix utilized and $(\eta^* - 1)$ would play a similar role for output expansions which do not alter the output mix.

In the literature of DEA, as well as economics, this proportion is referred to as "technical inefficiency" as contrasted with "scale inefficiencies," "allocative (or price) inefficiencies" and other type of inefficiencies that will be discussed in subsequent chapters.

With all such technical inefficiencies accounted for in the BCC and CCR models, the second stage optimization is directed to maximize the slacks in order to see whether further inefficiencies are present. Altering any nonzero slack obtained in the second stage optimization must then necessarily alter the mix.

As is now apparent, the CCR and BCC models distinguish between technical and mix inefficiencies and this is a source of trouble in trying to arrange for a single measure of efficiency. The allocative model makes no such distinction, however, and so in Section 4.4 we will use the opportunity provided by these models to develop a measure that comprehends *all* of the inefficiencies that the model can identify.

As noted above, there is a distinction to be made between the technical and mix inefficiencies identified by the BCC and CCR models. The DEA literature (much of it focused on weak efficiency) does not make this distinction and we will follow the literature in its rather loose usage of "technical efficiency" to cover both types of inefficiency. When it is important to do so, we should refine this terminology by using "purely technical inefficiency" for θ^* and η^* and distinguish this from the mix inefficiencies given in Definition 4.4.

4.3.2 Translation Invariance of the Additive Model

In many applications it may be necessary (or convenient) to be able to handle negative data in some of the inputs or outputs. For instance, in order to determine whether mutual insurance companies are more (or less) efficient than their stock-ownership counterparts, it was necessary to be able to go beyond the assumption of semipositive data (as defined at the start of Chapter 3) in order to handle losses as well as profits treated as output.[6] This was dealt with by employing a property of (ADD_o) known as "translation invariance" which we now develop from the following definition.

Definition 4.5 (Translation Invariance)
Given any problem, a DEA model is said to be translation invariant *if translating the original input and/or output data values results in a new problem that has the same optimal solution for the envelopment form as the old one.*

First we examine the input-oriented BCC model. In Figure 4.5, DMU D has the BCC-efficiency PR/PD. This ratio is invariant even if we shift the output value by changing the origin from O to O'. Thus, the BCC model is translation invariant with respect to *outputs* (but not *inputs*). Similar reasoning shows that the output oriented BCC model is invariant under the translation of *inputs* (but not *outputs*). (See Problem 4.3.)

Turning to the Additive model, Figure 4.6 shows that this model is translation invariant in both *inputs* and *outputs*, since the efficiency evaluation does not depend on the origin of the coordinate system when this model is used.

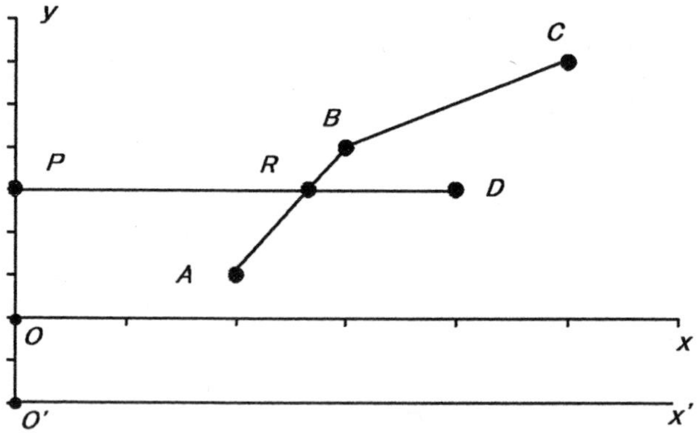

Figure 4.5. Translation in the BCC Model

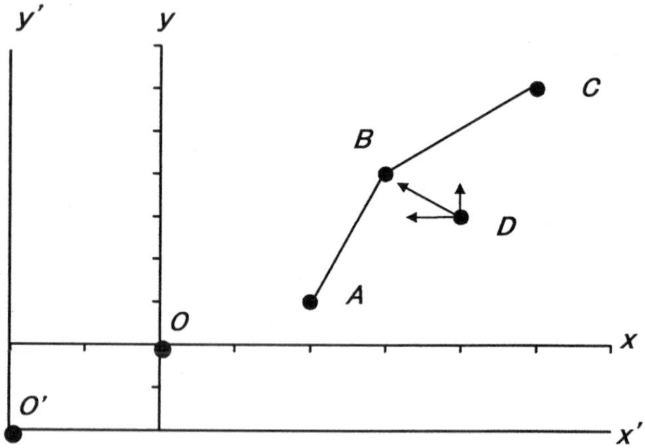

Figure 4.6. Translation in the Additive Model

We now develop this property of the Additive model in detail.

Let us translate the data set (X, Y) by introducing arbitrary constants $(\alpha_i : i = 1, \ldots, m)$ and $(\beta_r : r = 1, \ldots, s)$ to obtain new data

$$x'_{ij} = x_{ij} + \alpha_i \quad (i = 1, \ldots, m : j = 1, \ldots, n) \tag{4.46}$$

$$y'_{rj} = y_{rj} + \beta_r, \quad (r = 1, \ldots, s : j = 1, \ldots, n) \tag{4.47}$$

To show that this model is invariant under this arbitrary translation we observe that the x values (4.35) become

$$\sum_{j=1}^{n}(x'_{ij} - \alpha_i)\lambda_j + s_i^- = \sum_{j=1}^{n} x'_{ij}\lambda_j + s_i^- - \alpha_i = x'_{io} - \alpha_i$$

so that

$$\sum_{j=1}^{n} x'_{ij}\lambda_j + s_i^- = x'_{io} \quad (i = 1, \ldots, m)$$

which are the same λ_j, s_i^- that satisfy

$$\sum_{j=1}^{n} x_{ij}\lambda_j + s_i^- = x_{io}. \quad (i = 1, \ldots, m)$$

Similarly, the same λ_j, s_r^+ that satisfy

$$\sum_{j=1}^{n} y_{rj}\lambda_j - s_r^+ = y_{ro}. \quad (r = 1, \ldots, s)$$

will also satify

$$\sum_{j=1}^{n} y'_{rj}\lambda_j - s_r^+ = y'_{ro}. \quad (r = 1, \ldots, s)$$

Notice that the convexity condition $\boldsymbol{e\lambda} = 1$ is a key factor in deriving the above relations. The above equalities show that if $(\boldsymbol{\lambda}^*, \boldsymbol{s}^{-*}, \boldsymbol{s}^{+*})$ is an optimal solution of the original problem, then it is also optimal for the translated problem and *vice versa*. Finally, we also have

$$s_i^- = \sum_{j=1}^{n} x_{ij}\lambda_j - x_{io} = \sum_{j=1}^{n} x'_{ij}\lambda_j - x'_{io}$$

$$s_r^+ = \sum_{j=1}^{n} y_{rj}\lambda_j - y_{ro} = \sum_{j=1}^{n} y'_{rj}\lambda_j - y'_{ro}$$

so the value of the objective is also not affected and we therefore have the following theorem for the Additive model (4.34)-(4.38) from Ali and Seiford (1990).[7]

Theorem 4.6 (Ali and Seiford (1990)) *The Additive model given by (4.34)-(4.38) is translation invariant.*

4.4 A SLACKS-BASED MEASURE OF EFFICIENCY (SBM)

We now augment the Additive models by introducing a measure that makes its efficiency evaluation, as effected in the objective, invariant to the units of measure used for the different inputs and outputs. That is, we would like this summary measure to assume the form of a scalar that yields the same efficiency value when distances are measured in either kilometers or miles. More generally, we want this measure to be the same when x_{io} and x_{ij} are replaced by $k_i x_{io} = \widehat{x}_{io}$, $k_i x_{ij} = \widehat{x}_{ij}$ and y_{ro} and y_{rj} are replaced by $c_r y_{ro} = \widehat{y}_{ro}$, $c_r y_{rj} = \widehat{y}_{rj}$ where the k_i and c_r are arbitrary positive constants, $i = 1, \ldots, m$; $r = 1, \ldots, s$.

This property is known by names such as "dimension free"[8] and "units invariant." In this section we introduce such a measure for Additive models in the form of a single scalar called "SBM," (Slacks-Based Measure) which was introduced by Tone (1997, 2001) and has the following important properties:

1. (P1) The measure is invariant with respect to the unit of measurement of each input and output item. (Units invariant)
2. (P2) The measure is monotone decreasing in each input and output slack. (Monotone)

4.4.1 *Definition of SBM*

In order to estimate the efficiency of a DMU $(\boldsymbol{x}_o,\ \boldsymbol{y}_o)$, we formulate the following fractional program in $\boldsymbol{\lambda}$, $\boldsymbol{s}^-$ and $\boldsymbol{s}^+$.

$$(SBM) \qquad \min_{\boldsymbol{\lambda},\boldsymbol{s}^-,\boldsymbol{s}^+} \quad \rho = \frac{1 - \frac{1}{m}\sum_{i=1}^{m} s_i^-/x_{io}}{1 + \frac{1}{s}\sum_{r=1}^{s} s_r^+/y_{ro}} \tag{4.48}$$

$$\text{subject to} \quad \boldsymbol{x}_o = X\boldsymbol{\lambda} + \boldsymbol{s}^-$$
$$\boldsymbol{y}_o = Y\boldsymbol{\lambda} - \boldsymbol{s}^+$$
$$\boldsymbol{\lambda} \geq \mathbf{0},\ \boldsymbol{s}^- \geq \mathbf{0},\ \boldsymbol{s}^+ \geq \mathbf{0}.$$

In this model, we assume that $X \geq O$. If $x_{io} = 0$, then we delete the term s_i^-/x_{io} in the objective function. If $y_{ro} \leq 0$, then we replace it by a very small positive number so that the term s_r^+/y_{ro} plays a role of penalty.

It is readily verified that the objective function value ρ satisfies (P1) because the numerator and denominator are measured in the same units for every item in the objective of (4.48). It is also readily verified that an increase in either s_i^- or s_r^+, all else held constant, will decrease this objective value and, indeed, do so in a strictly monotone manner.

Furthermore, we have

$$0 \leq \rho \leq 1. \tag{4.49}$$

To see that this relation holds we first observe that $s_i^- \leq x_{io}$ for every i so that $0 \leq s_i^-/x_{io} \leq 1$ $(i = 1, \ldots, m)$ with $s_i^-/x_{io} = 1$ only if the evidence shows that only a zero amount of this input was required. It follows that

$$0 \leq \frac{\sum_{i=1}^{m} s_i^-/x_{io}}{m} \leq 1.$$

This same relation does not hold for outputs since an output shortfall represented by a nonzero slack can exceed the corresponding amount of output produced. In any case, however, we have

$$0 \leq \frac{\sum_{r=1}^{s} s_r^+/y_{ro}}{s}.$$

Thus these represent ratios of average input and output mix inefficiencies with the upper limit, $\rho = 1$, reached in (4.48) only if slacks are zero in all inputs and outputs.

4.4.2 Interpretation of SBM as a Product of Input and Output Inefficiencies

The formula for ρ in (4.48) can be transformed into

$$\rho = \left(\frac{1}{m}\sum_{i=1}^{m}\frac{x_{io}-s_i^-}{x_{io}}\right)\left(\frac{1}{s}\sum_{r=1}^{s}\frac{y_{ro}+s_r^+}{y_{ro}}\right)^{-1}.$$

The ratio $(x_{io} - s_i^-)/x_{io}$ evaluates the relative reduction rate of input i and, therefore, the first term corresponds to the mean proportional reduction rate of inputs or *input mix inefficiencies.* Similarly, in the second term, the ratio $(y_{ro} + s_r^+)/y_{ro}$ evaluates the relative proportional expansion rate of output r and $(1/s)\sum(y_{ro} + s_r^+)/y_{ro}$ is the mean proportional rate of output expansion. Its inverse, the second term, measures *output mix inefficiency.* Thus, SBM ρ can be interpreted as the ratio of mean input and output mix inefficiencies. Further, we have the theorem:

Theorem 4.7 *If DMU A dominates DMU B so that $\boldsymbol{x}_A \leq \boldsymbol{x}_B$ and $\boldsymbol{y}_A \geq \boldsymbol{y}_B$, then $\rho_A^* \geq \rho_B^*$.*

4.4.3 Solving SBM

(SBM) as formulated in (4.48) can be transformed into the program below by introducing a positive scalar variable t. See Problem 3.1 at the end of the preceding chapter.

$$(SBMt) \qquad \min_{t,\boldsymbol{\lambda},\boldsymbol{s}^-,\boldsymbol{s}^+} \quad \tau = t - \frac{1}{m}\sum_{i=1}^{m} ts_i^-/x_{io} \tag{4.50}$$

$$\text{subject to} \quad 1 = t + \frac{1}{s}\sum_{r=1}^{s} ts_r^+/y_{ro}$$
$$\boldsymbol{x}_o = X\boldsymbol{\lambda} + \boldsymbol{s}^-$$
$$\boldsymbol{y}_o = Y\boldsymbol{\lambda} - \boldsymbol{s}^+$$
$$\boldsymbol{\lambda} \geq \mathbf{0},\ \boldsymbol{s}^- \geq \mathbf{0},\ \boldsymbol{s}^+ \geq \mathbf{0},\ t > 0.$$

Now let us define

$$\boldsymbol{S}^- = t\boldsymbol{s}^-, \boldsymbol{S}^+ = t\boldsymbol{s}^+, \text{and } \boldsymbol{\Lambda} = t\boldsymbol{\lambda}.$$

Then $(SBMt)$ becomes the following linear program in t, $\boldsymbol{S}^-$, $\boldsymbol{S}^+$, and $\boldsymbol{\Lambda}$:

$$(LP) \quad \min \quad \tau = t - \frac{1}{m}\sum_{i=1}^{m} S_i^-/x_{io} \tag{4.51}$$

$$\text{subject to} \quad 1 = t + \frac{1}{s}\sum_{r=1}^{s} S_r^+/y_{ro}$$
$$t\boldsymbol{x}_o = X\boldsymbol{\Lambda} + \boldsymbol{S}^-$$
$$t\boldsymbol{y}_o = Y\boldsymbol{\Lambda} - \boldsymbol{S}^+$$
$$\boldsymbol{\Lambda} \geq \mathbf{0},\ \boldsymbol{S}^- \geq \mathbf{0},\ \boldsymbol{S}^+ \geq \mathbf{0},\ t > 0.$$

Note that $t > 0$ by virtue of the first constraint. This means that the transformation is reversible. Thus let an optimal solution of (LP) be

$$(\tau^*, t^*, \boldsymbol{\Lambda}^*, \boldsymbol{S}^{-*}, \boldsymbol{S}^{+*}).$$

We then have an optimal solution of (SBM) defined by,

$$\rho^* = \tau^*, \ \boldsymbol{\lambda}^* = \boldsymbol{\Lambda}^*/t^*, \ \boldsymbol{s}^{-*} = \boldsymbol{S}^{-*}/t^*, \ \boldsymbol{s}^{+*} = \boldsymbol{S}^{+*}/t^*. \tag{4.52}$$

From this optimal solution, we can decide whether a DMU is *SBM-efficient* as follows:

Definition 4.6 (SBM-efficient) *A DMU* $(\boldsymbol{x}_o, \boldsymbol{y}_o)$ *is SBM-efficient if and only if* $\rho^* = 1$.

This condition is equivalent to $\boldsymbol{s}^{-*} = \boldsymbol{0}$ *and* $\boldsymbol{s}^{+*} = \boldsymbol{0}$, i.e., no input excess and no output shortfall in an optimal solution.

For an SBM-inefficient DMU $(\boldsymbol{x}_o, \boldsymbol{y}_o)$, we have the expression:

$$\boldsymbol{x}_o = X\boldsymbol{\lambda}^* + \boldsymbol{s}^{-*}$$
$$\boldsymbol{y}_o = Y\boldsymbol{\lambda}^* - \boldsymbol{s}^{+*}.$$

The DMU $(\boldsymbol{x}_o, \boldsymbol{y}_o)$ can be improved and becomes efficient by deleting the input excesses and augmenting the output shortfalls. This is accomplished by the following formulae (SBM-projection):

$$\widehat{\boldsymbol{x}}_o \Leftarrow \boldsymbol{x}_o - \boldsymbol{s}^{-*} \tag{4.53}$$
$$\widehat{\boldsymbol{y}}_o \Leftarrow \boldsymbol{y}_o + \boldsymbol{s}^{+*}, \tag{4.54}$$

which are the same as for the Additive model. See (4.44) and (4.45).

Based on $\boldsymbol{\lambda}^*$, we define the reference set for $(\boldsymbol{x}_o, \boldsymbol{y}_o)$ as:

Definition 4.7 (Reference set) *The set of indices corresponding to positive* λ_j^**s is called the* reference set *for* $(\boldsymbol{x}_o, \boldsymbol{y}_o)$.

For multiple optimal solutions, the reference set is not unique. We can, however, choose any one for our purposes.

Let R_o be the reference set designated by

$$R_o = \{\, j \mid \lambda_j^* > 0\} \quad (j \in \{1, \ldots, n\}). \tag{4.55}$$

Then using R_o, we can also express $(\widehat{\boldsymbol{x}}_o, \widehat{\boldsymbol{y}}_o)$ by,

$$\widehat{\boldsymbol{x}}_o = \sum_{j \in R_o} \boldsymbol{x}_j \lambda_j^* \tag{4.56}$$

$$\widehat{\boldsymbol{y}}_o = \sum_{j \in R_o} \boldsymbol{y}_j \lambda_j^*. \tag{4.57}$$

This means that $(\widehat{\boldsymbol{x}}_o, \widehat{\boldsymbol{y}}_o)$, a point on the efficient frontier, is expressed as a positive combination of the members of the reference set, R_o, each member of which is also efficient. See Definition 4.2.

4.4.4 SBM and the CCR Measure

In its weak efficiency form, the CCR model can be formulated as follows:

$$(CCR) \qquad \min_{\theta,\boldsymbol{\mu},\boldsymbol{t}^-,\boldsymbol{t}^+} \theta$$

$$\text{subject to} \quad \theta \boldsymbol{x}_o = X\boldsymbol{\mu} + \boldsymbol{t}^- \tag{4.58}$$

$$\boldsymbol{y}_o = Y\boldsymbol{\mu} - \boldsymbol{t}^+ \tag{4.59}$$

$$\boldsymbol{\mu} \geq \boldsymbol{0}, \ \boldsymbol{t}^- \geq \boldsymbol{0}, \ \boldsymbol{t}^+ \geq \boldsymbol{0}.$$

Now let an optimal solution of (CCR) be $(\theta^*, \boldsymbol{\mu}^*, \boldsymbol{t}^{-*}, \boldsymbol{t}^{+*})$. From (4.58), we can derive

$$\boldsymbol{x}_o = X\boldsymbol{\mu}^* + \boldsymbol{t}^{-*} + (1-\theta^*)\boldsymbol{x}_o \tag{4.60}$$

$$\boldsymbol{y}_o = Y\boldsymbol{\mu}^* - \boldsymbol{t}^{+*}. \tag{4.61}$$

Let us define

$$\boldsymbol{\lambda} = \boldsymbol{\mu}^* \tag{4.62}$$

$$\boldsymbol{s}^- = \boldsymbol{t}^{-*} + (1-\theta^*)\boldsymbol{x}_o \tag{4.63}$$

$$\boldsymbol{s}^+ = \boldsymbol{t}^{+*}. \tag{4.64}$$

Then, $(\boldsymbol{\lambda}, \boldsymbol{s}^-, \boldsymbol{s}^+)$ is feasible for (SBM) and, by inserting the definitions in (4.63) and (4.64), its objective value can be expressed as:

$$\rho = \frac{1 - \frac{1}{m}\{\sum_{i=1}^m t_i^{-*}/x_{io} + m(1-\theta^*)\}}{1 + \frac{1}{s}\sum_{r=1}^s t_r^{+*}/y_{ro}} = \frac{\theta^* - \frac{1}{m}\sum_{i=1}^m t_i^{-*}/x_{io}}{1 + \frac{1}{s}\sum_{r=1}^s t_r^{+*}/y_{ro}}. \tag{4.65}$$

Evidently, the last term is not greater than θ^*. Thus, we have:

Theorem 4.8 *The optimal SBM ρ^* is not greater than the optimal CCR θ^*.*

This theorem reflects the fact that SBM accounts for *all* inefficiencies whereas θ^* accounts only for "purely technical" inefficiencies. Notice that the coefficient $1/(m\, x_{io})$ of the input excesses s_i^- in ρ plays a crucial role in validating Theorem 4.8.

Conversely, for an optimal solution $(\rho^*, \boldsymbol{\lambda}^*, \boldsymbol{s}^{-*}, \boldsymbol{s}^{+*})$ to SBM, let us transform the constraints from (4.48) into

$$\theta \boldsymbol{x}_o = X\boldsymbol{\lambda}^* + (\theta - 1)\boldsymbol{x}_o + \boldsymbol{s}^{-*} \tag{4.66}$$

$$\boldsymbol{y}_o = Y\boldsymbol{\lambda}^* - \boldsymbol{s}^{+*}. \tag{4.67}$$

Further, we add the constraint

$$(\theta - 1)\boldsymbol{x}_o + \boldsymbol{s}^{-*} \geq \boldsymbol{0}. \tag{4.68}$$

Then, $(\theta, \boldsymbol{\lambda}^*, \boldsymbol{t}^- = (\theta-1)\boldsymbol{x}_o + \boldsymbol{s}^{-*}, \boldsymbol{t}^+ = \boldsymbol{s}^{+*})$ is feasible for (CCR).

The relationship between CCR-efficiency and SBM-efficiency is given in the following theorem:

Theorem 4.9 (Tone (1997)) *A DMU $(\boldsymbol{x}_o, \boldsymbol{y}_o)$ is CCR-efficient if and only if it is SBM-efficient.*

Proof. Suppose that $(\boldsymbol{x}_o, \boldsymbol{y}_o)$ is CCR-inefficient. Then, we have either $\theta^* < 1$ or ($\theta^* = 1$ and $(\boldsymbol{t}^{-*}, \boldsymbol{t}^{+*}) \neq (\boldsymbol{0}, \boldsymbol{0})$). From (4.65), in both cases, we have $\rho < 1$ for a feasible solution of (SBM). Therefore, $(\boldsymbol{x}_o, \boldsymbol{y}_o)$ is SBM-inefficient.

On the other hand, suppose that $(\boldsymbol{x}_o, \boldsymbol{y}_o)$ is SBM-inefficient. Then, from Definition 4.6, $(\boldsymbol{s}^{-*}, \boldsymbol{s}^{+*}) \neq (\boldsymbol{0}, \boldsymbol{0})$. By (4.66) and (4.67), $(\theta, \boldsymbol{\lambda}^*, \boldsymbol{t}^- = (\theta - 1)\boldsymbol{x}_o + \boldsymbol{s}^{-*}, \boldsymbol{t}^+ = \boldsymbol{s}^{+*})$ is feasible for (CCR), provided $(\theta - 1)\boldsymbol{x}_o + \boldsymbol{s}^{-*} \geq \boldsymbol{0}$. There are two cases:
(Case 1) $\theta = 1$ and $(\boldsymbol{t}^- = \boldsymbol{s}^{-*}, \boldsymbol{t}^+ = \boldsymbol{s}^{+*}) \neq (\boldsymbol{0}, \boldsymbol{0})$. In this case, an optimal solution for (CCR) is CCR-inefficient.
(Case 2) $\theta < 1$. In this case, $(\boldsymbol{x}_o, \boldsymbol{y}_o)$ is CCR-inefficient.

Therefore, CCR-inefficiency is equivalent to SBM-inefficiency. Since the definitions of *efficient* and *inefficient* are mutually exclusive and collectively exhaustive, we have proved the theorem. □

4.4.5 *The Dual Program of the SBM Model*

The dual program of the problem (*LP*) in (4.50) can be expressed as follows, with the dual variables $\xi \in R$, $\boldsymbol{v} \in R^m$ and $\boldsymbol{u} \in R^s$:

$$(DP) \qquad \max_{\xi, \boldsymbol{v}, \boldsymbol{u}} \ \xi \tag{4.69}$$

$$\text{subject to} \qquad \xi + \boldsymbol{v}\boldsymbol{x}_o - \boldsymbol{u}\boldsymbol{y}_o = 1 \tag{4.70}$$

$$-\boldsymbol{v}X + \boldsymbol{u}Y \leq \boldsymbol{0} \tag{4.71}$$

$$\boldsymbol{v} \geq \frac{1}{m}[1/\boldsymbol{x}_o] \tag{4.72}$$

$$\boldsymbol{u} \geq \frac{\xi}{s}[1/\boldsymbol{y}_o], \tag{4.73}$$

where the notation $[1/\boldsymbol{x}_o]$ designates the row vector $(1/x_{1o}, \ldots, 1/x_{mo})$.

Using (4.70) we can eliminate ξ and we have an equivalent program:

$$(DP') \qquad \max \qquad \boldsymbol{u}\boldsymbol{y}_o - \boldsymbol{v}\boldsymbol{x}_o \tag{4.74}$$

$$\text{subject to} \qquad -\boldsymbol{v}X + \boldsymbol{u}Y \leq \boldsymbol{0} \tag{4.75}$$

$$\boldsymbol{v} \geq \frac{1}{m}[1/\boldsymbol{x}_o] \tag{4.76}$$

$$\boldsymbol{u} \geq \frac{1 - \boldsymbol{v}\boldsymbol{x}_o + \boldsymbol{u}\boldsymbol{y}_o}{s}[1/\boldsymbol{y}_o]. \tag{4.77}$$

The dual variables $\boldsymbol{v} \in R^m$ and $\boldsymbol{u} \in R^s$ can be interpreted as the "virtual" costs and prices of inputs and outputs, respectively. The dual program aims to find the optimal virtual costs and prices for the DMU $(\boldsymbol{x}_o, \boldsymbol{y}_o)$ so that the virtual profit $\boldsymbol{u}\boldsymbol{y}_j - \boldsymbol{v}\boldsymbol{x}_j$ does not exceed zero for any DMU (including $(\boldsymbol{x}_o, \boldsymbol{y}_o)$), and maximizes the "virtual" profit $\boldsymbol{u}\boldsymbol{y}_o - \boldsymbol{v}\boldsymbol{x}_o$ for the DMU $(\boldsymbol{x}_o, \boldsymbol{y}_o)$ concerned. Apparently, the optimal profit is at best zero and hence $\xi = 1$ for the SBM efficient DMUs. Constraints (4.76) and (4.77) restrict the feasible $\boldsymbol{v}$ and $\boldsymbol{u}$ to the positive orthant.

4.4.6 Oriented SBM Models

The input (output)-oriented SBM model can be defined by neglecting the denominator (numerator) of the objective function (4.48) of (SBM). Thus, the efficiency values ρ_I^* and ρ_O^* can be obtained as follows:

[Input-oriented SBM Model]

$$(SBM-I) \qquad \rho_I^* = \min_{\boldsymbol{\lambda}, \boldsymbol{s}^-} \ 1 - \frac{1}{m}\sum_{i=1}^{m} s_i^-/x_{io} \tag{4.78}$$

$$\begin{aligned} \text{subject to} \quad & \boldsymbol{x}_o = X\boldsymbol{\lambda} + \boldsymbol{s}^- \\ & \boldsymbol{y}_o \leq Y\boldsymbol{\lambda} \\ & \boldsymbol{\lambda} \geq \boldsymbol{0}, \ \boldsymbol{s}^- \geq \boldsymbol{0}. \end{aligned}$$

[Output-oriented SBM Model]

$$(SBM-O) \qquad \rho_O^* = \min_{\boldsymbol{\lambda}, \boldsymbol{s}^+} \ \frac{1}{1 + \frac{1}{s}\sum_{r=1}^{s} s_r^+/y_{ro}} \tag{4.79}$$

$$\begin{aligned} \text{subject to} \quad & \boldsymbol{x}_o \geq X\boldsymbol{\lambda} \\ & \boldsymbol{y}_o = Y\boldsymbol{\lambda} - \boldsymbol{s}^+ \\ & \boldsymbol{\lambda} \geq \boldsymbol{0}, \ \boldsymbol{s}^+ \geq \boldsymbol{0}. \end{aligned}$$

This leads to:

Theorem 4.10

$$\rho_I^* \geq \rho^* \quad and \quad \rho_O^* \geq \rho^*, \tag{4.80}$$

where ρ^* is the optimal value for (4.48).

The input-oriented SBM is substantially equivalent to what is called the "Russell input measure of technical efficiency."[9] See Section 4.5, below.

4.4.7 A Weighted SBM Model

We can assign weights to inputs and outputs corresponding to the relative importance of items as follows:

$$\rho = \frac{1 - \sum_{i=1}^{m} w_i^- s_i^-/x_{io}}{1 + \sum_{r=1}^{s} w_r^+ s_r^+/y_{ro}}, \tag{4.81}$$

with

$$\sum_{i=1}^{m} w_i^- = 1 \text{ and } \sum_{r=1}^{s} w_r^+ = 1. \tag{4.82}$$

The weights should reflect the intentions of the decision-maker. If all outputs are in the same unit, e.g., dollars, a conventional scheme is that:

$$w_r^+ = \sum_{j=1}^{n} y_{rj} \Big/ \sum_{j=1}^{n}\sum_{k=1}^{s} y_{kj}. \tag{4.83}$$

This weight selection reflects the importance of the output r being proportional to its contribution to the total magnitude. The input weights can be determined analogously.

4.4.8 Numerical Example (SBM)

We illustrate SBM using an example. Table 4.3 exhibits data for eight DMUs using two inputs (x_1, x_2) to produce a single output $(y = 1)$, along with CCR, SBM scores, slacks and reference set. Although DMUs F and G have full CCR-score $(\theta^* = 1)$, they have slacks compared to C and this is reflected by drops in the SBM scores to $\rho_F^* = 0.9$ and $\rho_G^* = 0.83333$. Also, the SBM scores of inefficient DMUs A, B and H dropped slightly from the CCR scores due to their slacks. Thus, the SBM measure reflects not only the weak efficiency values in θ^* but also the other (slack) inefficiencies as well.

Table 4.3. Data and Results of CCR and SBM

	Data			CCR	SBM					Mix
DMU	x_1	x_2	y	θ^*	$\rho^* = \rho_I^*$	Ref.	s_1^{-*}	s_2^{-*}	s^{+*}	Eff.[a]
A	4	3	1	0.857	0.833	D	0	1	0	0.972
B	7	3	1	0.632	0.619	D	3	1	0	0.98
C	8	1	1	1	1	C	0	0	0	1
D	4	2	1	1	1	D	0	0	0	1
E	2	4	1	1	1	E	0	0	0	1
F	10	1	1	1	0.9	C	2	0	0	0.9
G	12	1	1	1	0.833	C	4	0	0	0.833
H	10	1.5	1	0.75	0.733	C	2	0.5	0	0.978

[a]: Mix Eff. $= \rho^*/\theta^*$. See Section 5.8.2 for detail.

4.5 RUSSELL MEASURE MODELS

We now introduce a model described as the "Russell Measure Model." Actually it was introduced and named by Färe and Lovell(1978).[10] Their formulation is difficult to compute, however, so we turn to a more recent development due to Pastor, Ruiz and Sirvent.[11] This model is

$$R(\boldsymbol{x}_o, \boldsymbol{y}_o) = \min_{\theta,\phi,\lambda} \frac{\sum_{i=1}^{m} \theta_i/m}{\sum_{r=1}^{s} \phi_r/s} \tag{4.84}$$

subject to

$$\theta_i x_{io} \geq \sum_{j=1}^{n} x_{ij}\lambda_j, \; i = 1, \ldots, m$$

$$\phi_r y_{ro} \leq \sum_{j=1}^{n} y_{rj}\lambda_j, \; r = 1, \ldots, s$$

$$0 \leq \lambda_j \; \forall j$$

$$0 \leq \theta_i \leq 1; \; 1 \leq \phi_r \; \forall i, \; r.$$

Pastor *et al.* refer to this as the "Enhanced Russell Graph Measure of Efficiency" but we shall refer to it as ERM (Enhanced Russell Measure). See Färe, Grosskopf and Lovell (1985)[12] for the meaning of "graph measure." Such measures are said to be "closed" so $R(\boldsymbol{x}_o, \boldsymbol{y}_o)$ includes all inefficiencies that the model can identify. In this way we avoid limitations of the radial measures which cover only some of the input or output inefficiencies and hence measure only "weak efficiency."

The "inclusive" (=closure) property is shared by SBM. In fact SBM and ERM are related as in the following theorem,

Theorem 4.11 *ERM as formulated in (4.84) and SBM as formulated in (4.48) are equivalent in that λ_j^* values that are optimal for one are also optimal for the other.*

Proof : As inspection makes clear, a necessary condition for optimality of ERM is that the constraints in (4.84) must be satisfied as equalities. Hence we can replace those constraints with

$$\theta_i = \sum_{j=1}^{n} x_{ij}\lambda_j \Big/ x_{io}, i = 1, \ldots, m \tag{4.85}$$

$$\phi_r = \sum_{j=1}^{n} y_{rj}\lambda_j \Big/ y_{ro}, r = 1, \ldots, s.$$

Following Pastor *et al.* or Bardhan *et al.*, [13] we next set

$$\theta_i = \frac{x_{io} - s_i^-}{x_{io}} = 1 - \frac{s_i^-}{x_{io}}, \; i = 1, \ldots, m \tag{4.86}$$

$$\phi_r = \frac{y_{ro} + s_r^+}{y_{ro}} = 1 + \frac{s_r^+}{y_{ro}}, \; r = 1, \ldots, s.$$

Substituting these values in (4.85) produces

$$1 = \frac{\sum_{j=1}^{n} x_{ij}\lambda_j}{x_{io}} + \frac{s_i^-}{x_{io}}, \; i = 1, \ldots, m \tag{4.87}$$

$$1 = \frac{\sum_{j=1}^{n} y_{rj}\lambda_j}{y_{ro}} + \frac{s_r^+}{y_{ro}}, \; r = 1, \ldots, s$$

or

$$x_{io} = \sum_{j=1}^{n} x_{ij}\lambda_j + s_i^-,\ i = 1, \ldots, m \tag{4.88}$$
$$y_{ro} = \sum_{j=1}^{n} y_{rj}\lambda_j - s_r^+,\ r = 1, \ldots, s$$

which are the same as the constraints for SBM in (4.48).

Turning to the additional conditions,$0 \leq \theta_i \leq 1$ and $1 \leq \phi_r$ we again substitute from (4.86) to obtain $0 \leq s_i^- \leq x_{io}$ and $0 \leq s_r^+$. The condition $s_i^- \leq x_{io}$ is redundant since it is satisfied by the first set of inequalities in (4.84). Hence, we only have non-negativity for all slacks, the same as in (4.48).

Now turning to the objective in (4.84) we once more substitute from (4.86) to obtain

$$\frac{s}{m}\frac{\sum_{i=1}^{m}\theta_i}{\sum_{r=1}^{s}\phi_r} = \frac{s}{m}\frac{\sum_{i=1}^{m}\left(1 - \frac{s_i^-}{x_{io}}\right)}{\sum_{r=1}^{s}\left(1 + \frac{s_r^+}{y_{ro}}\right)} = \frac{1 - \frac{1}{m}\sum_{i=1}^{m}\frac{s_i^-}{x_{io}}}{1 + \frac{1}{s}\sum_{r=1}^{s}\frac{s_r^+}{y_{ro}}} \tag{4.89}$$

which is the same as the objective for SBM. Using the relations in (4.86) we may therefore use SBM to solve ERM or *vice versa.* □

4.6 SUMMARY OF THE BASIC DEA MODELS

In Table 4.4, we summarize some important topics for consideration in choosing between basic DEA models. In this table, 'Semi-p' (=semipositive) means nonnegative with at least one positive element in the data for each DMU, and 'Free' permits negative, zero or positive data. Although we have developed some DEA models under the assumption of positive data set, this assumption can be relaxed as exhibited in the table. For example, in the BCC-I (-O) model, outputs (inputs) are free due to the translation invariance theorem. In the case of SBM, nonpositive outputs can be replaced by a very small positive number and nonpositive input terms can be neglected for consideration in the objective function. The θ^* of the output oriented model (CCR-O) is the reciprocal of $\eta^*(\geq 1)$. 'Tech. or Mix' indicates whether the model measures 'technical efficiency' or 'mix efficiency'. 'CRS' and 'VRS' mean *constant* and *variable* returns to scale, respectively. The returns to scale of ADD and SBM depends on the added convexity constraint $e\boldsymbol{\lambda} = 1$.

Model selection is one of the problems to be considered in DEA up to, and including, choices of multiple models to test whether or not a result is dependent on the models (or methods) used.[14] Although other models will be developed in succeeding chapters, we will mention here some of the considerations to be taken into account regarding model selection.

1. The Shape of the Production Possibility Set.

The CCR model is based on the assumption that *constant* returns to scale

Table 4.4. Summary of Model Characteristics

Model		CCR-I	CCR-O	BCC-I	BCC-O	ADD	SBM
Data	X	Semi-p	Semi-p	Semi-p	Free	Free	Semi-p
	Y	Free	Free	Free	Semi-p	Free	Free
Trans.	X	No	No	No	Yes	Yes[a]	No
Invariance	Y	No	No	Yes	No	Yes[a]	No
Units invariance		Yes	Yes	Yes	Yes	No	Yes
θ^*		[0, 1]	[0, 1]	(0, 1]	(0, 1]	No	[0, 1]
Tech. or Mix		Tech.	Tech.	Tech.	Tech.	Mix	Mix
Returns to Scale		CRS	CRS	VRS	VRS	C(V)RS[b]	C(V)RS

[a]: The Additive model is translation invariant only when the convexity constraint is added.
[b]: C(V)RS means Constant or Variable returns to scale according to whether or not the convexity constraint is included.

prevails at the efficient frontiers, whereas the BCC and Additive models assume variable returns to scale frontiers, i.e., *increasing, constant* and *decreasing* returns to scale.[15] If preliminary surveys on the production functions identify a preferable choice by, say, such methods as linear regression analysis, (e.g., a Cobb-Douglas type) or expert opinions, then we can choose a DEA model that fits the situation. However, we should bear in mind that conventional regression-based methods deal with single output and multiple input cases, while DEA models analyze multiple outputs and multiple inputs correspondences.

2. Input or Output Oriented?
One of the main purposes of a DEA study is to project the inefficient DMUs onto the production frontiers, e.g., the CCR-projection and the BCC-projection, among others. There are three directions, one called *input-oriented* that aims at reducing the input amounts by as much as possible while keeping at least the present output levels, and the other, called *output-oriented*, maximizes output levels under at most the present input consumptions. There is a third choice, represented by the Additive and SBM models that deal with the input excesses and output shortfalls simultaneously in a way that jointly maximizes both. If achievement of efficiency, or failure to do so, is the only topic of interest, then these different models will all yield the same result insofar as technical and mix inefficiency is concerned. However, we need to note that the Additive and BCC models may give different estimates when inefficiencies are present. Moreover, as we shall see in the next chapter, the CCR and BCC models differ in that the former evaluates scale as well as technical inefficiencies simultaneously whereas the latter evaluates the two in a fashion that identifies them separately.

3. **Translation Invariance.**
As is seen from Table 4.4, we can classify the models according to whether or not they use the efficiency measure θ^*. It is to be noted that θ^* is measured in a manner that depends on the coordinate system of the data set. On the other hand, the θ^*-free models, such as the Additive model, are essentially coordinate-free and translation invariant. They evaluate the efficiency of a DMU by the l_1-metric distance from the efficient frontiers and are invariant with respect to the translation of the coordinate system. Although they supply information on the projection of the inefficient DMUs onto the production frontiers, they lack a one-dimensional efficiency measure like θ^*. The SBM model developed in Section 4.4 is designed to overcome such deficiencies in both the CCR-type and Additive models. However, SBM as represented in (4.48) is not translation invariant.

4. **Number of Input and Output Items.**
Generally speaking, if the number of DMUs (n) is less than the combined number of inputs and outputs ($m + s$), a large portion of the DMUs will be identified as efficient and efficiency discrimination among DMUs is questionable due to an inadequate number of degrees of freedom. See the opening discussion in Chapter 9. Hence, it is desirable that n exceed $m+s$ by several times. A rough rule of thumb in the envelopment model is to choose n (= the number of DMUs) equal to or greater than $\max\{m \times s, 3 \times (m + s)\}$. The selection of input and output items is crucial for successful application of DEA. We therefore generally recommend a process of selecting a small set of input and output items at the beginning and gradually enlarging the set to observe the effects of the added items. In addition, other methods, e.g., the assurance region method, the cone ratio model and others (that will be presented in the succeeding chapters) lend themselves to a sharper discrimination among DMUs.

5. **Try Different Models.**
If we cannot identify the characteristics of the production frontiers by preliminary surveys, it may be risky to rely on only one particular model. If the application has important consequences it is wise to try different models and methods, compare results and utilize expert knowledge of the problem, and possibly try other devices, too, before arriving at a definitive conclusion. See the discussion in Chapter 9 dealing with the use of statistical regressions and DEA to cross check each other.

4.7 SUMMARY OF CHAPTER 4

In this chapter, we introduced several DEA models.

1. The CCR and BCC models which are radial measures of efficiency that are either input oriented or output oriented.

2. The Additive models which identify input excesses and output shortfalls simultaneously.

3. The slacks-based measure of efficiency (SBM) which uses the Additive model and provides a scalar measure ranging from 0 to 1 that encompasses all of the inefficiencies that the model can identify.

4. We investigated the problem of model choice. Chapters 6, 7 and 9 will serve as a complement in treating this subject.

5. The translation invariance of the Additive model was introduced. This property of the Additive model (with convexity constraint) was identified by Ali and Seiford (1990) and was extended by Pastor (1996).[16] See, however, Thrall (1996)[17] for limitations involved in relying on this invariance property.

4.8 NOTES AND SELECTED BIBLIOGRAPHY

Translation invariance was first shown by Ali and Seiford (1990) and extended by Pastor *et al.* (1999). The SBM model was introduced by Tone (1997, 2001).

4.9 APPENDIX: FREE DISPOSAL HULL (FDH) MODELS

Another model which has received a considerable amount of research attention is the FDH (Free Disposal Hull) model as first formulated by Deprins, Simar and Tulkens (1984)[18] and developed and extended by Tulkens and his associates at the University of Louvain in Belgium.[19] The basic motivation is to ensure that efficiency evaluations are effected from only actually observed performances. Points like Q in Figure 3.2, for example, are not allowed because they are derived and not actually observed performances. Hence they are hypothetical.

Figure 4.7 provides an example of what is referred to as the Free Disposal Hull for five DMUs using two inputs in amounts x_1 and x_2 to produce a single output in amount $y = 1$.

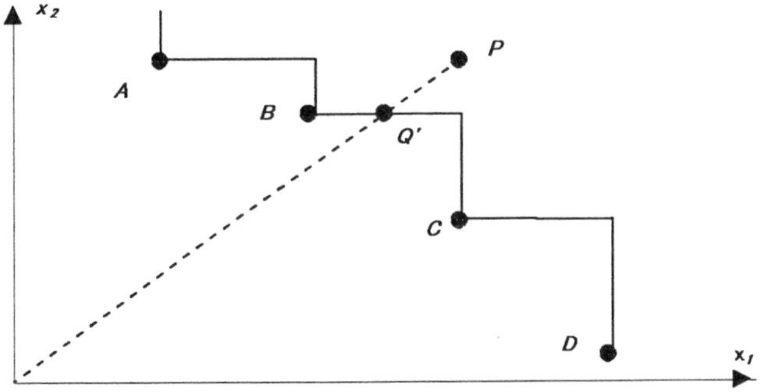

Figure 4.7. FDH Representation

The boundary of the set and its connection represents the "hull" defined as the "smallest set" that encloses all of the production possibilities that can be generated from the observations. Formally,

$$P_{FDH} = \{(\boldsymbol{x}, \boldsymbol{y}) | \boldsymbol{x} \geq \boldsymbol{x}_j, \boldsymbol{y} \leq \boldsymbol{y}_j,\ \boldsymbol{x},\ \boldsymbol{y} \geq \mathbf{0},\ j = 1, \ldots, n\}$$

where $\boldsymbol{x}_j (\geq \mathbf{0}), \boldsymbol{y}_j (\geq \mathbf{0})$ are actually observed performances for $j = 1, \ldots, n$ DMUs. In words, a point is a member of the production possibility set if all of its input coordinates are at least as large as their corresponds in the vector of observed values $\boldsymbol{x}_j$ for any $j = 1, \ldots, n$ and if their output coordinates are no greater than their corresponds in the vectors $\boldsymbol{y}_j$ of observed values for this same j.

This gives rise to the staircase (or step) function which we have portrayed by the solid line in the simple two-input one-output example of Figure 4.7. No point below this solid line has the property prescribed for P_{FDH}. Moreover this boundary generates the smallest set with these properties. For instance, connecting points B and C in the manner of Figure 3.2 in Chapter 3 would generate the boundary of a bigger production possibility set. Tulkens and his associates use an algorithm that eliminates all dominated points as candidates for use in generating the FDH. This algorithm proceeds in pairwise comparison fashion as follows: Let DMU$_k$ with coordinate $\boldsymbol{x}_k$, $\boldsymbol{y}_k$ be a candidate. If for any DMU$_j$ we have $\boldsymbol{x}_j \leq \boldsymbol{x}_k$ or $\boldsymbol{y}_j \geq \boldsymbol{y}_k$ with either $\boldsymbol{x}_j \neq \boldsymbol{x}_k$ or $\boldsymbol{y}_j \neq \boldsymbol{y}_k$ then DMU$_k$ is dominated (strictly) and removed from candidacy. Actually this can be accomplished more simply by using the following mixed integer programming formulation,

$$\begin{aligned} \min \quad & \theta \\ \text{subject to} \quad & \theta \boldsymbol{x}_o - X\boldsymbol{\lambda} \geq \mathbf{0} \\ & \boldsymbol{y}_o - Y\boldsymbol{\lambda} \leq \mathbf{0} \\ & \boldsymbol{e}\boldsymbol{\lambda} = 1, \quad \lambda_j \in \{0,\ 1\} \end{aligned} \tag{4.90}$$

where X and Y contain the given input and output matrices and $\lambda_j \in \{0,\ 1\}$ means that the components of $\boldsymbol{\lambda}$ are constrained to be bivalent. That is, they must all have values of zero or unity so that together with the condition $\boldsymbol{e}\boldsymbol{\lambda} = 1$ one and only one of the performances actually observed can be chosen. This approach was first suggested in Bowlin *et al.*(1984)[20] where it was coupled with an additive model to ensure that the "most dominant" of the non-dominated DMUs was chosen for making the indicated efficiency evaluation. In the case of Figure 4.7, the designation would thus have been A rather than B or C in order to maximize the sum of the slacks s_1^- and s_2^- when evaluating P. However, this raises an issue because the results may depend on the units of measure employed. One way to avoid this difficulty is to use the radial measure represented by $\min \theta = \theta^*$. This course, as elected by Tulkens and associates, would yield the point Q' shown in Figure 4.7. This, however, leaves the slack in going from Q' to B unattended, which brings into play the assumption of "Free Disposal" that is incorporated in the name "Free Disposal Hull." As noted earlier this means that nonzero slacks are ignored or, equivalently, weak efficiency suffices because the slacks do not appear in the objective — or, equivalently, they are present in the objective with zero coefficients assigned to them.

One way to resolve all of these issues is to return to our slacks-based measure (SBM) as given in (4.48) or its linear programming (LP) equivalent in (4.51) with the conditions $\lambda_j \in \{0, 1\}$ and $e\boldsymbol{\lambda} = 1$ adjoined. This retains the advantage of the additive model which (a) selects an "actually observed performance" and (b) provides a measure that incorporates all of the thus designated inefficiencies. Finally, it provides a measure which is units invariant as well. See R.M. Thrall (1999)[21] for further discussions of FDH and its limitations.

4.10 RELATED DEA-SOLVER MODELS FOR CHAPTER 4

BCC-I (The input-oriented Banker-Charnes-Cooper model).

This code solves the BCC model expressed by (4.2)-(4.6). The data format is the same as for the CCR model. The "Weight" sheet includes the optimal value of the multiplier u_0 corresponding to the constraint $\sum_{j=1}^{n} \lambda_j = 1$, as well as $\boldsymbol{v}^*$ and $\boldsymbol{u}^*$. In addition to the results which are similar to the CCR case, this model finds the returns-to-scale characteristics of DMUs in the "RTS" sheet. For inefficient DMUs, we identify returns to scale with the projected point on the efficient frontier. BCC-I uses the projection formula (4.17)-(4.18).

BCC-O (The output-oriented BCC model).

This code solves the output-oriented BCC model expressed in (4.22)-(4.26). The optimal expansion rate η_B^* is displayed in "Score" by its inverse in order to facilitate comparisons with the input-oriented case. Usually, the efficiency score differs in both cases for inefficient DMUs. This model also finds the returns-to-scale characteristics of DMUs in the "RTS" sheet. For inefficient DMUs, we identify returns to scale with the projected point on the efficient frontier. BCC-O uses the BCC version of the formula (3.74)-(3.15). The returns-to-scale characteristics of inefficient DMUs may also change from the input-oriented case.

SBM-C(V or GRS) (The slacks-based measure of efficiency under the constant (variable or general) returns-to-scale assumption.)

This code solves the SBM model. The format of the input data and the output results is the same as for the CCR case. In the GRS (general returns-to-scale) case, L (the lower bound) and U (the upper bound) of the sum of the intensity vector $\boldsymbol{\lambda}$ must be supplied through keyboard. The defaults are $L = 0.8$ and $U = 1.2$.

SBM-I-C(V or GRS) (The input-oriented slacks-based measure of efficiency under the constant (variable or general) returns-to-scale assumption).

This code solves the SBM model in input-orientation. Therefore, output slacks (shortfalls) are not accounted for in this efficiency measure.

SBM-O-C(V or GRS) (The output-oriented SBM model under constant (variable or general) returns-to-scale assumption).

This is the output-oriented version of the SBM model, which puts emphasis on the output shortfalls. Therefore, input slacks (excesses) are not accounted for in this efficiency measure.

FDH (The free disposal hull model).

This code solves the FDH model introduced in Section 4.8 and produces the worksheets "Score," "Projection," "Graph1" and "Graph2." The data format is the same as for the CCR model.

4.11 PROBLEM SUPPLEMENT FOR CHAPTER 4

Problem 4.1

It is suggested on occasion that the weak (or Farrell) efficiency value, θ^*, be used to rank DMU performances as determined from CCR or BCC models.

Assignment : Discuss possible shortcomings in this approach.

Suggested Response : There are two main shortcomings that θ^* possesses for the proposed use. First, the value of θ^* is not a complete measure and, instead, the nonzero slacks may far outweigh the value of $(1-\theta^*)$. Second, the θ^* values will generally be determined relative to different reference groups. Note, therefore, that a value of $\theta^* = 0.9$ for DMU A means that its purely technical efficiency is 90% of the efficiency generated by members of its (efficient) reference group. Similarly a value of $\theta^* = 0.8$ for DMU B refers to its performance relative to a different reference group. This does *not* imply that DMU B is less efficient than DMU A.

Ranking can be useful, of course, so when this is wanted recourse should be had to an explicitly stated principle of ranking. This was done by the Texas Public Utility Commission, for example, which is required by law to conduct "efficiency audits" of the 75 electric cooperatives which are subject to its jurisdiction in Texas. See Dennis Lee Thomas "Auditing the Efficiency of Regulated Companies through the Use of Data Envelopment Analysis: An Application to Electric Cooperatives," Ph.D. Thesis, Austin, TX: University of Texas at Austin, Graduate School of Business, 1985.[22] As discussed in Thomas (subsequently Chairman of the Texas PUC), the ranking principle was based on dollarizing the inefficiencies. This included assigning dollar values to the slacks as well as to the purely technical inefficiencies. For instance, line losses in KWH were multiplied by the revenue charges at prices that could otherwise be earned from sales of delivered power. Fuel excesses, however, were costed at the purchase price per unit. Thomas then also remarks that the reference groups designated by DEA were used by PUC auditors to help in their evaluation by supplying comparison DMUs for the cooperative being audited. See also A. Charnes, W.W. Cooper, D. Divine, T.W. Ruefli and D. Thomas "Comparisons of DEA and Existing Ratio and Regression Systems for Effecting Efficiency Evaluations of Regulated Electric Cooperatives in Texas," *Research in Government and Nonprofit Accounting* 5, 1989, pp.187-210.

Problem 4.2

Show that the CCR model is not translation invariant.

Suggested Answer : Let us examine the CCR model via a simple one input and one output case, as depicted in Figure 4.8. The efficient frontier is the line connecting the origin O and the point B. The CCR-efficiency of A is evaluated by PQ/PA. If we translate the output value by a given amount, say 2, in the manner shown in Figure 4.8, the origin will move to O' and the efficient frontier will shift to the dotted line extending beyond B from O'. Thus, the CCR-efficiency of A is evaluated, under the new coordinate system, by PR/PA, which differs from the old value PQ/PA. A similar observation of the translation of the input item shows that the CCR model is *not* invariant for translation of both *input* and *output*. From this we conclude that the condition $\sum_{j=1}^{n} \lambda_j = 1$ plays a crucial role which is exhibited in limited fashion in the BCC model and more generally in Additive models.

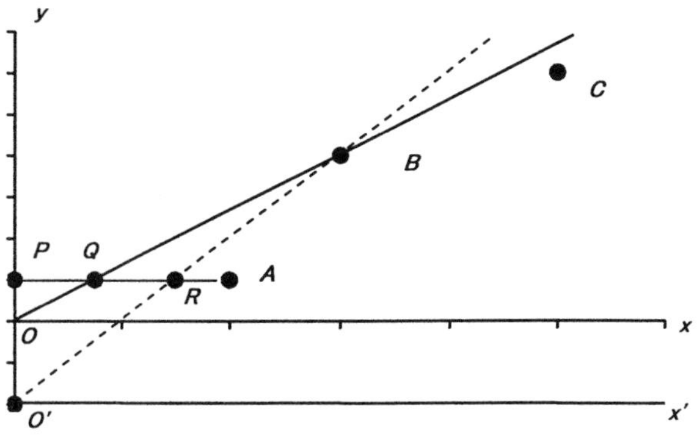

Figure 4.8. Translation in the CCR Model

Problem 4.3

As noted in the text, the critical condition for translation invariance is $\boldsymbol{e\lambda} = 1$. This suggests that parts of the BCC model may be translation invariant. Therefore,

Assignment : Prove that the input-oriented (output-oriented) BCC model is translation invariant with respect to the coordinates of outputs (inputs).

Answer : The input-oriented BCC model is:

$$
\begin{aligned}
\min \quad & \theta_B \\
\text{subject to} \quad & \theta_B \boldsymbol{x}_o - X\boldsymbol{\lambda} - \boldsymbol{s}^- = \boldsymbol{0} \\
& Y\boldsymbol{\lambda} - \boldsymbol{s}^+ = \boldsymbol{y}_o \\
& \boldsymbol{e\lambda} = 1 \\
& \boldsymbol{\lambda} \geq \boldsymbol{0}, \quad \boldsymbol{s}^- \geq \boldsymbol{0}, \quad \boldsymbol{s}^+ \geq \boldsymbol{0}.
\end{aligned}
\tag{4.91}
$$

If we translate the output data Y by some constants $(\beta_r : r = 1, \dots, s)$ to obtain

$$y'_{rj} = y_{rj} + \beta_r, \quad (r = 1, \dots, s : j = 1, \dots, n) \tag{4.92}$$

then (4.91) becomes, in terms of the components of Y',

$$\sum_{j=1}^{n}(y'_{rj} - \beta_r)\lambda_j - s_r^+ = \sum_{j=1}^{n} y'_{rj}\lambda_j - \beta_r - s_r^+ = y'_{ro} - \beta_r, \quad (r = 1, \dots, s) \tag{4.93}$$

so subtracting β_r from both sides gives

$$\sum_{j=1}^{n} y'_{rj}\lambda_j - s_r^+ = y'_{ro}, \quad (r = 1, \dots, s) \tag{4.94}$$

which is the same as the original expression for this output constraint. Notice that the convexity condition $\boldsymbol{e\lambda} = 1$ is used to derive the above relation. Notice also that this result extends to inputs *only* in the special case of $\theta_B^* = 1$.

It should now be readily apparent that if $(\theta_B^*, \ \lambda^*, \boldsymbol{s}^{-*}, \boldsymbol{s}^{+*})$ is an optimal solution of the original BCC model, then it is also optimal for the translated problem.

By the same reasoning used for the input-oriented case, we conclude that the output-oriented BCC model is translation invariant with respect to the input data X but *not* to the output data.

Comment and References : The property of translation invariance for Additive models was first demonstrated in Ali and Seiford (1990), as noted in the text. It has been carried further in J.T Pastor (1996).[23] See also R.M. Thrall (1996)[24] who shows that this property does not carry over to the dual of Additive models.

Problem 4.4

Statement : Consider the following version of an Additive model

$$\begin{aligned} \max \quad & \sum_{i=1}^{m} s_i^- + \sum_{r=1}^{s} s_r^+ \\ \text{subject to} \quad & \sum_{j=1}^{n} \widehat{x}_{ij}\lambda_j + s_i^- = \widehat{x}_{io}, \ i = 1, \dots, m \\ & \sum_{j=1}^{n} \widehat{y}_{rj}\lambda_j - s_r^+ = \widehat{y}_{ro}, \ r = 1, \dots, s \end{aligned} \tag{4.95}$$

where the variables are also constrained to be nonnegative and the symbol "^" means that the data are stated in natural logarithmic units. (Unlike (ADD_o), as given in (4.34)-(4.38), we have omitted the condition $\boldsymbol{e\lambda} = 1$ because this restricts results to the case of constant returns to scale in the models we are now considering.

Assignment : Part I: Use anti-logs (base e) to derive yet another DEA model called the "multiplicative model," and relate the resulting formulation to the Cobb-Douglas forms of production frontiers that have played prominent roles in many econometric and statistical studies.
Part II: Relate the conditions for full (100%) efficiency to Definition 5.1 for the Additive model.

Suggested Response: : Part I: Let an optimal solution be represented by $\lambda_j^*, s_i^{-*}, s_r^{+*}$. Then, taking anti-logs of (4.95) we obtain

$$x_{io} = \prod_{j=1}^{n} x_{ij}^{\lambda_j^*} e^{s_i^{-*}} = a_i^* \prod_{j=1}^{n} x_{ij}^{\lambda_j^*}, \; i = 1, \ldots, m \tag{4.96}$$

$$y_{ro} = \prod_{j=1}^{n} y_{rj}^{\lambda_j^*} e^{-s_r^{+*}} = b_r^* \prod_{j=1}^{n} y_{rj}^{\lambda_j^*}, \; r = 1, \ldots, s$$

where $a_i^* = e^{s_i^{-*}}$, $b_r^* = e^{-s_r^{+*}}$. This shows that each x_{io} and y_{ro} is to be considered as having been generated by Cobb-Douglas (=log-linear) processes with estimated parameters indicated by the starred values of the variables.

Part II: To relate these results to ratio forms we proceed in a manner analogous to what was done for (4.7)-(4.10) and turn to the dual (multiplier) form of (4.95),

$$\begin{aligned} \min \quad & \sum_{i=1}^{m} v_i \widehat{x}_{io} - \sum_{r=1}^{s} u_r \widehat{y}_{ro} \\ \text{subject to} \quad & \sum_{i=1}^{m} v_i \widehat{x}_{ij} - \sum_{r=1}^{s} u_r \widehat{y}_{rj} \geq 0, \; j = 1, \ldots, n \\ & v_i \geq 1, \; i = 1, \ldots, m \\ & u_r \geq 1, \; r = 1, \ldots, s. \end{aligned} \tag{4.97}$$

This transformation to logarithm values gives a result that is the same as in the ordinary Additive model. Hence, no new software is required to solve this multiplicative model. To put this into an easily recognized efficiency evaluation form we apply anti-logs to (4.97). We change the objective in (4.97) and reverse the constraints to obtain:

$$\begin{aligned} \max \quad & \sum_{r=1}^{s} u_r \widehat{y}_{ro} - \sum_{i=1}^{m} v_i \widehat{x}_{io} \\ \text{subject to} \quad & \sum_{r=1}^{s} u_r \widehat{y}_{rj} - \sum_{i=1}^{m} v_i \widehat{x}_{ij} \leq 0, \; j = 1, \ldots, n \\ & v_i \geq 1, \; i = 1, \ldots, m \\ & u_r \geq 1, \; r = 1, \ldots, s. \end{aligned}$$

We then obtain:

$$\begin{aligned}
\max \quad & \prod_{r=1}^{s} y_{ro}^{u_r} \Big/ \prod_{i=1}^{m} x_{io}^{v_i} \\
\text{subject to} \quad & \prod_{r=1}^{s} y_{rj}^{u_r} \Big/ \prod_{i=1}^{m} x_{ij}^{v_i} \le 1, \; j = 1, \ldots, n \\
& v_i, u_r \ge 1, \forall i, \; r.
\end{aligned} \tag{4.98}$$

To obtain conditions for efficiency we apply antilogs to (4.95) and use the constraints in (4.97) to obtain

$$\max \frac{\prod_{r=1}^{s} e^{s_r^+}}{\prod_{i=1}^{m} e^{-s_i^-}} = \frac{\prod_{r=1}^{s} \prod_{j=1}^{n} y_{rj}^{\lambda_j^*} \Big/ y_{ro}}{\prod_{i=1}^{m} \prod_{j=1}^{n} x_{ij}^{\lambda_j^*} \Big/ x_{io}} \ge 1. \tag{4.99}$$

The lower bound represented by the unity value on the right is obtainable if and only if all slacks are zero. Thus, the conditions for efficiency of the multiplicative model are the same as for the Additive model where the latter has been stated in logarithmic units. See Definition 5.1. We might also note that the expression on the left in (4.99) is simpler and easier to interpret and the computation via (4.95) is straightforward with the $\widehat{x}_{ij}$ and $\widehat{y}_{rj}$ stated in logarithmic units.

In conclusion we might note that (4.99) is not units invariant unless $\sum_{j=1}^{n} \lambda_j^* = 1$. This "constant-returns-to-scale condition" can, however, be imposed on (4.95) if desired. See also A. Charnes, W.W. Cooper, L.M. Seiford and J. Stutz "Invariant Multiplicative Efficiency and Piecewise Cobb-Douglas Envelopments" *Operations Research Letters* 2, 1983, pp.101-103.

Comments : The class of multiplicative models has not seen much used in DEA applications. It does have potential for uses in extending returns-to-scale treatments to situations that are not restricted to concave cap functions like those which are assumed to hold in the next chapter. See R.D. Banker and A. Maindiratta "Piecewise Loglinear Estimation of Efficient Production Surfaces" *Management Science* 32, 1986, pp.126-135. Its properties suggest possible uses not only in its own right but also in combination with other models and, as will be seen in the next chapter, it can be used to obtain scale elasticities that are not obtainable with other DEA models.

Problem 4.5

Work out the dual version of the weighted SBM model in Section 4.4.8 and interpret the results.

Suggested Response:

$$\max \quad \boldsymbol{u}\boldsymbol{y}_o - \boldsymbol{v}\boldsymbol{x}_o \tag{4.100}$$

$$\text{subject to} \quad -\boldsymbol{v}X + \boldsymbol{u}Y \le \boldsymbol{0} \tag{4.101}$$

$$v_i x_{io} \geq \frac{1}{m} w_i^- \ (i = 1, \ldots, m) \tag{4.102}$$

$$u_r y_{ro} \geq \frac{1 - \boldsymbol{v}\boldsymbol{x}_o + \boldsymbol{u}\boldsymbol{y}_o}{s} w_r^+. \ (r = 1, \ldots, s) \tag{4.103}$$

Thus, each $v_i x_{io}$ is bounded below by a value proportional to $1/m$ times the selected weight w_i^- for input and the value of unity increment for the $\boldsymbol{y}_o$ and decrement for the $\boldsymbol{x}_o$ combined in the numerator on s, the number of outputs in the denominator.

Problem 4.6

Assignment: Suppose the objective for ERM in (4.84) is changed from its fractional form to the following

$$\max \frac{\sum_{r=1}^{s} \phi_r}{s} - \frac{\sum_{i=1}^{m} \theta_i}{m}. \tag{4.104}$$

Using the relations in (4.86) transform this objective to the objective for the corresponding Additive model and discuss the properties of the two measures.

Suggested Response: Transforming (4.104) in the suggested manner produces the following objective for the corresponding Additive model.

$$\max z = \frac{\sum_{r=1}^{s}\left(1 + \frac{s_r^+}{y_{ro}}\right)}{s} - \frac{\sum_{i=1}^{m}\left(1 - \frac{s_i^-}{x_{io}}\right)}{m} = \sum_{r=1}^{s} \frac{s_r^+}{y_{ro}} + \sum_{i=1}^{m} \frac{s_i^-}{x_{io}}. \tag{4.105}$$

The ERM constraints are similarly transformed to the constraints of an Additive model as described in the proof of Theorem 4.15.

Discussion: The objective in (4.104) jointly maximizes the "inefficiencies" measured by the ϕ_r and minimizes the "efficiencies" measured by the θ_i. This is analogous to the joint minimization represented by the objective in (4.84) for ERM but its z^* values, while non-negative, may exceed unity. The objective stated on the right in (4.105) jointly maximizes the "inefficiencies" in both the s_r^+/y_{ro} and s_i^-/x_{io}. This is analogous to the joint maximization of both of these measures of inefficiency in SBM except that, again, its value may exceed unity.

Notes

1. R. D. Banker, A. Charnes and W.W. Cooper (1984), "Some Models for Estimating Technical and Scale Inefficiencies in Data Envelopment Analysis," *Management Science* 30, pp.1078-1092

2. SBM was first proposed by K. Tone (1997), "A Slacks-based Measure of Efficiency in Data Envelopment Analysis," Research Reports, Graduate School of Policy Science, Saitama University and subsequently published in *European Journal of Operational Research* 130 (2001), pp.498-509.

3. For a discussion of this constant-returns-to-scale Additive model see A.I. Ali and L.M. Seiford, "The Mathematical Programming Approach to Efficiency Measurement," in *The Measurement of Productive Efficiency: Techniques and Applications*, H. Fried, C. A. Knox Lovell, and S. Schmidt (editors), Oxford University Press, London, (1993).

4. This distance is measured in what is called an l_1-metric for the slack vectors s^- and s^+. Also called the "city-block" metric, the objective in (4.34) is directed to maximizing the distance as measured by the sum of the input plus output slacks.

5. T. Ahn, A. Charnes and W.W. Cooper (1988), "Efficiency Characterizations in Different DEA Models," *Socio-Economic Planning Sciences*, 22, pp.253-257.

6. P.L. Brockett, W.W. Cooper, L.L. Golden, J.J. Rousseau and Y. Wang, "DEA Evaluation of Organizational Forms and Distribution System in the U.S. Property and Liability Insurance Industry," *International Journal of Systems Science*, 1998.

7. A.I. Ali and L.M. Seiford (1990), "Translation Invariance in Data Envelopment Analysis," *Operations Research Letters* 9, pp.403-405.

8. This term is taken from the literature on dimensional analysis. See pp.123-125, in R.M. Thrall "Duality, Classification and Slacks in DEA," *Annals of Operations Research* 66, 1996.

9. See R.S. Färe and C.A.K. Lovell (1978), "Measuring the Technical Efficiency of Production," *Journal of Economic Theory* 19, pp.150-162. See also R.R. Russell (1985) "Measures of Technical Efficiency," *Journal of Economic Theory* 35, pp.109-126.

10. See the Note 9 references.

11. J.T. Pastor, J.L. Ruiz and I. Sirvent (1999), "An Enhanced DEA Russell Graph Efficiency Measure," *European Journal of Operational Research* 115, pp.596-607.

12. R.S. Färe, S. Grosskopf and C.A.K Lovell (1985), *The Measurement of Efficiency of Production* Boston: Kluwer-Nijhoff.

13. I. Bardhan, W.F. Bowlin, W.W. Cooper and T. Sueyoshi (1996), "Models and Measures for Efficiency Dominance in DEA. Part II: Free Disposal Hull (FDI) and Russell Measure (RM) Approaches," *Journal of the Operations Research Society of Japan* 39, pp.333-344.

14. See the paper by T. Ahn and L.M. Seiford "Sensitivity of DEA Results to Models and Variable Sets in a Hypothesis Test Setting: The Efficiency of University Operations," in Y. Ijiri, ed., *Creative and Innovative Approaches to the Science of Management* (New York: Quorum Books, 1993). In this paper Ahn and Seiford find that U.S. public universities are more efficient than private universities when student outputs are emphasized but the reverse is true when research is emphasized. The result, moreover, is found to be invariant over all of the DEA models used in this study. For "methods" cross-checking, C.A.K. Lovell, L. Walters and L. Wood compare statistical regressions with DEA results in "Stratified Models of Education Production Using Modified DEA and Regression Analysis," in *Data Envelopment Analysis: Theory, Methodology and Applications* (Norwell, Mass.: Kluwer Academic Publishers, 1994.)

15. See Chapter 5 for details.

16. J.T. Pastor (1996), "Translation Invariance in DEA: A Generalization," *Annals of Operations Research* 66, pp.93-102.

17. R.M. Thrall (1996), "The Lack of Invariance of Optimal Dual Solutions Under Translation," *Annals of Operations Research* 66, pp.103-108.

18. Deprins D., L. Simar and H. Tulkens (1984), "Measuring Labor Efficiency in Post Offices," in M. Marchand, P. Pestieau and H. Tulkens, eds. *The Performance of Public Enterprises: Concepts and Measurement* (Amsterdam, North Holland), pp.243-267.

19. See the review on pp. 205-210 in C.A.K. Lovell "Linear Programming Approaches to the Measurement and Analysis of Productive Efficiency," *TOP* 2, 1994, pp. 175-243.

20. Bowlin, W.F., J. Brennan, A. Charnes, W.W. Cooper and T. Sueyoshi (1984), "A Model for Measuring Amounts of Efficiency Dominance," Research Report, The University of Texas at Austin, Graduate School of Business. See also Bardhan I., W.F. Bowlin, W.W. Cooper and T. Sueyoshi "Models and Measures for Efficiency Dominance in DEA," *Journal of the Operations Research Society of Japan* 39, 1996, pp.322-332.

21. R.M. Thrall (1999), "What is the Economic Meaning of FDH?" *Journal of Productivity Analysis,* 11, pp.243-250.

22. Also available in microfilm form from University Microfilms, Inc., Ann Arbor, Michigan.

23. See the Note 17 reference.

24. See the Note 18 reference.

5 RETURNS TO SCALE

5.1 INTRODUCTION

This chapter deals with returns to scale along with modifications and extensions that are needed to use these concepts in actual applications. We start with conceptual formulations obtained from economics for which we use Figure 5.1. The function $y = f(x)$ in the upper portion of the diagram is a "production function" defined so that the value of y is maximal for every x. This definition assumes that technical efficiency is always achieved. Therefore, points like P, which lie inside the production possibility set, do not find a place in the concept that we are presently considering. Only points on the frontier defined by $y = f(x)$ are of interest.

The lower diagram portrays the behavior of average productivity ($a.p.= y/x$) and marginal productivity ($m.p.= dy/dx$) which are obtained from the upper diagram by noting that y/x corresponds to the slope of the ray from the origin to y and dy/dx is the derivative of $f(x)$ at this same point.

As can be seen, the slopes of the rays increase with x until x_0 is reached after which the slopes of the corresponding rays (and hence $a.p.$) begin to decrease. In a similarly visualized manner, the derivative $dy/dx = m.p.$ increases with x until the inflection point at $f(x_1)$ is reached after which it begins to decrease. As shown in the lower diagram, $m.p.$ lies above $a.p.$ to the left of x_0 and below it to the right. This means that output is changing proportionally faster than input to the left of x_0 while the reverse situation occurs to the right of x_0.

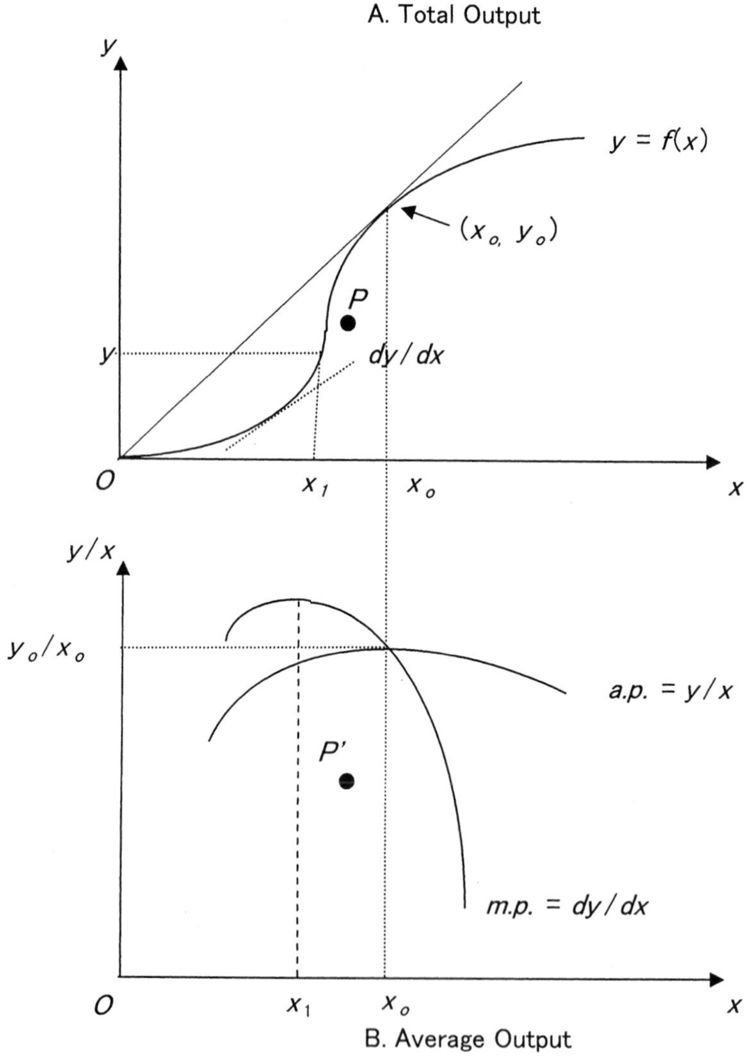

Figure 5.1. Returns to Scale

We can formalize this more precisely by determining where the maximum value of *a.p.* occurs by differentiating *a.p.* with respect to x via

$$\frac{d(y/x)}{dx} = \frac{xdy/dx - y}{x^2} = 0,$$

which at $x = x_0$ gives

$$e(x) = \frac{x}{y}\frac{dy}{dx} = \frac{d\ln y}{d\ln x} = 1. \tag{5.1}$$

Here $e(x)$ is defined as a logarithmic derivative to reflect the fact that proportional changes are being considered. Called an "elasticity" in economics it measures the *relative* change in output compared to the *relative* change in input. At x_0, where *a.p.* is maximal, we have $e(x) = 1$ so that $dy/dx = y/x$ and the rate of change of output to input does not change.

Stated differently, $e(x) = 1$ represents constant returns-to-scale because the proportional increase in y given by dy/y is the same as the proportional increase in x given by dx/x. To the left of x_0 we will have $dy/dx > y/x$ so $e(x) > 1$ and returns to scale is increasing. To the right of x_0 we have $dy/dx < y/x$ so $e(x) < 1$ and returns to scale is decreasing with $dy/y < dx/x$.

We now extend the above example to the case of multiple inputs while retaining the assumption that output is a scalar represented by the variable y. To avoid confusing changes in scale with changes in mix, we write

$$y = f(\theta x_1, \theta x_2, \ldots, \theta x_m)$$

and use the scalar θ to represent increase in scale when $\theta > 1$. Thus, keeping input mix constant, — we increase all input by the common scale θ — we can represent "elasticity of scale" by

$$\varepsilon(\theta) = \frac{\theta}{y}\frac{dy}{d\theta}. \tag{5.2}$$

Then, at least locally, we have increasing returns-to-scale when $\varepsilon(\theta) > 1$ and decreasing returns-to-scale when $\varepsilon(\theta) < 1$. Finally $\varepsilon(\theta) = 1$ when returns to scale is constant. All analyses are thus conducted from a point $(x_1, \ldots, x_m)$ on the efficient frontier without changing the input proportion. That is, the mix is held constant.

To treat realistic applications we will need to modify these formulations. For instance, to deal with multiple outputs and inputs simultaneously we need to replace the "maximal value of y" which characterizes the production frontier in the single output case. Because this "maximum output" becomes ambiguous in multiple output situations we have recourse to the Pareto-Koopmans definition of efficiency given in Chapter 3. We also need to allow for occurrences of points like P and P' in Figure 5.1 which are not on the frontier. We must therefore develop methods that enable us to distinguish adjustments that eliminate such inefficiencies and adjustments associated with returns-to-scale inefficiencies. That is, we will need to be able to distinguish increasing inputs that achieve more (or less) than proportional increase in outputs at points on the efficient frontier and separate them from output increases resulting from the elimination of Pareto-Koopmans inefficiencies.

It is possible that such more-than-proportional increases in outputs are *less* valuable than the input increases required to obtain them. In economics this issue is solved by recourse to cost, profit or revenue functions — such as we treat later in this text. However, in many cases the unit price or cost data needed for such analyses are not available. A knowledge of returns to scale is worthwhile in any case and the means for attaining this knowledge by DEA occupies the rest of this chapter.

5.2 GEOMETRIC PORTRAYALS IN DEA

Figure 5.2 is a modification of Figure 5.1 to provide a simplified version of situations likely to be encountered in actual applications. To start with some of the differences, we note that A in Figure 5.2 represents the smallest of the observed input values and thereby reflects the fact that it is usually necessary to commit at least a minimum of resources — i.e., a positive threshold input value must be achieved — before any output is observed.

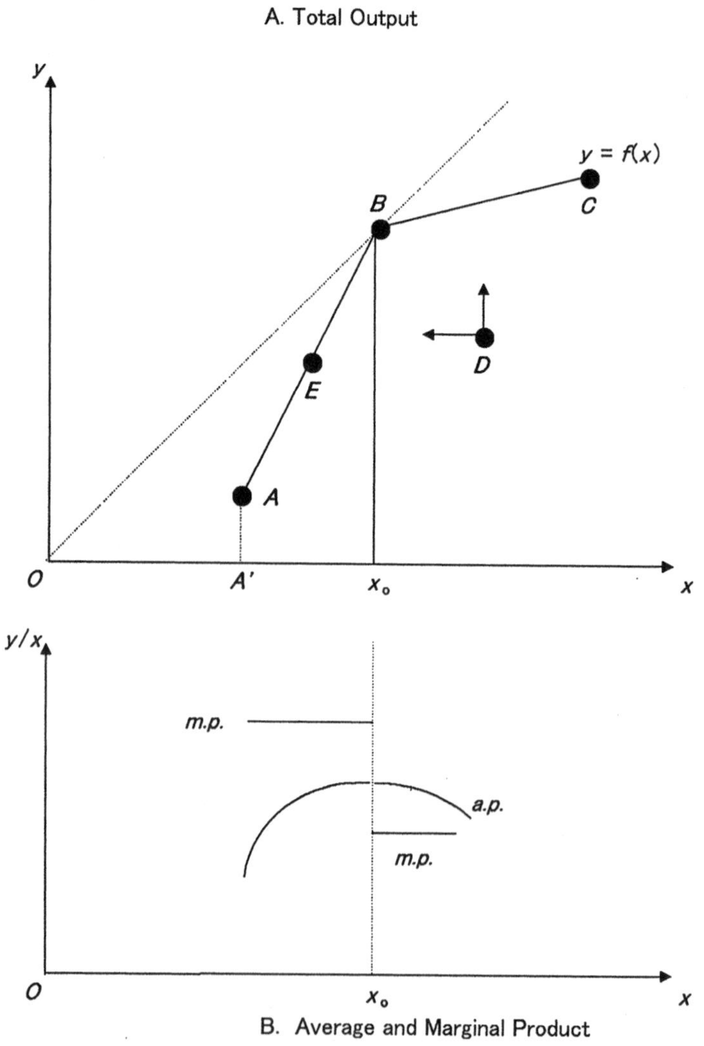

Figure 5.2. Returns to Scale: One Output-One Input

The smooth (everywhere differentiable) function used in Figure 5.1 now gives way to a piecewise linear function shown at the top of Figure 5.2, with consequences like the failure of $m.p.$ to intersect $a.p.$ at its maximum point in the bottom figure. The failure of the derivative to exist at x_0 is there reflected in a jump associated with the "staircase-like form" that occurs at the gap exhibited for $m.p.$ in Figure 5.2.

The concept of a tangent plane (or line) for dy/dx in Figure 5.1 now needs to be extended to the more general concept of a "supporting hyperplane" which we develop as follows. A hyperplane H_o in an $(m+s)$ dimensional input-output space passing through the point represented by the vectors $(\boldsymbol{x}_o,\ \boldsymbol{y}_o)$ can be expressed by the equation:

$$H_o: \qquad \boldsymbol{u}(\boldsymbol{y}-\boldsymbol{y}_o)-\boldsymbol{v}(\boldsymbol{x}-\boldsymbol{x}_o)=0, \tag{5.3}$$

where $\boldsymbol{u} \in R^s$ and $\boldsymbol{v} \in R^m$ are coefficient vectors. Now we define u_0, a scalar, by

$$u_0 = \boldsymbol{u}\boldsymbol{y}_o - \boldsymbol{v}\boldsymbol{x}_o. \tag{5.4}$$

Thus, the hyperplane in (5.3) can be expressed as:

$$\boldsymbol{u}\boldsymbol{y} - \boldsymbol{v}\boldsymbol{x} - u_0 = 0. \tag{5.5}$$

Generally, a hyperplane divides the space into two halfspaces. If the hyperplane H_o contains the production possibility set P in only one of the halfspaces, then it is said to be a *supporting hyperplane* of P *at the point* $(\boldsymbol{x}_o,\ \boldsymbol{y}_o)$. That is, a supporting hyperplane touches the production possibility set at this point.

More concretely, for every $(\boldsymbol{x},\ \boldsymbol{y}) \in P$, associated with any DMU,

$$\boldsymbol{u}\boldsymbol{y} - \boldsymbol{v}\boldsymbol{x} - u_0 \leq 0 \tag{5.6}$$

holds. From the definition of P for the BCC model and the above property of a supporting hyperplane, it can be demonstrated that

$$\boldsymbol{u} \geq \boldsymbol{0}, \quad \boldsymbol{v} \geq \boldsymbol{0}, \tag{5.7}$$

because the production possibility set is on only one side of the hyperplane.

In addition, a linear equation is invariant under multiplication by a nonzero number, so, to eliminate this source of indeterminacy, we add the constraint

$$\boldsymbol{v}\boldsymbol{x}_o = 1, \tag{5.8}$$

which we refer to as a *normalization* condition. P is generated by the data set represented in the matrices $(X,\ Y)$ and hence (5.6) can be replaced by

$$\boldsymbol{u}Y - \boldsymbol{v}X - u_0\boldsymbol{e} \leq \boldsymbol{0}, \tag{5.9}$$

which represents this relation for *all* DMUs. Also, from (5.4) and (5.8),

$$\boldsymbol{u}\boldsymbol{y}_o - u_0 = 1. \tag{5.10}$$

Thus, it is concluded that the vector of coefficients $(\boldsymbol{v},\ \boldsymbol{u},\ u_0)$ of the supporting hyperplane H_o is a (dual) optimal solution of the BCC model. Hence we have,

Theorem 5.1 *If a DMU* $(\boldsymbol{x}_o,\ \boldsymbol{y}_o)$ *is BCC-efficient, the normalized coefficient* $(\boldsymbol{v},\ \boldsymbol{u},\ u_0)$ *of the supporting hyperplane to* P *at* $(\boldsymbol{x}_o,\ \boldsymbol{y}_o)$ *gives an optimal solution of the BCC model and vice versa.*

We will make the sense (and use) of this theorem more concrete in the next section. Here we simply pause to note that the supporting hyperplane at a point like B in Figure 5.2 is not unique. The support property applies when the hyperplane (here a line) coincides with the line from A to B. It continues to apply, however, as the line is rotated around B until coincidence with the line from B to C is achieved. Thus the uniqueness associated with the assumption that $f(x)$ is analytic, as in Figure 5.1, is lost. This lack of uniqueness needs to be reflected in the returns-to-scale characterizations that we now supply.

5.3 BCC RETURNS TO SCALE

In (4.7)-(4.10) of the preceding chapter, the dual (multiplier) problem of the BCC model was expressed as:[1]

$$
\begin{aligned}
(DBCC_o)\quad \max \quad & z = \boldsymbol{u}\boldsymbol{y}_o - u_0 & (5.11)\\
\text{subject to} \quad & \boldsymbol{v}\boldsymbol{x}_o = 1 & (5.12)\\
& -\boldsymbol{v}X + \boldsymbol{u}Y - u_0\boldsymbol{e} \le \boldsymbol{0} & (5.13)\\
& \boldsymbol{v} \ge \boldsymbol{0},\ \boldsymbol{u} \ge \boldsymbol{0},\ u_0 \text{ free in sign}, & (5.14)
\end{aligned}
$$

where the free-in-sign variable, u_0, is associated with the condition $\boldsymbol{e\lambda} = 1$ in the corresponding envelopment model given as (BCC_o) in (4.2)-(4.6). See also Table 4.1.

When a DMU is BCC-efficient we can invoke Theorem 5.1 and employ the sign of u_0^* to portray the situation for returns to scale. This can be done in a simple manner by returning to A in the picture portrayed in the top diagram of Figure 5.2. It is apparent that returns to scale is increasing at A because the slopes of the rays going from the origin to $f(x)$ are increasing at this point. The supports for all rotations at A starting with coincidence at $A'A$ and continuing toward the line AB will all have negative intercepts which we can associate with $u_0^* < 0$. At E the support which has this same negative intercept property is unique. Thus, all of the supports have this same negative intercept property — which, we might note, is also shared by the tangent line at (x_0, y_0) in Figure 5.1. Jumping over to C in Figure 5.2 we note that the supports at this point will all have positive intercepts which are associated with $u_0^* > 0$. All such supports can thus be associated with decreasing returns to scale at C. Finally, at B we note that the rotations include the support represented by the broken line through the origin which we can associate with $u_0^* = 0$ as the intercept where returns to scale is constant — as was the case in Figure 5.1 where the tangent line also goes through the origin.

Using $(\boldsymbol{x}_o, \boldsymbol{y}_o)$ to represent a point with coordinate values corresponding to the multiple inputs and outputs recorded for DMU_o we formalize the above in the following theorem for use with the Model (BCC_o) — from Banker and Thrall (1992)[2] —

Theorem 5.2 *Assuming that $(\boldsymbol{x}_o, \boldsymbol{y}_o)$ is on the efficient frontier the following conditions identify the situation for returns to scale at this point,*

(i) *Increasing returns-to-scale prevails at $(\boldsymbol{x}_o, \boldsymbol{y}_o)$ if and only if $u_0^* < 0$ for all optimal solutions.*

(ii) *Decreasing returns-to-scale prevails at $(\boldsymbol{x}_o, \boldsymbol{y}_o)$ if and only if $u_0^* > 0$ for all optimal solutions.*

(iii) *Constant returns-to-scale prevails at $(\boldsymbol{x}_o, \boldsymbol{y}_o)$ if and only if $u_0^* = 0$ in any optimal solution.*

Chasing down *all* optimal solutions can be onerous. We can avoid this task, however, and also eliminate the efficiency assumption that is embedded in this theorem (so that we can treat points like D in Figure 5.2). This can be accomplished as follows. Assume that a solution is available from the BCC model represented in (4.2)-(4.6). This gives the information needed to project $(\boldsymbol{x}_o, \boldsymbol{y}_o)$ into $(\widehat{\boldsymbol{x}}_o, \widehat{\boldsymbol{y}}_o)$, a point on the efficient frontier, via

$$\begin{aligned} \widehat{x}_{io} &= \theta_B^* x_{io} - s_i^{-*}, \quad i = 1, \ldots, m \\ \widehat{y}_{ro} &= y_{ro} + s_r^{+*}, \quad r = 1, \ldots, s \end{aligned} \tag{5.15}$$

where $\theta_B^*, s_i^{-*}, s_r^{+*}$ are part of a solution to (4.2)-(4.6) — which identifies this as a BCC model.

From this solution to the BCC envelopment model, most computer codes report the value of u_0^* as part of the optimal solution for the dual (multiplier) problem. Suppose therefore that our solution yields $u_0^* < 0$. To take advantage of this information in deciding whether (i) or (iii) applies from the above theorem, we utilize a path opened by Banker, Bardhan and Cooper (1996)[3] and replace the customary dual (multiplier) problem as given in (5.11)-(5.14) with the following,

$$\begin{aligned} \max \quad & u_0 && (5.16) \\ \text{subject to} \quad & -\sum_{i=1}^m v_i x_{ij} + \sum_{r=1}^s u_r y_{rj} - u_0 \le 0, \quad j = 1, \ldots, n; j \neq o \\ & -\sum_{i=1}^m v_i \widehat{x}_{io} + \sum_{r=1}^s u_r \widehat{y}_{ro} - u_0 \le 0, \quad j = o \\ & \sum_{i=1}^m v_i \widehat{x}_{io} = 1, \quad \sum_{r=1}^s u_r \widehat{y}_{ro} - u_0 = 1 \\ & u_0 \le 0 \\ & v_i \ge 0, \; u_r \ge 0 \end{aligned}$$

where the $\hat{x}_{io}$ and $\hat{y}_{ro}$ are obtained from (5.15), and the added constraint $\sum_{r=1}^{s} u_r \hat{y}_{ro} - u_0 = 1$ is necessary for the solution to lie on the efficient frontier. Complemented by Definition 2.1, the condition becomes sufficient. Because $(\hat{\boldsymbol{x}}_o, \hat{\boldsymbol{y}}_o)$ is on the efficiency frontier for the BCC model with (5.15) there is no need to introduce the non-Archimedean element $\varepsilon > 0$.

The constraint $u_0 \leq 0$ means that the maximum of (5.16) cannot exceed zero. If zero is achieved the returns to scale is constant by (iii) of Theorem 5.2. If it is not achieved – i.e., the maximum is achieved with $u_0^* < 0$ – then condition (i) is applicable and returns to scale is decreasing.

The formulation (5.16) therefore forms part of a two-stage solution procedure in which one first solves (5.11)-(5.14) to obtain a solution with $u_0^* < 0$. This is then followed by the use of (5.16) to see whether an alternative optimum is present which can achieve $u_0^* = 0$. This corresponds to the treatment we accorded to D in Figure 5.2 where (a) one first eliminated the technical and mix inefficiencies and then (b) determined whether output increases (or decreases) could be obtained which are more, less, or exactly proportional to input variations that might then be undertaken along the efficient frontier.

We have formulated (5.16) for use when $u_0^* < 0$ occurred in the initial stage. If $u_0^* > 0$ had occurred then the inequality $u_0 \leq 0$ would be replaced by $u_0 \geq 0$ and the objective in (5.16) would be reoriented to "min u_0." (This will be illustrated by a numerical example in the next section.)

5.4 CCR RETURNS TO SCALE

In keeping with the DEA literature, we have used "constant returns-to-scale" to characterize the CCR model. This is technically correct but somewhat misleading because this model can also be used to determine whether returns to scale is increasing or decreasing. This is accomplished by slightly sharpening the conditions discussed in Section 5.3 to obtain the disjunction in the following theorem by Banker and Thrall (1992).

Theorem 5.3 *Let $(\boldsymbol{x}_o, \boldsymbol{y}_o)$ be a point on the efficient frontier. Employing a CCR model in envelopment form to obtain an optimal solution $(\lambda_1^*, \ldots, \lambda_n^*)$, returns to scale at this point can be determined from the following conditions,*

(i) *If $\sum_{j=1}^{n} \lambda_j^* = 1$ in any alternate optimum then constant returns-to-scale prevails.*

(ii) *If $\sum_{j=1}^{n} \lambda_j^* > 1$ for all alternate optima then decreasing returns-to-scale prevails.*

(iii) *If $\sum_{j=1}^{n} \lambda_j^* < 1$ for all alternate optima then increasing returns-to-scale prevails.*

Banker and Thrall (1992) proved this theorem on the assumption that $(\boldsymbol{x}_o, \boldsymbol{y}_o)$ is on the efficient frontier. However, we now utilize the route provided by Banker, Chang and Cooper (1996)[4] to eliminate the need for this assumption. This is accomplished by proceeding in a two-stage manner analogous to the preceding section. That is, we utilize the information available from the first stage solution to formulate the following problem,

$$\min \quad \sum_{j=1}^{n} \widehat{\lambda}_j - \varepsilon \left(\sum_{i=1}^{m} \widehat{s}_i^{\,-} + \sum_{r=1}^{s} \widehat{s}_r^{\,+} \right) \tag{5.17}$$

$$\text{subject to} \quad \theta_o^* \boldsymbol{x}_o = \sum_{j=1}^{n} \boldsymbol{x}_j \widehat{\lambda}_j + \widehat{\boldsymbol{s}}^{\,-}$$

$$\boldsymbol{y}_o = \sum_{j=1}^{n} \boldsymbol{y}_j \widehat{\lambda}_j - \widehat{\boldsymbol{s}}^{\,+}$$

$$1 \leq \sum_{j=1}^{n} \widehat{\lambda}_j$$

$$0 \leq \widehat{\lambda}_j, \quad \forall j,$$

where $\boldsymbol{x}_j, \boldsymbol{y}_j$ are vectors of input and output values with components corresponding to $\boldsymbol{x}_o, \boldsymbol{y}_o$ representing the observation for the DMU_o to be evaluated. The slack vectors $\boldsymbol{s}^-, \boldsymbol{s}^+$ as well as the vector $\widehat{\boldsymbol{\lambda}}$ are constrained to be nonnegative.

Here θ_o^* is obtained from the first stage solution, as in (5.15). However, we do not utilize the slack values, as is done in (5.15), because these are automatically provided by the way the objective is formulated in (5.17). Thus,

$$\theta_o^* \boldsymbol{x}_o - \widehat{\boldsymbol{s}}^{\,-*} = \sum_{j=1}^{n} \boldsymbol{x}_j \widehat{\lambda}_j^* \tag{5.18}$$

$$\boldsymbol{y}_o + \widehat{\boldsymbol{s}}^{\,+*} = \sum_{j=1}^{n} \boldsymbol{y}_j \widehat{\lambda}_j^*$$

exhibits the coordinates of an efficient point obtained from the expressions on the left in terms of a nonnegative combination of data from the other DMUs on the right. As noted in Definition 3.4, the DMUs on the right for which $\widehat{\lambda}_j^* > 0$ constitute efficient members of the reference set used to evaluate DMU_o. The efficient point represented on the left of (5.18) is therefore expressed in terms of the DMUs that define this portion (a facet) of the efficient frontier.

We have formulated (5.17) for use after a first stage solution with $\sum_{j=1}^{n} \lambda_j^* > 1$ has been achieved. The case where $\sum_{j=1}^{n} \lambda_j^* = 1$ needs no further treatment since by (i) of Theorem 5.3 constant returns to scale will then prevail at the efficient point represented in (5.18). The remaining case — viz., $\sum_{j=1}^{n} \lambda_j^* <$

1 — is addressed by writing $\sum_{j=1}^{n} \widehat{\lambda}_j \leq 1$ and replacing the objective with max $\sum_{j=1}^{n} \widehat{\lambda}_j + \varepsilon(\sum_{i=1}^{m} \widehat{s}_i^{\,-} + \sum_{r=1}^{s} \widehat{s}_r^{\,+})$.

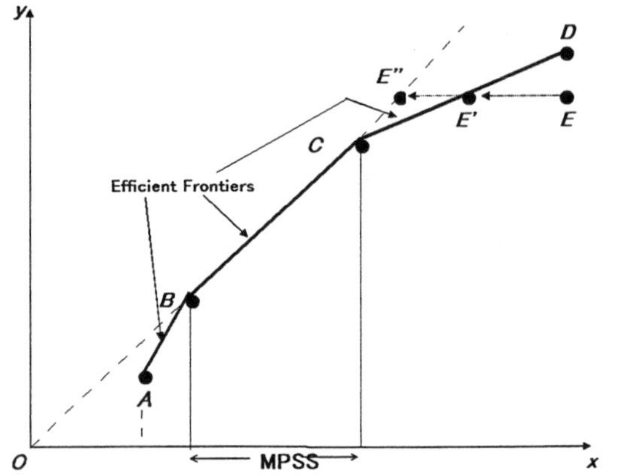

Figure 5.3. Most Productive Scale Size

For a numerical illustration using (5.17) we start with E in Figure 5.3. With coordinates $(x, y) = (4, 9/2)$ it is the only point that is not on the efficient frontier for either the BCC model (represented by the solid lines connecting A, B, C and D) or the efficient frontier of the CCR model (represented by the dotted line that passes through points B and C). In this case the first stage solution yields $\theta_o^* = 27/32$ with an alternate optimum for either $\lambda_B^* = 9/4$ or $\lambda_C^* = 9/8$ and all other variables $= 0$. Hence in both optima we have $\sum_{j=1}^{n} \lambda_j^* > 1$. To check whether condition (i) or (ii) in Theorem 5.3 is applicable we apply (5.17) to the (x, y) values located at the bottom of Figure 5.3, to obtain

$$
\begin{aligned}
\min \quad & (\widehat{\lambda}_A + \widehat{\lambda}_B + \widehat{\lambda}_C + \widehat{\lambda}_D + \widehat{\lambda}_E) - \varepsilon(\widehat{s}^{\,-} + \widehat{s}^{\,+}) \\
\text{subject to} \quad & \theta_o^* x_o = 27/8 = 1\widehat{\lambda}_A + 3/2\widehat{\lambda}_B + 3\widehat{\lambda}_C + 4\widehat{\lambda}_D + 4\widehat{\lambda}_E + \widehat{s}^{\,-} \\
& y_o = 9/2 = 1\widehat{\lambda}_A + 2\widehat{\lambda}_B + 4\widehat{\lambda}_C + 5\widehat{\lambda}_D + 9/2\widehat{\lambda}_E - \widehat{s}^{\,+} \\
& 1 \leq \widehat{\lambda}_A + \widehat{\lambda}_B + \widehat{\lambda}_C + \widehat{\lambda}_D + \widehat{\lambda}_E \\
& 0 \leq \widehat{\lambda}_A, \widehat{\lambda}_B, \widehat{\lambda}_C, \widehat{\lambda}_D, \widehat{\lambda}_E, \widehat{s}^{\,-}, \widehat{s}^{\,+}.
\end{aligned}
$$

This has an optimum at $\widehat{\lambda}_C^* = 9/8$ with all other variables equal to zero. Because this is the minimizing solution, it follows that $\sum_{j=1}^{n} \widehat{\lambda}_j^* > 1$ in *all* alternate optima so (ii) – as the applicable condition in Theorem 5.3 – means that decreasing returns to scale prevails.

To illustrate the opposite case we turn to A, a point on the BCC-efficient frontier in Figure 5.3 where returns to scale is increasing. For this case we report a first stage solution with $\min \theta_o = \theta_o^* = 3/4$ which is associated with alternate optima consisting of $\lambda_B^* = 1/2$ or $\lambda_C^* = 1/4$, and all other variables zero. To deal with this case, for which $\sum_{j=1}^n \lambda_j^* < 1$, we modify (5.17) in the manner indicated by the discussion following (5.18) to obtain

$$
\begin{aligned}
\max \quad & (\widehat{\lambda}_A + \widehat{\lambda}_B + \widehat{\lambda}_C + \widehat{\lambda}_D + \widehat{\lambda}_E) + \varepsilon(\widehat{s}^{\,-} + \widehat{s}^{\,+}) \\
\text{subject to} \quad & 3/4 = 1\widehat{\lambda}_A + 3/2\widehat{\lambda}_B + 3\widehat{\lambda}_C + 4\widehat{\lambda}_D + 4\widehat{\lambda}_E + \widehat{s}^{\,-} \\
& 1 = 1\widehat{\lambda}_A + 2\widehat{\lambda}_B + 4\widehat{\lambda}_C + 5\widehat{\lambda}_D + 9/2\widehat{\lambda}_E - \widehat{s}^{\,+} \\
& 1 \geq \widehat{\lambda}_A + \widehat{\lambda}_B + \widehat{\lambda}_C + \widehat{\lambda}_D + \widehat{\lambda}_E \\
& 0 \leq \widehat{\lambda}_A,\ \widehat{\lambda}_B,\ \widehat{\lambda}_C,\ \widehat{\lambda}_D,\ \widehat{\lambda}_E,\ \widehat{s}^{\,-},\ \widehat{s}^{\,+}.
\end{aligned}
$$

This has an optimum solution with $\widehat{\lambda}_B^* = 1/2$ and all other values zero so that condition (iii) of Theorem 5.3 is applicable and returns to scale is found to be increasing at A.

We next turn to E and now evaluate it by the BCC model first with (5.11)-(5.14) and then with (5.16). This will allow us to show how to modify (5.16) to treat the case $u_0^* > 0$ because the first stage solution gives $\theta_B^* = 7/8$ with $\lambda_C^* = \lambda_D^* = 1/2$ and the corresponding multiplier values are $u_0^* = u^* = v^* = 1/4$.

Reorienting (5.16) for use with this first stage $u_0^* > 0$ replaces $\max u_0$ with $\min u_0$ and $u_0 \leq 0$ with $u_0 \geq 0$ as in the following example.

$$
\begin{aligned}
\min \quad & u_0 \\
\text{subject to} \quad & -v + u - u_0 \leq 0 \\
& -3/2v + 2u - u_0 \leq 0 \\
& -3v + 4u - u_0 \leq 0 \\
& -4v + 5u - u_0 \leq 0 \\
& -7/2v + 9/2u - u_0 \leq 0 \\
& 7/2v = 1 \\
& 9/2u - u_0 = 1 \\
& v,\ u \geq 0,\ u_0 \geq 0.
\end{aligned}
$$

A minimizing solution to this problem is $v^* = u^* = u_0^* = 2/7$. This confirms the results we earlier found with the CCR model — viz., returns to scale is decreasing at E', the point with coordinate $(\widehat{x}_o, \widehat{y}_o) = (27/8, 9/2)$ in Figure 5.3. We now note that this coordinate is on the efficient frontier of the BCC model, but not for the CCR model. This is confirmed by noting that $\theta_B^* = 7/8 = 28/32$ for the BCC model applied to E whereas for the CCR model we get $\theta_o^* = 27/32$. Hence, a still further input contraction is needed to attain the CCR-efficiency available at E'', even after E' is attained in Figure 5.3.

We are now in a position to interpret the two models in the following manner. The BCC model separates the analysis in a twofold manner by (i) evaluating technical (and mix) efficiency in its envelopment model and (ii) evaluating returns-to-scale efficiency in its dual (multiplier) model. The CCR model, on the other hand, simultaneously evaluates both types of efficiency using only the envelopment form in the manner set forth in our treatment of Theorem 5.3.

The term "scale efficiency" in DEA involves recourse to constant returns-to-scale in a manner that generalizes the situation in Figure 5.1 where *a.p.* (average product) reached a maximum at (x_o, y_o). In addition to treating multiple outputs and inputs simultaneously it is necessary in DEA to allow for cases in which constant returns-to-scale prevails over an entire region such as the very simple example of the segment stretching from B to C in Figure 5.3.

In the next section we refer to regions like this as MPSS (Most Productive Scale Size) and use this as a gauge of scale efficiency. Here we note that we can access this region by invoking the following projection formulas from Banker and Morey (1986),[5]

$$\begin{aligned} \widehat{x}^*_{io} &\Leftarrow \frac{\theta^*_o x_{io} - \widehat{s}_i^{-*}}{\sum_{j=1}^n \widehat{\lambda}^*_j}, \quad i = 1, \dots, m \\ \widehat{y}^*_{ro} &\Leftarrow \frac{y_{ro} + \widehat{s}_r^{+*}}{\sum_{j=1}^n \widehat{\lambda}^*_j}, \quad r = 1, \dots, s \end{aligned} \tag{5.19}$$

where the $\widehat{\lambda}^*_j, \widehat{s}_i^{-*}, \widehat{s}_r^{+*}$ are obtained from an optimal solution for (5.17). In fact the numerators of (5.19) are obtained from (5.15), and division by the denominators thus yields the expressions

$$\begin{aligned} \widehat{\boldsymbol{x}}^*_o &= \sum_{j=1}^n \boldsymbol{x}_j \widetilde{\lambda}^*_j \\ \widehat{\boldsymbol{y}}^*_o &= \sum_{j=1}^n \boldsymbol{y}_j \widetilde{\lambda}^*_j \\ \text{with } 1 &= \sum_{j=1}^n \widetilde{\lambda}^*_j. \end{aligned} \tag{5.20}$$

To illustrate we return to the initial example in which E was evaluated with $\sum_{j=1}^n \widehat{\lambda}^*_j = 9/4$. Applying (5.19) to $(x_o, y_o) = (27/8, 9/2)$ therefore gives

$$\begin{aligned} \widehat{x}^* &= \frac{27}{8}\frac{4}{9} = \frac{3}{2} \\ \widehat{y}^* &= \frac{9}{2}\frac{4}{9} = 2 \end{aligned}$$

which are the coordinates of B, a point of constant returns-to-scale that is in the MPSS region of Figure 5.3.

We carry the discussion of MPSS further in the next section. Here we conclude by returning to the remark with which we began this chapter — viz., characterizing CCR as a constant returns-to-scale model is somewhat misleading. In fact use of the CCR model might be preferred to use of the BCC model for returns-to-scale analyses for reasons like the following:

First the CCR model generally involves fewer constraints and hence is likely to be slightly more efficient computationally. Second the solutions from this model directly provide more of the pertinent information in forms for use in projective formulas like (5.15) and (5.18) - to which we now add (5.19).

Relations between the CCR and BCC models are studied further in the next section. Here we conclude with the following theorem — due to Banker and Thrall (1992) — which relates the returns-to-scale characterizations from these two models as follows,

Theorem 5.4

(i) $u_o^* > 0$ *for all optimal solutions to a BCC model if and only if* $\sum_{j=1}^n \widehat{\lambda}_j^* > 1$ *for all optimal solutions to the corresponding CCR model.*

(ii) $u_o^* < 0$ *for all optimal solutions to a BCC model if and only if* $\sum_{j=1}^n \widehat{\lambda}_j^* < 1$ *for all optimal solutions to the corresponding CCR model.*

(iii) $u_o^* = 0$ *for some optimal solution to a BCC model if and only if* $\sum_{j=1}^n \widehat{\lambda}_j^* = 1$ *for some optimal solution to the corresponding CCR model.*

This is conveniently summarized in the following,

Corollary 5.1 (Corollary to Theorem 5.4)

(i) $u_o^* > 0$ *for all optimal solutions to a BCC model if and only if* $\sum_{j=1}^n \lambda_j^* - 1 > 0$ *in all optimal solutions to the corresponding CCR model.*

(ii) $u_o^* < 0$ *for all optimal solutions to a BCC model if and only if* $\sum_{j=1}^n \lambda_j^* - 1 < 0$ *in all optimal solutions to the corresponding CCR model.*

(iii) $u_o^* = 0$ *for some optimal solution to a BCC model if and only if* $\sum_{j=1}^n \lambda_j^* - 1 = 0$ *in some optimal solution to the corresponding CCR model.*

5.5 MOST PRODUCTIVE SCALE SIZE

To further examine relations between BCC and CCR models we start with the following theorem due to Ahn, Charnes and Cooper (1989),[6]

Theorem 5.5 *A* DMU_o *found to be efficient with a CCR model will also be found to be efficient with the corresponding BCC model and constant returns-to-scale prevails at* DMU_o.

This theorem, which is simple to prove, is critically important for the topics covered in this section. It is also important to note that the converse is not necessarily true. As exemplified by E' in Figure 5.3, a DMU can be simultaneously characterized as efficient by a BCC model and inefficient by a CCR model with $\theta_o^* < \theta_B^*$. However, if $\theta_o^* = \theta_B^*$ then there will be at least one alternate optimum for this θ_o^* for which $\sum_{j=1}^n \lambda_j^* = 1$.

This theorem is directed to mathematical properties of the two types of models. To obtain interpretations that we can relate to our earlier discussions of returns to scale and to open possibilities for managerial uses we utilize

R.D. Banker's concept of Most Productive Scale Size (MPSS),[7] which we initiate with the following expression

$$(\boldsymbol{x}_o\alpha,\ \boldsymbol{y}_o\beta). \tag{5.21}$$

Here $(\boldsymbol{x}_o, \boldsymbol{y}_o)$ are vectors with components corresponding to coordinates of the point being evaluated and (α, β) are scalars representing expansion or contraction factors according to whether $\alpha, \beta > 1$ or $\alpha, \beta < 1$ are applied to the inputs and outputs represented by these coordinates.

Our discussion in Section 1 of this chapter was confined to a single output so a scale factor, applied to all inputs, provided everything needed to study scale variations in a manner that would avoid confusing variations in mix with economies or diseconomies of scale. Here α plays the role previously assigned to θ for inputs and the scalar β plays a similar role for outputs.

We also need to allow for inequality relations in place of the equalities used in (5.1) and (5.2). We then want a measure of scale which is "dimensionless" (i.e., it does not depend on the units of measure used) so we utilize a ratio of these two scalars, each of which is dimensionless.

To accomplish these (and other) purposes we use the following model

$$\begin{aligned} \max \quad & \beta/\alpha \\ \text{subject to} \quad & \beta\boldsymbol{y}_o \le \sum_{j=1}^{n} \boldsymbol{y}_j\lambda_j \\ & \alpha\boldsymbol{x}_o \ge \sum_{j=1}^{n} \boldsymbol{x}_j\lambda_j \\ & 1 = \sum_{j=1}^{n} \lambda_j \\ & 0 \le \beta,\ \alpha. \end{aligned} \tag{5.22}$$

This is a fractional programming problem. By the procedures described in Problem 3.1 (in Chapter 3)[8] we can transform this to an ordinary linear programming problem to obtain solutions $\widehat{\boldsymbol{x}}_o = \alpha\boldsymbol{x}_o, \widehat{\boldsymbol{y}}_o = \beta\boldsymbol{y}_o$ which we can associate with points which are MPSS by means of the following theorem,

Theorem 5.6 *A necessary condition for a DMU_o with output and input vectors $\boldsymbol{y}_o$ and $\boldsymbol{x}_o$ to be at MPSS is $\beta^*/\alpha^* = \max\beta/\alpha = 1$ in (5.22), in which case $\beta^* = \alpha^*$ and returns to scale will be constant.*

We illustrate for D the point with coordinates (4, 5) in Figure 5.3. Thus, applying (5.22) we obtain the following problem,

$$\begin{aligned} \max \quad & \beta/\alpha \\ \text{subject to} \quad & 5\beta \le 1\lambda_A + 2\lambda_B + 4\lambda_C + 5\lambda_D + 9/2\lambda_E \\ & 4\alpha \ge 1\lambda_A + 3/2\lambda_B + 3\lambda_C + 4\lambda_D + 4\lambda_E \\ & 1 = \lambda_A + \lambda_B + \lambda_C + \lambda_D + \lambda_E \\ & 0 \le \lambda_A, \ldots, \lambda_E. \end{aligned}$$

This has an optimum with $\lambda_B^* = 1$ and $\beta^* = 2/5, \alpha^* = 3/8$. It also has an alternate optimum with $\lambda_C^* = 1$ and $\beta^* = 4/5, \alpha^* = 3/4$. In either case we have $\beta^*/\alpha^* = 16/15$ so, via Theorem 5.6, D is not at MPSS.

The information needed to bring D into MPSS is at hand in both cases. Thus, utilizing the first solution we have $\alpha^* x_o = 3/8 \times 4 = 3/2$ and $\beta^* y_o = 2/5 \times 5 = 2$ which yields new coordinates $(\widehat{x}_o, \widehat{y}_o) = (3/2, 2)$ where $(\widehat{x}_o,\ \widehat{y}_o) = (\alpha^* x_o,\ \beta^* y_o)$ as in (5.21). Reference to Figure 5.3 shows these values to be the coordinates of B, a point which is MPSS in the constant-returns-to-scale region of Figure 5.3. Turning next to the alternate optimum we have $\beta^* y_o = 4/5 \times 5 = 4$ and $\alpha^* x_o = 3/4 \times 4 = 3$, which are the coordinates of C, a point which is also MPSS in Figure 5.3. Indeed, every convex combination of these two solutions yields $\beta^*/\alpha^* = 16/15$ with components $(\widehat{x}_o,\ \widehat{y}_o) = (\alpha^* x_o,\ \beta^* y_o)$ corresponding to the coordinates of points in the region of constant-returns-to-scale in Figure 5.3 and, hence, each such point is MPSS.

To extend the relation $\alpha^* x_o = \widehat{x}_o$ and $\beta^* y_o = \widehat{y}_o$ to the case of multiple inputs and outputs we write these vector relations in the following form,

$$e\beta^* = \left(\frac{\widehat{y}_{1o}}{y_{1o}}, \ldots, \frac{\widehat{y}_{ro}}{y_{ro}}, \ldots, \frac{\widehat{y}_{so}}{y_{so}}\right) \tag{5.23}$$
$$e\alpha^* = \left(\frac{\widehat{x}_{1o}}{x_{1o}}, \ldots, \frac{\widehat{x}_{io}}{x_{io}}, \ldots, \frac{\widehat{x}_{mo}}{x_{mo}}\right),$$

where e is the vector with unity for all of its elements. To show that the proportions in all outputs and inputs are maintained at the values β^* and α^*, respectively, we simply note that $\beta^* = \widehat{y}_{ro}/y_{ro} = \widehat{y}_{ko}/y_{ko}$ implies $y_{ko}/y_{ro} = \widehat{y}_{ko}/\widehat{y}_{ro}$ and $\alpha^* = \widehat{x}_{io}/x_{io} = \widehat{x}_{lo}/x_{lo}$ implies $x_{io}/x_{lo} = \widehat{x}_{io}/\widehat{x}_{lo}$ so output and input mixes are both preserved.

We can also use these formulations to generalize the elasticity in Section 1 to the case of multiple-inputs and multiple-outputs by noting that

$$\frac{1-\beta^*}{1-\alpha^*} = \frac{y_{ro} - \widehat{y}_{ro}}{y_{ro}} \Bigg/ \frac{x_{io} - \widehat{x}_{io}}{x_{io}} = \frac{x_{io}}{y_{ro}} \frac{\Delta y_{ro}}{\Delta x_{io}}. \tag{5.24}$$

For each of these mix pairs we then have a measure of scale change associated with movement to the region of MPSS.

To see what these results mean we return to D in Figure 5.3 to note that the change from D to B can be associated with output-to-input ratios which have values $5/4 = 15/12$ at D and $2/(3/2) = 16/12$ at B. We can then use (5.21) to note that $\beta^*/\alpha^* = 16/15$ gives the change of output in ratio form when the inputs are set at $x = 12$ in both cases.

Returning to (5.24) we recall that an elasticity measures the proportional change in output relative to the proportional change in input. Hence when input decreases from $x = 4$ to $x = 3/2$ the proportional decrease is $(4 - 3/2)/4 = 5/8$ and $(1 - \beta^*)/(1 - \alpha^*) = 24/25$ which means that the proportional decrease in output will be $24/23 \times 5/8 = 3/5$ — which is the proportional output change in moving from D to B. As is to be expected in the case of decreasing returns-to-scale, the proportional decrease in output is less than the proportional decrease in input. This is confirmed by the movement from D to C which gives

$$\frac{x_o}{y_o}\frac{\Delta y_o}{\Delta x_o} = \frac{4}{5}\cdot\frac{5-4}{4-3} = \frac{4}{5} \text{ or } \frac{1}{5} = \frac{5-4}{5} = \frac{\Delta y_o}{y_o} < \frac{\Delta x_o}{x_o} = \frac{4-3}{4} = \frac{1}{4},$$

and the relative change in output is smaller than the relative change in input in this case as well.

The immediately preceding developments assume that the observations for the DMUs being evaluated are on the BCC-efficient frontiers. This is to be expected because, as previously emphasized, elasticity and returns to scale are both frontier concepts. However, this is not a limiting assumption because formula (5.15) may be invoked when these kinds of concepts (and their related measures) are to be utilized.

When only access to MPSS is of interest we can develop an approach from the following

Definition 5.1 (MPSS) *For DMU_o to be MPSS* both *of the following conditions need be satisfied: (i) $\beta^*/\alpha^* = 1$ and (ii) all slacks are zero.*

This is intended to mean that the slacks in *all* alternative optima are zero. This contingency is covered in a manner analogous to the two-stage procedure outlined in Chapter 3. That is, we employ (5.22) — or its associated linear programming problem — to obtain a solution β^*/α^*. If this has any nonzero slacks it is not necessary to proceed further. However, if all slacks are zero we check to see that no alternate optimum has nonzero slacks by moving to a second stage in which $\beta = \beta^*$ and $\alpha = \alpha^*$ are fixed and the slacks are maximized. The following formulas may then be used to obtain a point with coordinates y^*_{ro}, x^*_{io} which are MPSS,

$$\begin{aligned} \beta^* y_{ro} + s_r^{+*} &= y^*_{ro}, \quad r = 1,\ldots,s \\ \alpha^* x_{io} - s_i^{-*} &= x^*_{io}, \quad i = 1,\ldots,m \end{aligned} \tag{5.25}$$

This allows us to conclude this discussion on a note which is concerned with practical implementation of the concepts we have been considering. In a conference on the economics of the iron and steel industry one of the economic consultants, Harold H. Wein, complained that the returns-to-scale formulations in economics were of little use. He had never participated in a case where changes in scale were separated from changes in mix, so formulas like (5.2) were of no use either in (a) arriving at such decisions or (b) analyzing their consequences.

Dr. Wein might also have complained about the restriction to a single output and the assumed prevalence of technical efficiency. Here, however, we have addressed all of these concerns. Thus we have moved to the multiple output case and removed the assumption that technical efficiency has been achieved. We also have provided (5.25) to disentangle mix inefficiencies for both inputs and outputs (as represented in the nonzero slacks) and separated them from the scale inefficiencies represented by values of $\beta^*/\alpha^* \neq 1$. Hence we have allowed for inequalities in inputs and in outputs and we have treated both multiple inputs and multiple outputs in a way that distinguishes scale and mix inefficiencies.

5.6 FURTHER CONSIDERATIONS

Still another approach to returns to scale analysis may be used as given in Tone (1996),[9] which we develop as follows:

Theorem 5.7 (Tone(1996)) *In the BCC model a reference set to any BCC inefficient DMU does not include both increasing and decreasing returns-to-scale DMUs at the same time.*

Proof : Let the BCC inefficient DMU be $(\boldsymbol{x}_o, \boldsymbol{y}_o)$ and its projection onto the efficient frontier by (5.15) be $(\widehat{\boldsymbol{x}}_o, \widehat{\boldsymbol{y}}_o)$. Since $(\widehat{\boldsymbol{x}}_o, \widehat{\boldsymbol{y}}_o)$ is BCC-efficient, there exists an optimal weight $(\boldsymbol{v}_e, \boldsymbol{u}_e, u_{0e})$ that satisfies

$$\boldsymbol{u}_e \widehat{\boldsymbol{y}}_o - u_{0e} = 1 \tag{5.26}$$

$$\boldsymbol{v}_e \widehat{\boldsymbol{x}}_o = 1 \tag{5.27}$$

$$-\boldsymbol{v}_e X + \boldsymbol{u}_e Y - u_{0e}\boldsymbol{e} \leq \boldsymbol{0} \tag{5.28}$$

$$\boldsymbol{v}_e > \boldsymbol{0}, \quad \boldsymbol{u}_e > \boldsymbol{0}. \tag{5.29}$$

Hence, for $j \in E_o$ (a reference set to $(\boldsymbol{x}_o, \boldsymbol{y}_o)$), we have:

$$-\boldsymbol{v}_e \boldsymbol{x}_j + \boldsymbol{u}_e \boldsymbol{y}_j - u_{0e} \leq 0. \tag{5.30}$$

From (5.26) and (5.27),

$$-\boldsymbol{v}_e \widehat{\boldsymbol{x}}_o + \boldsymbol{u}_e \widehat{\boldsymbol{y}}_o - u_{0e} = 0. \tag{5.31}$$

Since

$$\widehat{\boldsymbol{x}}_o = \sum_{j \in E_o} \lambda_j^* \boldsymbol{x}_j \quad \text{and} \quad \widehat{\boldsymbol{y}}_o = \sum_{j \in E_o} \lambda_j^* \boldsymbol{y}_j,$$

we obtain:

$$-\boldsymbol{v}_e \sum_{j \in E_o} \lambda_j^* \boldsymbol{x}_j + \boldsymbol{u}_e \sum_{j \in E_o} \lambda_j^* \boldsymbol{y}_j - u_{0e} = 0. \tag{5.32}$$

This equation can be transformed, using $\sum_{j \in E_o} \lambda_j^* = 1$, into:

$$\sum_{j \in E_o} \lambda_j^* \left(-\boldsymbol{v}_e \boldsymbol{x}_j + \boldsymbol{u}_e \boldsymbol{y}_j - u_{0e} \right) = 0. \tag{5.33}$$

However, from (5.30), (5.33) we have some $\lambda_j^* > 0$ $(j \in E_o)$ to satisfy $\sum_{j \in E_o} \lambda_j^* = 1$. Hence, for each j we have the equation:

$$-\boldsymbol{v}_e \boldsymbol{x}_j + \boldsymbol{u}_e \boldsymbol{y}_j - u_{0e} = 0. \quad (\forall j \in E_o) \tag{5.34}$$

Thus, $(\boldsymbol{v}_e, \boldsymbol{u}_e, u_{0e})$ are coefficients of a supporting hyperplane at each $(\boldsymbol{x}_j, \boldsymbol{y}_j)$ for every $j \in E_o$. Let $t_j = 1/\boldsymbol{v}_e \boldsymbol{x}_j$ (> 0), then $t_j \boldsymbol{v}_e = \boldsymbol{v}'_e, t_j \boldsymbol{u}_e = \boldsymbol{u}'_e, t_j u_{0e} = u'_{0e}$

gives $u'_e y_j - u'_{0e} = 1$ which is an optimal solution for the multiplier form associated with $j \in E_o$.

Suppose that E_o contains an *increasing* returns-to-scale DMU a and a *decreasing* returns-to-scale DMU b. Then, u_{0e} must be negative, since returns to scale is increasing at DMU a. At the same time, u_{0e} must be positive, since DMU b has decreasing returns-to-scale characteristics. This leads to a contradiction. □

We also have a corollary from the above,

Corollary 5.2 *Let a reference set to a BCC-inefficient DMU $(\boldsymbol{x}_o, \boldsymbol{y}_o)$ be E_o. Then, E_o consists of one of the following combinations of BCC-efficient DMUs where IRS, CRS and DRS stand for increasing, constant and decreasing returns-to-scale, respectively.*

(*i*) *All DMUs have IRS.*

(*ii*) *Mixture of DMUs with IRS and CRS.*

(*iii*) *All DMUs have CRS.*

(*iv*) *Mixture of DMUs with CRS and DRS.*

(*v*) *All DMUs show DRS.*

Based on the above observations, we have

Theorem 5.8 (Characterization of Return to Scale)
Let the BCC-projected activity of a BCC-inefficient DMU $(\boldsymbol{x}_o, \boldsymbol{y}_o)$ be $(\widehat{\boldsymbol{x}}_o, \widehat{\boldsymbol{y}}_o)$ and a reference set to $(\boldsymbol{x}_o, \boldsymbol{y}_o)$ be E_o. Then, $(\widehat{\boldsymbol{x}}_o, \widehat{\boldsymbol{y}}_o)$ belongs to

1. *IRS, if E_o consists of DMUs in categories (i) or (ii) of Corollary 5.2, and*

2. *DRS, if E_o consists of DMUs in categories (iv) or (v).*

Proof: In the case of (*i*) or (*ii*), E_o contains at least one DMU with IRS and any supporting hyperplane at $(\widehat{\boldsymbol{x}}_o, \widehat{\boldsymbol{y}}_o)$ is also a supporting hyperplane at the IRS DMU, as shown in the proof of Theorem 5.7. Thus, the upper bound of u_{0e} must be negative. By the same reasoning, in the case of (*iv*) or (*v*), the projected activity has DRS. □

However, in the case of (*iii*) we cannot identify the returns to scale of $(\widehat{\boldsymbol{x}}_o, \widehat{\boldsymbol{y}}_o)$ directly from its reference set. In this case, $(\widehat{\boldsymbol{x}}_o, \widehat{\boldsymbol{y}}_o)$ may display IRS, CRS or DRS. We need further LP calculations for identifying its returns-to-scale characteristics as described below.

Using the preceding results a procedure for identifying the returns-to-scale characteristics of DMU$_o$ represented by vectors $(\boldsymbol{x}_o, \boldsymbol{y}_o)$ can be developed as follows.

First we solve the envelopment form of the (input-oriented) BCC model for $(\boldsymbol{x}_o,\ \boldsymbol{y}_o)$ via (4.2)-(4.6).

(i) If DMU$_o$ is found to be BCC-efficient then we can, at the same time, find an optimal u_0^* of the multiplier form (5.11)-(5.14) as a value of an optimal dual variable associated with the envelopment model constraint $\boldsymbol{e\lambda} = 1$.

(ia) If $u_0^* = 0$ then we can conclude that *constant* returns-to-scale prevails at $(\boldsymbol{x}_o,\ \boldsymbol{y}_o)$ by Theorem 5.2.

(ib) If $u_0^* < (>) 0$ then we solve the following LP using its dual (envelopment form), since the envelopment form is computationally much easier than the multiplier form.

$$\begin{array}{llll} \text{[Multiplier form]} & \max\ (\min) & u_0 & (5.35) \\ & \text{subject to} & -\boldsymbol{v}X + \boldsymbol{u}Y - u_0\boldsymbol{e} \leq \boldsymbol{0} & \\ & & \boldsymbol{v}\boldsymbol{x}_o = 1, \quad \boldsymbol{u}\boldsymbol{y}_o - u_0 = 1 & \\ & & \boldsymbol{v},\ \boldsymbol{u} \geq \boldsymbol{0}, \quad u_0 \ \text{free in sign.} & \end{array}$$

$$\begin{array}{llll} \text{[Envelopment form]} & \min\ (\max) & z = \theta - \lambda_0 \quad (\lambda_0 - \theta) & (5.36) \\ & \text{subject to} & \theta\boldsymbol{x}_o \geq X\boldsymbol{\lambda} & \\ & & \lambda_0\boldsymbol{y}_o \leq Y\boldsymbol{\lambda} & \\ & & -\boldsymbol{e\lambda} + \lambda_0 = 1 \quad (\boldsymbol{e\lambda} - \lambda_0 = 1) & \\ & & \boldsymbol{\lambda} \geq \boldsymbol{0},\ \theta,\ \lambda_0 \ \text{free in sign.} & \end{array}$$

Both LPs above have the same optimal value if [Envelopment form] has a finite minimum (maximum). If [Multiplier form] has an unbounded optimal value (∞ in the max case or $-\infty$ in the min cas) the [Envelopment form] has no feasible solution. Let the optimal objective value of [Multiplier form] be $\overline{u}_0^*$ ($\underline{u}_0^*$), including ∞ ($-\infty$). Then by Theorem 5.2 we have,

(a) If $\overline{u}_0^* \geq 0$ ($\underline{u}_0^* \leq 0$) we can conclude that *constant* returns-to-scale prevails at $(\boldsymbol{x}_o,\ \boldsymbol{y}_o)$.

(b) If $\overline{u}_0^* < 0$ ($\underline{u}_0^* > 0$) we can conclude that *increasing (decreasing)* returns-to-scale prevails at $(\boldsymbol{x}_o,\ \boldsymbol{y}_o)$.

(ii) If DMU$_o$ is found to be BCC inefficient then we can relate the returns-to-scale characteristics of its projected DMU $(\widehat{\boldsymbol{x}}_o, \widehat{\boldsymbol{y}}_o)$ to those of its reference set as represented in (5.15) in the following manner:

(iia) The projected DMU $(\widehat{\boldsymbol{x}}_o, \widehat{\boldsymbol{y}}_o)$ displays *increasing* returns-to-scale if the reference set of $(\boldsymbol{x}_o, \boldsymbol{y}_o)$ consists of either *increasing* returns-to-scale DMUs or a mixture of *increasing* and *constant* returns-to-scale DMUs.

(iib) The projected DMU $(\widehat{\boldsymbol{x}}_o, \widehat{\boldsymbol{y}}_o)$ displays *decreasing* returns-to-scale if the reference set of $(\boldsymbol{x}_o, \boldsymbol{y}_o)$ consists of either *decreasing* returns-to-scale DMUs or a mixture of *decreasing* and *constant* returns-to-scale DMUs.

(iic) If the reference set of $(\widehat{\boldsymbol{x}}_o, \widehat{\boldsymbol{y}}_o)$ consists of DMUs all belonging to *constant* returns to scale, then we apply the above procedure (i) to $(\widehat{\boldsymbol{x}}_o, \widehat{\boldsymbol{y}}_o)$ and identify its returns to scale characteristics.

A merit of the above procedure exists in the fact that no extra computations are needed for identifying returns-to-scale characteristics of BCC inefficient DMUs in the cases (iia) and (iib). This will reduce a fairly great amount of computations. Numerical examples are presented in Problem 5.4.

5.7 RELAXATION OF THE CONVEXITY CONDITION

We can extend the BCC envelopment model by relaxing the convexity condition $\boldsymbol{e\lambda} = 1$ to

$$L \leq \boldsymbol{e\lambda} \leq U, \tag{5.37}$$

where L $(0 \leq L \leq 1)$ and U $(1 \leq U)$ are upper and lower bounds for the sum of the λ_j. Notice that $L = 0$ and $U = \infty$ correspond to the CCR model and $L = U = 1$ corresponds to the BCC model. Some typical extensions are described below.

1. The Increasing Returns-to-Scale (IRS) Model
The case $L = 1$, $U = \infty$ is called the *Increasing Returns-to-Scale* (IRS) or *Non-Decreasing Returns-to-Scale* (NDRS) model. The constraint on $\boldsymbol{\lambda}$ is :

$$\boldsymbol{e\lambda} \geq 1. \tag{5.38}$$

The condition $L = 1$ assumes that we cannot reduce the scale of DMU but it is possible to expand the scale to infinity. Figure 5.4 shows the production possibility set of the IRS model for the single input and single output case.

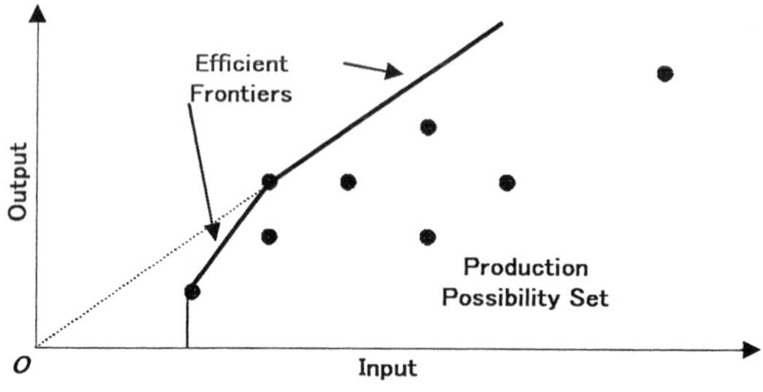

Figure 5.4. The IRS Model

The output/input ratio for any point on the efficient frontier is not decreasing with respect to input and the term NDRS is derived from this fact. That is, a proportional increase in output is always at least as great as the related proportional increase in input. In mathematical terms, we always have

$$\frac{\Delta y/y}{\Delta x/x} = \frac{x}{y}\frac{\Delta y}{\Delta x} \geq 1$$

so that

$$\frac{\Delta y}{y} \geq \frac{\Delta x}{x}$$

where Δy, Δx are the increases to be made from a frontier point with coordinate (x, y). This model focuses on the scale efficiencies of relatively small DMUs.

2. The Decreasing Returns-to-Scale (DRS) Model

The case $L = 0$, $U = 1$ is called the *Decreasing Returns-to-Scale* (DRS) or *Non-Increasing Returns-to-Scale* (NIRS) model. The constraints on $\boldsymbol{\lambda}$ are:

$$0 \leq \boldsymbol{e\lambda} \leq 1. \tag{5.39}$$

By the condition $U = 1$, scaling up of DMUs is interdicted, while scaling down is permitted. Figure 5.5 depicts the production possibility set. The output/input ratio of efficient frontier points is decreasing with respect to the input scale. That is, $\Delta y/y \leq \Delta x/x$ with equality (=constant returns-to-scale) for the first segment on the frontier and strict inequality holding thereafter, so this model puts the emphasis on larger DMUs where returns to scale is decreasing.

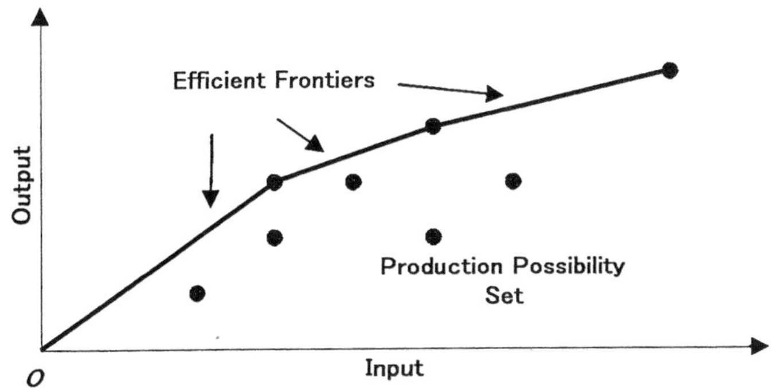

Figure 5.5. The DRS Model

3. The Generalized Returns-to-Scale (GRS) Model

The case $0 \leq L \leq 1$, $U \geq 1$ is called the *Generalized Returns-to-Scale* (GRS) model because it is possible to use these values to control the admissible range allowed for returns to scale. For instance $L = 0.8$ and $U = 1.2$ means that returns to scale can be decreasing at most in the proportion $L = 0.8$ and they can be increasing at most in the proportion $U = 1.2$. Figure 5.6 shows the production possibility set of the GRS model for the single-input and single-output case, which sharpens the BCC model. Efficiency measured by this model is not better than that measured by BCC.

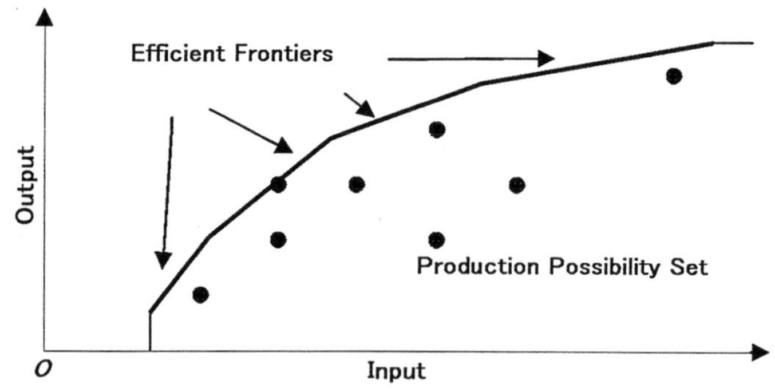

Figure 5.6. The GRS Model

It is logically true that for every DMU we have the relations

$$\theta^*_{CCR} \leq \theta^*_{IRS}, \theta^*_{DRS}, \theta^*_{GRS} \leq \theta^*_{BCC}.$$

5.8 DECOMPOSITION OF TECHNICAL EFFICIENCY

It is interesting to investigate the sources of inefficiency that a DMU might have. Are they caused by the inefficient operation of the DMU itself or by the disadvantageous conditions under which the DMU is operating?

For this purpose, comparisons of the (input-oriented) CCR and BCC scores deserve consideration. The CCR model assumes the constant returns-to-scale production possibility set, i.e., it is postulated that the radial expansion and reduction of all observed DMUs and their nonnegative combinations are possible and hence the CCR score is called *global technical* efficiency. On the other hand, the BCC model assumes that convex combinations of the observed DMUs form the production possibility set and the BCC score is called *local pure technical* efficiency. If a DMU is fully efficient (100%) in both the CCR and BCC scores, it is operating in the *most productive scale size* as pointed out in Section 5.5. If a DMU has full BCC efficiency but a low CCR score, then it is operating locally efficiently but not globally efficiently due to the scale size of the DMU. Thus, it is reasonable to characterize the *scale efficiency* of a DMU by the ratio of the two scores.

5.8.1 Scale Efficiency

Here we extend this approach to show how to decompose these inefficiencies into their component parts. Based on the CCR and BCC scores, we define *scale efficiency* as follows:

Definition 5.2 (Scale Efficiency) *Let the CCR and BCC scores of a DMU be θ^*_{CCR} and θ^*_{BCC}, respectively. The scale efficiency is defined by*

$$SE = \frac{\theta^*_{CCR}}{\theta^*_{BCC}}. \tag{5.40}$$

SE is not greater than one. For a BCC-efficient DMU with CRS characteristics, i.e., in the most productive scale size, its scale efficiency is one. The CCR score is called the (global) *technical* efficiency (TE), since it takes no account of scale effect as distinguished from PTE. On the other hand, BCC expresses the (local) *pure technical* efficiency (PTE) under variable returns-to-scale circumstances. Using these concepts, relationship (5.40) demonstrates a decomposition of efficiency as $\theta^*_{CCR} = \theta^*_{BCC} \times \text{SE}$ or,

$$[\text{Technical Eff. (TE)}] = [\text{Pure Technical Eff. (PTE)}] \times [\text{Scale Eff. (SE)}]. \tag{5.41}$$

This decomposition, which is unique, depicts the sources of inefficiency, i.e., whether it is caused by inefficient operation (PTE) or by disadvantageous conditions displayed by the scale efficiency (SE) or by both.

In the single input and single output case, the scale efficiency can be illustrated by Figure 5.7. For the BCC-efficient A with IRS, its scale efficiency is given by

$$\text{SE}(A) = \theta^*_{CCR}(A) = \frac{LM}{LA} < 1,$$

which denotes that A is operating locally efficient (PTE=1) and its overall inefficiency (TE) is caused by its failure to achieve scale inefficiency (SE) as represented by LM/LA. For DMUs B and C, their scale efficiency is one, i.e.,

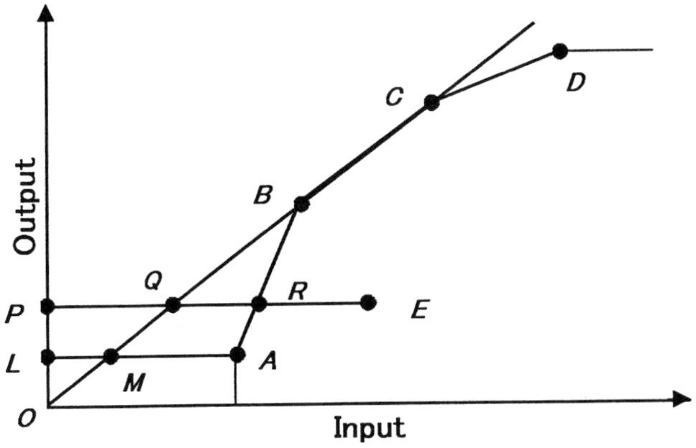

Figure 5.7. Scale Efficiency

they are operating at the most productive scale size. Their technical efficiency is also one so they are both scale and technically efficient for both the CCR and BCC models. For the BCC-inefficient DMU E, we have

$$\mathrm{SE}(E) = \frac{PQ}{PE}\frac{PE}{PR} = \frac{PQ}{PR},$$

which is equal to the scale efficiency of the input-oriented BCC projection R. The decomposition of E is

$$\mathrm{TE}(E) = \mathrm{PTE}(E) \times \mathrm{SE}(E) \quad \text{or}$$
$$\frac{PQ}{PE} = \frac{PR}{PE}\frac{PQ}{PR}.$$

Thus, E's overall inefficiency is caused by the technically inefficient operation of E and at the same time by the disadvantageous scale condition of E measured by PQ/PR.

Although the above scale efficiency is input-oriented, we can define the *output-oriented scale efficiency* using the output-oriented scores, as well.

5.8.2 Mix Efficiency

There are two types of efficiency measures, i.e., radial and non-radial. In Section 4.4 of Chapter 4, we introduced a non-radial measure of efficiency called SBM (slacks-based measure of efficiency). In order to turn SBM to an input-orientation, we modify the objective function (4.51) to deal only with the input slacks $\boldsymbol{s}^-$. Thus we have the input-oriented SBM model formulated as

$$(SBM_{in}) \quad \min \quad \rho_{in} = 1 - \frac{1}{m}\sum_{i=1}^{m} s_i^-/x_{io} \tag{5.42}$$
$$\text{subject to} \quad \boldsymbol{x}_o = X\boldsymbol{\lambda} + \boldsymbol{s}^-$$
$$\boldsymbol{y}_o = Y\boldsymbol{\lambda} - \boldsymbol{s}^+$$
$$\boldsymbol{\lambda} \geq \boldsymbol{0}, \; \boldsymbol{s}^- \geq \boldsymbol{0}, \; \boldsymbol{s}^+ \geq \boldsymbol{0}.$$

Let an optimal solution of (SBM_{in}) be $(\rho_{in}^*, \boldsymbol{\lambda}^*, \boldsymbol{s}^{-*}, \boldsymbol{s}^{+*})$. Then we have the relationship $\rho_{in}^* \leq \theta_{CCR}^*$ (see also Problem 4.5). Furthermore, we have the following theorem.

Theorem 5.9 *The equality $\rho_{in}^* = \theta_{CCR}^*$ holds if and only if the input-oriented CCR model has zero input-slacks for every optimal solution.*

Thus, the strict inequality $\rho_{in} < \theta_{CCR}^*$ holds if and only if the CCR solution reveals an input mix inefficiency. These observations lead us to the following definition of "mix efficiency."

Definition 5.3 (Mix Efficiency) *Let the input-oriented CCR and SBM scores of DMU_o be θ_{CCR}^* and ρ_{in}^*, respectively. The mix efficiency is defined by*

$$\mathrm{MIX} = \frac{\rho_{in}^*}{\theta_{CCR}^*}. \tag{5.43}$$

MIX is not greater than one and we have a decomposition of the non-radial efficiency into radial and mix efficiencies as

$$[\text{Input-oriented SBM}] = [\text{Radial Eff. (TE)}] \times [\text{Mix Eff.(MIX)}]. \tag{5.44}$$

Using (5.41) (the decomposition of TE), we have the decomposition of the non-radial technical efficiency ρ_{in}^* into MIX, pure technical efficiency (PTE) and scale efficiency (SE) as

$$\rho_{in}^* = [\text{MIX}] \times [\text{PTE}] \times [\text{SE}]. \tag{5.45}$$

The above decomposition is unique and will help to interpret sources of inefficiencies for each non-radial inefficient DMU.

5.8.3 An Example of Decomposition of Technical Efficiency

We applied the above decomposition to the hospital example in Table 1.5 in Chapter 1. The results are exhibited in Table 5.1. It is observed that F's

Table 5.1. Decomposition of Technical Efficiency

DMU	SBM ρ_{in}^*	CCR TE	BCC PTE	Mix Eff MIX	Scale Eff SE
A	1	1	1	1	1
B	1	1	1	1	1
C	0.852	0.883	0.896	0.965	0.985
D	1	1	1	1	1
E	0.756	0.763	0.882	0.99	0.866
F	0.704	0.835	0.939	0.843	0.889
G	0.895	0.902	1	0.992	0.902
H	0.774	0.796	0.799	0.972	0.997
I	0.905	0.96	0.989	0.942	0.971
J	0.781	0.871	1	0.896	0.871
K	0.866	0.955	1	0.907	0.955
L	0.936	0.958	1	0.977	0.958
Average	0.872	0.910	0.959	0.957	0.949

low SBM(0.704) is caused by MIX(0.843) and SE(0.889), and E's SBM(0.756) is mainly attributed to PTE(0.882) and MIX(0.866). Although H is roughly efficient with respect to MIX(0.972) and SE(0.997), its low PTE(0.799) forces it to be inefficient overall with SBM=0.774. These are main causes of their inefficiency as measured by the "non-radial" slacks-based model.

5.9 AN EXAMPLE OF RETURNS TO SCALE USING A BANK MERGER SIMULATION

Since the collapse of the bubble economy in 1990, Japan has been suffering from worsening recessions. Financial institutions may be one of the most affected of all industries. Although once regarded as the most stable and profitable sector, financial institutions in Japan have been greatly weakened. During the 1980s, banks and insurance companies loaned considerable amounts of money to builders and construction companies. However, the prices of land and buildings dropped by 70-90% from their highest levels in the 1980s. In order to protect against the ruin of the financial system, the Japanese Government began to assist these institutions by using public money to prevent them from becoming insolvent. At the same time, it was recommended that these institutions should restructure their organizations, with mergers being one method of restructuring. This section is concerned with this subject, i.e., how to check whether the merger policy is effective in the case of local banks.

5.9.1 Background

Ordinary Japanese banks can be classified roughly into three categories: "city," "regional" and "trust." City banks, of which there were nine in 1997, serve almost all the 47 prefectures in Japan as well as overseas. Regional banks, numbering more than one hundred, offer limited services to customers, mainly in one prefecture, with their branches concentrated more densely within the prefecture than the city banks.

Although each regional bank is small in scale compared with an average city bank, their total shares in deposits and loans are almost at the same levels as those of city banks. Hence, their robustness is one of the main concerns of bureaucrats, politicians and financial authorities in the Government. In this section we compare a set of regional banks with nine city banks with respect to efficiencies and returns to scale. Then we simulate the effects of merging in the case of two regional banks that are hypothetically merged into one under restructuring by reducing the number of branches and employees.

5.9.2 Efficiencies and Returns to Scale

Table 5.2 exhibits data for twenty banks in terms of the number of branches and employees, assets and operating net profits computed as an average over three years; 1995, 1996 and 1997. In this table the first eleven refer to regional banks and the last nine to city banks as noted by the symbols R and C, respectively. From the 104 regional banks, we chose the eleven in the northern part of Japan as typical rural area banks.

We employed 3 inputs (the number of branches and employees and assets) and one output (operating net profits) for evaluating the relative efficiency of the banks listed in the table. The CCR, BCC and Scale efficiencies and

Table 5.2. Data of 11 Regional and 9 City Banks*

No.	Bank	Branches	Employees	Assets**	Profits**
R1	Aomori	112	1,894	2,024,205	13,553
R2	Michinoku	111	1,727	1,702,017	16,540
R3	Iwate	123	1,843	2,048,014	14,760
R4	Tohoku	56	792	600,796	3,782
R5	Kita-Nippon	86	1,422	1,173,778	8,700
R6	Akita	111	1,841	2,290,274	18,590
R7	Hokuto	105	1,633	1,361,498	9,699
R8	Shonai	64	889	668,236	5,605
R9	Yamagata	90	1,611	1,563,930	9,993
R10	Yamagatashiawase	69	1,036	654,338	3,525
R11	Shokusan	73	1,011	656,794	3,177
C1	Dai-ichi Kangyo	407	17,837	53,438,938	411,149
C2	Sakura	529	18,805	52,535,262	302,901
C3	Fuji	363	15,188	51,368,976	371,364
C4	Tokyo Mitsubishi	419	20,235	78,170,694	504,422
C5	Asahi	436	13,149	29,531,193	170,835
C6	Sanwa	395	13,998	52,999,340	399,398
C7	Sumitomo	397	15,710	56,468,571	353,531
C8	Daiwa	218	8,671	16,665,573	112,121
C9	Tokai	305	11,546	31,376,180	186,640

* Source:*Analysis of Financial Statements of All Banks,*
Federation of Bankers Associations of Japan. 1995, 1996, 1997.
** Unit = One million yen ≈ U.S.$ 8300.

returns-to-scale characteristics of each bank are listed in Table 5.3. We used the input-oriented models in measuring efficiency.

The CCR results, listed in Column 2 of Table 5.3, show that the regional banks performed worse than the city banks when evaluated on the constant returns-to-scale assumption associated with this model, as evidenced by the fact that 8 out of 11 regional banks were below the average compared with 3 out of 9 city banks. However, regional bank R2 is one of 4 best performers, and furthermore it is the bank most frequently referenced for evaluating inefficient banks. It is used as a reference for all inefficient regional banks and serves as the most influential referent, i.e, with the largest λ value. It performs similarly for inefficient city banks. R2 is also the most efficient one in CCR measure. For confirmation, we might note that this bank is famous for its unique managerial strategies under the strong leadership of its owner.

The BCC scores provide efficiency evaluations using a local measure of scale, i.e. under variable returns-to-scale. In this model, two regional banks, R4 and R8, are accorded efficient status in addition to the four CCR efficient banks —

Table 5.3. Efficiencies and Returns to Scale

Bank	CCR		BCC		Scale
No.	Score	Reference	Score	RTS	Score
R1	0.715	R2 C1	0.765	Incr	0.934
R2	1		1	Const	1
R3	0.768	R2 C6	0.799	Incr	0.961
R4	0.648	R2	1	Incr	0.648
R5	0.763	R2	0.853	Incr	0.894
R6	0.892	R2 C1 C6	0.913	Incr	0.977
R7	0.733	R2	0.776	Incr	0.945
R8	0.863	R2	1	Incr	0.863
R9	0.676	R2 C1	0.777	Incr	0.870
R10	0.554	R2	0.918	Incr	0.604
R11	0.498	R2	0.915	Incr	0.544
C1	1		1	Const	1
C2	0.742	R2 C1 C6	0.747	Const	0.993
C3	0.981	C1 C4 C6	0.991	Incr	0.990
C4	1		1	Const	1
C5	0.728	R2 C1	0.743	Const	0.980
C6	1		1	Const	1
C7	0.854	C1 C4 C6	0.865	Incr	0.987
C8	0.853	R2 C1	0.854	Const	0.999
C9	0.766	R2 C1	0.767	Incr	0.999
Average	0.802		0.884		0.909

which retain their previous efficient status. R4's full efficiency with the BCC model is caused by its use of the smallest amount of inputs (refer to Theorem 4.3) even though it is the lowest in the CCR score. The BCC scores do not discriminate well between regional and city banks. The regional banks exhibit scores in which 5 out of 11 are below average, while the city banks have 5 out of 9. This means that both categories are positioned, on average, at about the same relative distance from the efficient BCC frontiers, although there are differences in efficiency within each category. Both groups are performing in a similar way with regard to efficiency under the managerial situations to which they belong. The scale efficiency as defined by the ratio, CCR/BCC (see Definition 5.2 and discussions following it), exhibits large differences between the two groups. All city banks are above average whereas 6 regional banks are below it. This may mean that banks in the city group are in an advantageous condition compared with those in the regional group, and their global inefficiencies (CCR) are mainly attributed to their inefficient operations or management. The cases of the regional banks are mixed. For example, R1 has a low BCC score and a relatively high scale efficiency among the group, meaning that the overall inefficiency (0.715) in the CCR column of R1 is caused by inefficient operations (0.765) rather than scale inefficiency (0.934). R4 has a fully efficient BCC score and a low scale efficiency (0.648). This can be interpreted to mean that the global inefficiency of this bank under CCR score is mainly attributed to disadvantageous regional conditions.

Now we turn to the returns to scale, as identified by the input-oriented BCC model and displayed in Table 5.3 under the heading "RTS." As was claimed in Theorem 5.5, banks with full efficiency in the CCR score are also efficient in the BCC model, and are in the MPSS, the region where constant returns-to-scale prevails. Banks R2, C1, C2, C4, C5, C6 and C8 have this status, while all other banks display increasing returns-to-scale. Interestingly, 10 of 11 regional banks belong to this class, which shows that they have a possibility to improve their efficiency by scaling up their activities. This observation leads us to hypothetically study the desirability of merging low ranked regional banks.

5.9.3 The Effects of a Merger

Confronted with the worst banking crisis in recent years, many banks have been forced to reconstruct their organizations and some have been nationalized under the control of the Japanese Financial Supervisory Agency. Mergers have always been thought of as a powerful tool for restructuring in competitive survival situations. In view of the above returns-to-scale characteristics of the inefficient regional banks, we will hypothetically merge two inefficient banks into one and evaluate changes in efficiency. The target banks are R9 and R10, which both belong to the same prefecture and have low CCR ranks (17th and 19th among 20). Their optimal weights and slacks for inputs, along with projections onto the CCR and the BCC efficient frontiers are listed in Table 5.4. Both banks have no slacks in output (operating net profits) and their projections remain at the status quo, so these values are not listed in the table. It is observed that both banks have positive weights in "Assets" (v_3^*), so a reduction in assets will contribute directly to improve their CCR - global - efficiencies. R10 has zero weights on "Branch" and "Employee," meaning that it needs to reduce these slacks in branch (15) and employee (206) before these inputs can have positive weights and hence contribute to improving efficiencies. (Refer to Problem 3.5 for discussions on these subjects.) The CCR projection requires drastic reductions in inputs, especially for R10, while the BCC projection is moderate compared with the CCR case, especially for R10.

First, we simulate the case in which the BCC projected version of R9 and R10 merge to form a new bank — called R12. Hence R12 has 125 (=69+56) branches, 1,887 (=1,095+792) employees and assets of 1,419,675 (=1,215,757+600,796) yen to produce operating net profits 13,518 (= 9,993 + 3,525) yen. See Table 5.2. Although reduction of branches and employees is a traditional way of restructuring, there may be problems in reducing assets of banks. However, we think of this merger as providing a means of strengthening both banks by eliminating bad loans. In any case, we added this new bank to

Table 5.4. Weights, Slacks and Projections

	CCR					
		Weights			Slacks	
Bank	Branch v_1^*	Employee v_2^*	Assets v_3^*	Branch s_1^{-*}	Employee s_2^{-*}	Assets s_3^{-*}
R9	2.38E-3	0	5.02E-7	0	110	0
R10	0	0	1.53E-7	15	206	0

	CCR Projection			BCC Projection		
	Branch	Employee	Assets	Branch	Employee	Assets
R9	60	978	1,056,915	69	1095	1,215,757
	(-29%)	(-39%)	(-32%)	(-22%)	(-32%)	(-22%)
R10	23	368	362,760	56	792	600,796
	(-66%)	(-64%)	(-45%)	(-19%)	(-24%)	(-8%)

the data set and evaluated its efficiencies. The results are displayed in Table 5.5 along with those of the projected values of R9 and R10 denoted by R9′ and R10′, respectively.

Table 5.5. Efficiency of Projected and Merged Banks

Bank	Branch	Employee	Assets	CCR	BCC	RTS	Scale
R9′	69	1,095	1,215,757	0.877	1	Incr	0.877
R10′	56	792	600,796	0.604	1	Incr	0.604
R12	125	1,887	1,419,675	0.766	0.780	Incr	0.982

As expected, R9′ and R10′ come to full BCC efficiency and also improve their global CCR efficiencies. However, the merged bank R12 has a BCC efficiency, 0.780, which is less than 1, and is identified as still having increasing returns-to-scale status. This result for R12 leads us to conclude that merging two locally (BCC) efficient banks which have increasing returns-to-scale status results in a locally inefficient one, if the merged bank exhibits increasing returns-to-scale characteristics. Thus, we have,

Proposition 5.1 (Tone (1999)) *When two locally (BCC) efficient DMUs merge to form a new DMU having the sum of inputs and outputs of the two DMUs as its inputs and outputs, the new DMU is neither locally (BCC) nor globally (CCR) efficient, if increasing returns-to-scale prevails at all three DMUs.*

We do not prove this proposition here. Instead, we illustrate it in the case of a single input and single output using Figure 5.8, where R9′, R10′, R12 and R2 are symbolically plotted in order to help understand the situations. In the

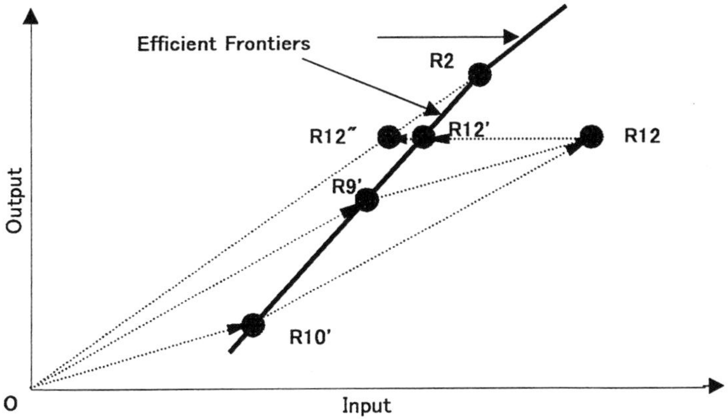

Figure 5.8. Merger is Not Necessarily Efficient.

figure, we merged R9′ and R10′ which when summed as a vector resulted in R12. This is a point inside the production possibility set of the BCC model and hence is locally inefficient. A further reduction in input is required to attain the BCC frontier point R12′.

This suggests that merging *and* restructuring may not be enough. Reduction of input resources is also necessary in order to arrive at a more efficient status. Otherwise, the merger will result in a worsened efficiency status.

To realize a globally efficient status is more difficult, as indicated by R12″ or R2 in the figure, the latter being attained by using the MPSS formula, as in (5.25).

As a second test, we used the data in Table 5.4 and merged the two banks R9 and R10 by projecting them onto the CCR frontiers. Comparing this result with R9 and R10 in Table 5.3 we find that these efficiency estimates result from a really drastic input reduction. The new bank has 83 (=60+23) branches, 1,346 (=978+368) employees, 1,419,675 (=1,056,915+362,760) assets to produce operating net profits of 13,518 (=9,993+3,525). After adding this new bank to the data set, we evaluated its efficiency and found that the new bank attains global full efficiency.

In conclusion to this very limited study, we might say that inefficient regional banks are in urgent need of reconstruction not only by merging but also by reducing input resources, as well as by directing their marketing activities to new areas which city banks may not be able to approach. R2 is an instructive example. Otherwise, regional banks will not survive in competition against other financial institutions under present competitive conditions, which include the challenge of "global" worldwide efficiency evaluations.

5.10 SUMMARY

We have now covered the topics of returns to scale in its technical and mix efficiency aspects by reference to the MPSS concept. Scale inefficiencies were identified with failure to achieve MPSS. In the process we have given implementable form to the basic concepts, as drawn from the literature of economics, and extended them for simultaneous treatment of multiple outputs and inputs. This is needed for many (if not most) managerial uses. We have also eliminated the need for assumptions associated with achievement of technical efficiency, as used in economics, which are not likely to hold in practice. Finally, we provided interpretations and examples of their use which are intended to facilitate managerial, engineering, and social policy uses as in our example of bank merger problems and possibilities. We next extend this to other types of models and their uses in the next few sections. Later in this text we will further extend our treatments of returns to scale for use when unit prices and costs are available.

5.11 ADDITIVE MODELS

The model (5.22) which we used for MPSS, avoids the problem of choosing between input and output orientations, but this is not the only type of model for which this is true. The additive models to be examined in this section also have this property. That is, these models simultaneously maximize outputs and minimize inputs, in the sense of vector optimizations. The additive model we select is

$$\begin{aligned}
&\max_{\boldsymbol{\lambda}, \boldsymbol{s}^-, \boldsymbol{s}^+} \sum_{i=1}^{m} g_i^- s_i^- + \sum_{r=1}^{s} g_r^+ s_r^+ \\
&\text{subject to} \\
&\sum_{j=1}^{n} x_{ij}\lambda_j + s_i^- = x_{io}, \ i = 1, 2, \ldots, m \qquad (5.46)\\
&\sum_{j=1}^{n} y_{rj}\lambda_j + s_r^+ = y_{ro}, \ r = 1, 2, \ldots, s \\
&\sum_{j=1}^{n} \lambda_j = 1 \\
&\lambda_j, s_i^-, s_r^+ \geq 0.
\end{aligned}$$

This model utilizes the "goal vector" approach of Thrall (1996a)[10] in which the slacks in the objective are accorded "goal weights" which may be subjective or objective in character. Here we want to use these "goal weights" to ensure that the units of measure associated with the slack variables do not affect the optimal solution choices.

Employing the language of "dimensional analysis," as in Thrall (1996a), we want these weights to be "contragradient" in order to insure that the resulting objective will be "dimensionless." That is, we want the solutions to be free of the units in which the inputs and outputs are stated. An example is the use of the input and output ranges in Cooper, Park and Pastor (1999)[11] to obtain $g_i = 1/R_i^-$, $g_r = 1/R_r^+$, where R_i^- is the range for the ith input and R_r^+ is the range for the rth output. This gives each term in the objective of (5.46) a contragradient weight. The resulting value of the objective is dimensionless, as follows from the fact that the s_i^- and s_r^+ in the numerators are measured in the same units as the R_i^- and R_r^+ in the denominators. Hence the units of measure cancel.

The condition for efficiency given in Chapter 1 for the CCR model is now replaced by the following simpler condition,

Definition 5.4 *A DMU$_o$ evaluated by (5.46) is efficient if and only if all slacks are zero.*

To start our returns-to-scale analyses for these additive models we first replace the projections in (5.15) with

$$\begin{aligned} \widehat{x}_{io} &= x_{io} - s_i^{-*} \; i = 1, \ldots, m \\ \widehat{y}_{ro} &= y_{ro} - s_r^{+*}, \; r = 1, \ldots, s \end{aligned} \tag{5.47}$$

where s_i^{-*} and s_r^{+*} are optimal slacks obtained from (5.46). Then we turn to the dual (multiplier) model associated with (5.46) which we write as follows,

$$\begin{aligned} &\min_{\boldsymbol{v},\boldsymbol{u},u_0} \sum_{i=1}^{m} v_i x_{io} - \sum_{r=1}^{s} u_r y_{ro} + u_0 \\ &\text{subject to} \\ &\sum_{i=1}^{m} v_i x_{ij} - \sum_{r=1}^{s} u_r y_{rj} + u_0 \geq 0, \; j = 1, \ldots, n \\ &v_i \geq g_i^-, \; u_r \geq g_r^+; \; u_0 \; free, \end{aligned} \tag{5.48}$$

where, as was true for the BCC model, we will use the variable u_0 to evaluate returns to scale.

We are thus in position to use Theorem 5.2 for "additive" as well "radial measures" as reflected in the BCC models discussed in earlier parts of this chapter. Hence we again have recourse to this theorem where, however, we note the difference in objectives between (5.11) and (5.48), including the change from $-u_0$ to $+u_0$. As a consequence of these differences we also modify (5.16) to the following,

$$
\begin{aligned}
&\max_{\boldsymbol{v},\boldsymbol{u},\widehat{u}_0} \ \widehat{u}_0 \\
&\text{subject to} \\
&\sum_{r=1}^{s} u_r y_{rj} - \sum_{i=1}^{m} v_i x_{ij} - \widehat{u}_0 \leq 0, \ j = 1,\ldots,n; \ j \neq o \qquad (5.49)\\
&\sum_{r=1}^{s} u_r \widehat{y}_{ro} - \sum_{i=1}^{m} v_i \widehat{x}_{ij} - \widehat{u}_0 = 0 \\
&u_r \geq g_r^+, \ v_i \geq g_i^-, \ \widehat{u}_0 \leq 0.
\end{aligned}
$$

Here we have assumed that $u_0^* < 0$ was achieved in a first-stage use of (5.48). Hence, if $\widehat{u}_0^* < 0$ is maximal in (5.49) then returns to scale are increasing at $(\widehat{\boldsymbol{x}}_o, \widehat{\boldsymbol{y}}_o)$ in accordance with (i) in Theorem 5.2 whereas if $\widehat{u}_0^* = 0$ then (iii) applies and returns to scale are constant at this point $(\widehat{\boldsymbol{x}}_o, \widehat{\boldsymbol{y}}_o)$ on the efficient frontier.

For $\widehat{u}_0^* > 0$ in stage one, the objective and the constraint on $\widehat{u}_0$ are simply reoriented in the manner we now illustrate by using (5.48) to evaluate E in Figure 5.3 via

$$
\begin{aligned}
&\max_{\boldsymbol{\lambda}, s^-, s^+} \ s^- + s^+ \\
&\text{subject to} \\
&\lambda_A + \frac{3}{2}\lambda_B + 3\lambda_C + 4\lambda_D + 4\lambda_E + s^- = 4 \\
&\lambda_A + 2\lambda_B + 4\lambda_C + 5\lambda_D + \frac{9}{2}\lambda_E - s^+ = \frac{9}{2} \\
&\lambda_A + \lambda_B + \lambda_C + \lambda_D + \lambda_E = 1 \\
&s^-, s^+, \lambda_A, \lambda_B, \lambda_C, \lambda_D, \lambda_E \geq 0,
\end{aligned}
$$

where we have used unit weights for the g_i^-, g_r^+, in (5.46), to obtain the usual additive model formulation. (See Thrall (1996b)[12] for a discussion of the applicable condition for a choice of such "unity" weights.) This has an optimal solution with $\lambda_C = \lambda_D = s^{-*} = 1/2$ and all other variables zero. To check that this is optimal we turn to the corresponding dual (multiplier) form for the above envelopment model which is

$$
\begin{aligned}
&\min \ 4v - \frac{9}{2}u + u_0 \\
&\text{subject to} \\
&v - u + u_0 \geq 0 \\
&\frac{3}{2}v - 2u + u_0 \geq 0 \\
&3v - 4u + u_0 \geq 0 \\
&4v - 5u + u_0 \geq 0
\end{aligned}
$$

$$4v - \frac{9}{2}u + u_0 \geq 0$$
$$v, u \geq 1, \; u_0 : free.$$

The solution $v^* = u^* = u_0^* = 1$ satisfies all constraints and gives $4v^* - \frac{9}{2}u^* + u_0^* = \frac{1}{2}$. This is the same value as in the preceding envelopment model so that, by the duality theorem of linear programming, both solutions are optimal.

To determine the conditions for returns to scale we use (5.47) to project E into E′ with coordinates $(\widehat{x}, \widehat{y}) = (\frac{7}{2}, \frac{9}{2})$ in Figure 5.3. Then we utilize the following reorientation of (5.49),

$$\begin{aligned}
&\min \; \widehat{u}_0 \\
&\text{subject to} \\
&v - u + \widehat{u}_0 \geq 0 \\
&\frac{3}{2}v - 2u + \widehat{u}_0 \geq 0 \\
&3v - 4u + \widehat{u}_0 \geq 0 \\
&4v - 5u + \widehat{u}_0 \geq 0 \\
&\frac{7}{2}v - \frac{9}{2}u + u_0 \geq 0 \\
&v, u \geq 1, \; \widehat{u}_0 \geq 0.
\end{aligned}$$

This also gives $v^* = u^* = \widehat{u}_0^* = 1$ so the applicable condition is (ii) in Theorem 5.2. Thus returns to scale are decreasing at E′, the point on the BCC efficient frontier which is shown in Figure 5.3.

5.12 MULTIPLICATIVE MODELS AND "EXACT" ELASTICITY

The treatments to this point have been confined to "qualitative" characterizations in the form of identifying whether RTS are "increasing," "decreasing," or "constant." There is a literature — albeit a relatively small one — which is directed to "quantitative" estimates of RTS in DEA. Examples are Banker, Charnes and Cooper (1984),[13] Førsund (1996)[14] and Banker and Thrall (1992). [15] However, there are problems in using the standard DEA models, as is done in these studies, to obtain scale elasticity estimates. Førsund (1996), for instance, lists a number of such problems. Also, the elasticity values in Banker and Thrall (1992) are determined only within upper and lower bounds. This is an inherent limitation that arises from the piecewise linear character of the frontiers for these models. Finally, attempts to extend the Färe, Grosskopf and Lovell (1985, 1994)[16] approaches to the determination of "exact" scale elasticities have not been successful. See the criticisms in Førsund (1996, p.296) and Fukuyama (2000, p.105). [17] (Multiple output-multiple input production and cost functions which meet the sub- and super-additivity requirements in economics are dealt with in Panzar and Willig (1977). [18] See also Baumol, Panzar and Willig (1982). [19]

This does not, however, exhaust the possibilities. The class of "multiplicative models" opens additional opportunities. Introduced by this name in Charnes *et al.* (1982)[20] — see also Banker *et al.* (1981)[21] — and extended in Charnes *et al.* (1983)[22] to accord these models non-dimensional (=units invariance) properties. Although not used very much in applications these multiplicative models can provide advantages for extending the range of potential uses for DEA. For instance, they are not confined to efficiency frontiers which are concave. They can be formulated to allow the efficiency frontiers to be concave in some regions and non-concave elsewhere. See Banker and Maindiratta (1986). [23] They can also be used to obtain "exact" estimates of elasticities in manners that we now describe.

The models we use for this discussion are due to Banker and Maindiratta (1986) — where analytical characterizations are supplied along with confirmation in controlled-experimentally designed simulation studies.

For this development we use an output oriented version of the multiplicative model which places us in consonance with the one in Banker and Maindiratta (1986) — viz.,

$$
\begin{aligned}
&\max_{\gamma_0, \boldsymbol{\lambda}} \ \gamma_0 \\
&\text{subject to} \\
&\prod_{j=1}^{n} x_{ij}^{\lambda_j} \le x_{io} \quad i = 1, \ldots, m \\
&\prod_{j=1}^{n} y_{rj}^{\lambda_j} \ge \gamma_0 y_{ro} \quad r = 1, \ldots, s \\
&\sum_{j=1}^{n} \lambda_j = 1 \\
&\gamma_0, \lambda_j \ge 0,
\end{aligned}
\tag{5.50}
$$

where "$\prod$" means "product of" and the variables with values to be determined are λ_j and γ_0.

We convert these inequalities to equalities as follows.

$$
\begin{aligned}
&e^{s_i^-} \prod_{j=1}^{n} x_{ij}^{\lambda_j} = x_{io} \quad i = 1, \ldots, m \\
&\text{and} \\
&e^{s_r^+} \prod_{j=1}^{n} y_{rj}^{\lambda_j} = \gamma_0 y_{ro} \quad r = 1, \ldots, s.
\end{aligned}
\tag{5.51}
$$

We also replace the objective in (5.50) with $\gamma_0 e^{\varepsilon \left(\sum_{i=1}^{m} s_i^- + \sum_{r=1}^{s} s_r^+\right)}$, where s_i^-, $s_r^+ \ge 0$ represent slacks. Employing (5.51) and taking logarithms convert this model to

$$
\min_{\boldsymbol{\lambda}, \widetilde{\gamma}_0, \boldsymbol{s}^-, \boldsymbol{s}^+} \quad -\widetilde{\gamma}_0 - \varepsilon \Big(\sum_{r=1}^{s} s_r^+ + \sum_{i=1}^{m} s_i^-\Big)
$$

$$
\begin{aligned}
&\text{subject to} \\
&\widetilde{x}_{io} = \sum_{j=1}^{n} \widetilde{x}_{ij}\lambda_j + s_i^- \;\; i = 1, \ldots, m \qquad (5.52)\\
&\widetilde{\gamma}_0 + \widetilde{y}_{ro} = \sum_{j=1}^{n} \widetilde{y}_{rj}\lambda_j - s_r^+ \;\; r = 1, \ldots, s \\
&1 = \sum_{j=1}^{n} \lambda_j \\
&\lambda_j,\; s_r^+,\; s_i^- \geq 0, \forall j, r, i,
\end{aligned}
$$

where "$\sim$" denotes "logarithm" so the $\widetilde{x}_{ij}$, $\widetilde{y}_{rj}$, $\widetilde{\gamma}_0$, $\widetilde{x}_{io}$ and $\widetilde{y}_{ro}$ are expressed in logarithmic units.

The dual to (5.52) is

$$
\begin{aligned}
&\max_{\alpha,\beta,\alpha_0} \sum_{r=1}^{s} \beta_r\widetilde{y}_{ro} - \sum_{i=1}^{m} \alpha_i\widetilde{x}_{io} - \alpha_0 \\
&\text{subject to} \\
&\sum_{r=1}^{s} \beta_r\widetilde{y}_{rj} - \sum_{i=1}^{m} \alpha_i\widetilde{x}_{ij} - \alpha_0 \leq 0, \; j = 1, \ldots, n \qquad (5.53)\\
&\sum_{r=1}^{s} \beta_r = 1 \\
&\alpha_i \geq \varepsilon, \beta_r \geq \varepsilon; \quad \alpha_0 : free.
\end{aligned}
$$

Using α_i^*, β_r^* and α_0^* for optimal values, $\sum_{r=1}^{s} \beta_r^*\widetilde{y}_{ro} - \sum_{i=1}^{m} \alpha_i^*\widetilde{x}_{io} - \alpha_0^* = 0$ represents a supporting hyperplane (in logarithmic coordinates) for DMU$_o$, where efficiency is achieved. We may rewrite this log-linear supporting hyperplane in terms of the original input/output values:

$$
\prod_{r=1}^{s} y_{ro}^{\beta_r^*} = e^{\alpha_0^*} \prod_{i=1}^{m} x_{io}^{\alpha_i^*}. \qquad (5.54)
$$

Then referring to (5.54) we introduce

Theorem 5.10 *Multiplicative Model RTS,*
(i) *RTS are increasing if and only if* $\sum_{i=1}^{m} \alpha_i^* > 1$ *for all optimal solutions,*
(ii) *RTS are decreasing if and only if* $\sum_{i=1}^{m} \alpha_i^* < 1$ *for all optimal solutions,*
(iii) *RTS are constant if and only if* $\sum_{i=1}^{m} \alpha_i^* = 1$ *for some optimal solutions.*

To see what this means we revert to the discussion of (5.21) and introduce scalars a, b in $(a\boldsymbol{x}_o, b\boldsymbol{y}_o)$. In conformation with (5.54) this means

$$
e^{\alpha_0^*} \prod_{i=1}^{m} (ax_{io})^{\alpha_i^*} = \prod_{r=1}^{s} (by_{ro})^{\beta_r^*} \qquad (5.55)
$$

so that the thus altered inputs and outputs satisfy this extension of the usual Cobb-Douglas single output formulations.

The problem now becomes: given a specific expansion $a > 1$, contraction $a < 1$, or neither, i.e., $a = 1$, for application to all inputs, what is the value of b that positions the solution in the supporting hyperplane at this point? The answer is given by the following

Theorem 5.11 *If* $(a\boldsymbol{x}_o, b\boldsymbol{y}_o)$ *lies in the supporting hyperplane then* $b = a^{\sum_{i=1}^{m} \alpha_i^*}$

This theorem, as proved in Banker *et al.*(2003), [24] gives the precise result for the exact elasticity. Thus if $\sum_{i=1}^{m} \alpha_i^* = 1$ then $b = a$ and the outputs are expanded in the same proportion as the value of a which has been selected for the inputs. For $\sum_{i=1}^{m} \alpha_i^* > 1$ the proportional expansion in outputs exceed the proportion in which the inputs were expanded. Finally, for $\sum_{i=1}^{m} \alpha_i^* < 1$, the case of decreasing returns to scale applies since the outputs expansion is less than proportional to the proportion in which the inputs were expanded. Note that with any a specified for the inputs we then have a numerical value for the elasticity given by $\sum_{i=1}^{m} \alpha_i^*$.

There may be alternative optimal solutions for (5.53) so the values for the α_i^* components need not be unique. For dealing with alternate optima, we return to (5.50) and note that a necessary condition for efficiency is $\gamma_0^* = 1$. For full efficiency we must also have all slacks at zero in (5.51). Adaptation of (2.3), our previous approaches, projects to point on the efficient frontier via

$$\prod_{j=1}^{n} x_{ij}^{\lambda_j^*} = e^{-s_i^{-*}} x_{io} = x'_{io}, \ i = 1, \ldots, m \tag{5.56}$$

$$\prod_{j=1}^{n} y_{rj}^{\lambda_j^*} = e^{s_r^{+*}} y_{ro} = y'_{ro}, \ r = 1, \ldots, s$$

and x'_{io}, y'_{ro} are the coordinates of the point on the efficiency frontier used to evaluate DMU_o.

Thus, we can extend the preceding models in a manner that is now familiar. Suppose we have obtained an optimal solution for (5.53) with $\sum_{i=1}^{m} \alpha_i^* < 1$. We then utilize (5.56) to form the following problem

$$\begin{aligned}
&\max_{\alpha,\beta,\alpha_0} \sum_{i=1}^{m} \alpha_i \\
&\text{subject to} \\
&\sum_{r=1}^{s} \beta_r \tilde{y}_{rj} - \sum_{i=1}^{m} \alpha_i \tilde{x}_{ij} - \alpha_0 \leq 0, \ j = 1, \ldots, n; \ j \neq o \\
&\sum_{r=1}^{s} \beta_r \tilde{y}_{ro} - \sum_{i=1}^{m} \alpha_i \tilde{x}_{io} - \alpha_0 = 0 \qquad (5.57) \\
&\sum_{r=1}^{s} \beta_r = 1
\end{aligned}$$

$$\sum_{i=1}^{m} \alpha_i \leq 1$$

$$\alpha_i \geq \varepsilon,\ \beta_r \geq \varepsilon;\ \alpha_0 \text{ free in sign.}$$

If $\sum_{i=1}^{m} \alpha_i^* = 1$ in (5.57), then returns to scale are constant by (iii) of Theorem 5.5. If the maximum is achieved with $\sum_{i=1}^{m} \alpha_i^* < 1$, however, condition (ii) of Theorem 5.4 is applicable and returns to scale are decreasing at the point $x'_{io},\ y'_{ro}; i = 1, \ldots, m; r = 1, \ldots, s$.

If we initially have $\sum_{i=1}^{m} \alpha_i^* > 1$ in (5.53), we replace $\sum_{i=1}^{m} \alpha_i \leq 1$ with $\sum_{i=1}^{m} \alpha_i \geq 1$ in (5.57) and also change the objective to minimize $\sum_{i=1}^{m} \alpha_i$. If the optimal value is greater than one, then (i) of Theorem 5.4 is applicable and RTS are increasing. On the other hand, if we attain $\sum_{i=1}^{m} \alpha_i^* = 1$ then condition (iii) applies and returns to scale are constant.

We can also derive pertinent scale elasticities in a straightforward manner. Thus, using the standard logarithmic derivative formulas for elasticities we obtain

$$\frac{d \ln b}{d \ln a} = \frac{a}{b}\frac{db}{da} = \sum_{i=1}^{m} \alpha_i^*. \tag{5.58}$$

Consisting of a sum of component elasticities, one for each input, this overall measure of elasticity is applicable to the value of the multiplicative expression with which DMU$_o$ is associated.

The derivation in (5.58) holds only for points where this derivative exists. However, we can bypass this possible source of difficulty by noting that Theorem 5.10 allows us to obtain this elasticity estimate via

$$\frac{\ln b}{\ln a} = \sum_{i=1}^{m} \alpha_i^*. \tag{5.59}$$

Further, as discussed in Cooper, Thompson and Thrall (1996), [25] it is possible to extend these concepts to the case in which all of the components of $\boldsymbol{y}_o$ are allowed to increase by at least the factor b. That is, we need not assume that all outputs are scaled by the same values of a. However, we cannot similarly treat the constant, a, as providing an upper bound for the inputs since mix alterations are not permitted in the treatment of returns to scale in economics. See Varian (1984, p.20)[26] for RTS characterizations in economics.

In conclusion we turn to properties of units invariance for these multiplicative models. Thus we note that $\sum_{i=1}^{m} \alpha_i^*$ is units invariant by virtue of the relation expressed in (5.58). The property of units invariance is also exhibited in (5.59) since a and b are both dimension free. Finally, we also have

Theorem 5.12 *The models given in (5.50) and (5.51) are dimension free. That is, changes in the units used to express the input quantities x_{ij} or the output quantities y_{rj} in (5.50) will not affect the solution set or alter the value of* $\max \gamma_0 = \gamma_0^*$.

This theorem, which is proved in Banker *et al.* (2003), has the following corollary.

Corollary 5.3 *The restatement of (5.51) in logarithmic form yields a model which is translation invariant.*

5.13 SUMMARY OF CHAPTER 5

Although we have now covered all of the presently available models, we have not covered all of the orientations in each case. Except for the multiplicative models we have not covered output-oriented objectives for a variety of reasons. There are no real problems with the mathematical development but further attention must be devoted to how changes in input scale and input mix should be treated when all outputs are to be scaled up in the same proportions. See the discussion in Cooper, Thompson and Thrall (1996).

As also noted in Cooper, Thompson and Thrall (1996), the case of increasing returns to scale can be clarified by using Banker's most productive scale size to write $(\alpha \boldsymbol{x}_o, \beta \boldsymbol{y}_o)$. The case $1 < \beta/\alpha$ means that all outputs are increased by at least the factor β and returns to scale are increasing as long as this condition holds. The case $1 > \beta/\alpha$ has the opposite meaning — *viz.*, no output is increasing at a rate that exceeds the rate at which all inputs are increased. Only for constant returns to scale do we have $1 = \beta/\alpha$, in which case all outputs and all inputs are required to be increasing (or decreasing) at the same rate so no mix change in involved for the inputs.

The results in this chapter (as in the literature to date) are restricted to this class of cases. This leaves unattended a wide class of cases. One example involves the case where management interest is centered on only subsets of the outputs and inputs. A direct way to deal with this situation is to partition the inputs and outputs of interest and designate the conditions to be considered by $(\boldsymbol{x}_o^I, \boldsymbol{x}_o^N, \boldsymbol{y}_o^I, \boldsymbol{y}_o^N)$ where I designates the inputs and outputs that are of interest to management and N designates those which are not of interest (for such scale returns studies). Proceeding as described in the present chapter and treating $\boldsymbol{x}_o^N$ and $\boldsymbol{y}_o^N$ as "exogenously fixed," in the spirit of Banker and Morey (1986), [27] would make it possible to determine the situation for returns to scale with respect to the thus designated subsets. Also, as noted in Tone (2001), [28] the status of returns to scale may vary by the addition of the assurance region which will be introduced in the next chapter. Other cases involve treatments with unit costs and prices as in FGL (1994), Sueyoshi (1999), [29] and Tone and Sahoo. [30]

The developments covered in this chapter have been confined to technical aspects of production. Our discussions follow a long-standing tradition in economics which distinguishes scale from input mix changes by not allowing the latter to vary when scale changes are being considered. This permits the latter (i.e., scale changes) to be represented by a single scalar — hence the name. However, this can be far from actual practice, where scale and mix are likely to be varied simultaneously when determining the size and scope of an operation.

See the discussion in Ray (2005)[31] who uses the concept of "size" as distinct from "scale" as introduced in Maindiratta (1990)[32] to study the efficiency of the banks in India.

There are, of course, many other aspects to be considered in treating returns to scale besides those attended to in the present chapter. Management efforts to maximize profits, even under conditions of certainty, require simultaneous determination of scale, scope and mix magnitudes with prices and costs known, as well as the achievement of the technical efficiency which is always to be achieved with *any* set of positive prices and costs. The topics treated in this chapter do not deal with such price-cost information. Moreover, the focus is on *ex-post facto* analysis of already effected decisions. This can have many uses, especially in the control aspects of management where evaluations of performance are required. Left unattended in this chapter, and in much of the DEA literature, is the *ex ante* (planning) problem of how to use this knowledge in order to determine how to blend scale and scope with mix and other efficiency considerations when efficiency future-oriented decisions. See Bogetoft and Wang (2005)[33] which extends DEA for use in evaluating proposed mergers in an *ex ante* fashion.

5.14 APPENDIX: FGL TREATMENT AND EXTENSIONS

In this Appendix, we first present the approach due to Färe, Grosskopf and Lovell (1985 and 1994). See also Färe and Grosskopf (1994). We then present a simple RTS approach without the need for checking the multiple optimal solutions as in Zhu and Shen (1995)[34] and Seiford and Zhu (1999)[35] where only the BCC and CCR models are involved. This approach will substantially reduce the computational burden, because it relies on the standard CCR and BCC computational codes (see Zhu (2002)[36] for a detailed discussion).

To start, we add to the BCC and CCR models by the following DEA model whose frontier exhibits non-increasing returns to scale (NIRS), as in Färe, Grosskopf and Lovell (FGL, 1985, 1994)

$$
\begin{aligned}
&\theta^*_{NIRS} = \min \theta_{NIRS} \\
&\text{subject to} \\
&\theta_{NIRS} x_{io} = \sum_{j=1}^{n} x_{ij}\lambda_j + s_i^- \quad i = 1, \ldots, m \\
&y_{ro} = \sum_{j=1}^{n} y_{rj}\lambda_j - s_r^+ \quad r = 1, \ldots, s \\
&1 \geq \sum_{j=1}^{n} \lambda_j \\
&0 \leq \lambda_j,\ s_i^-,\ s_r^+ \ \forall i, r, j
\end{aligned}
\tag{5.60}
$$

The FGL development rests on the following relation

$$\theta^*_{CCR} \le \theta^*_{NIRS} \le \theta^*_{BCC}$$

where "*" refers to an optimal value and θ^*_{NIRS} is defined in (5.60) while θ^*_{CCR} and θ^*_{BCC} refer to the BCC and CCR models.

FGL utilize this relation to form ratios that provide measures of RTS. However, we turn to the following tabulation which relates their RTS characterization to Theorems 2.3 and 2.4 (and accompanying discussion). See also Färe and Grosskopf (1994), Banker, Chang and Cooper (1996), and Seiford and Zhu (1999)

	FGL Model	RTS	CCR Model
Case 1	If $\theta^*_{CCR} = \theta^*_{BCC}$	Constant	$\sum \lambda^*_j = 1$
Case 2	If $\theta^*_{CCR} < \theta^*_{BCC}$ then		
Case 3	If $\theta^*_{CCR} = \theta^*_{NIRS}$	Increasing	$\sum \lambda^*_j < 1$
Case 4	If $\theta^*_{CCR} < \theta^*_{NIRS}$	Decreasing	$\sum \lambda^*_j > 1$

It should be noted that the problem of non-uniqueness of results in the presence of alternative optima is not encountered in the FGL approach (unless output-oriented as well as input-oriented models are used) whereas they do need to be considered as in Theorem 5.4. However, Zhu and Shen (1995) and Seiford and Zhu (1999) develop an alternative approach that is not troubled by the possibility of such alternative optima.

We here present their results. See also Zhu (2002).

	Seiford and Zhu (1999)	RTS	CCR Model
Case 1	If $\theta^*_{CCR} = \theta^*_{BCC}$	Constant	$\sum \lambda^*_j = 1$
Case 2	If $\theta^*_{CCR} \neq \theta^*_{BCC}$ then		
Case 3	If $\sum \lambda_j < 1$ in any CCR outcome	Increasing	$\sum \lambda^*_j < 1$
Case 4	If $\sum \lambda_j > 1$ in any CCR outcome	Decreasing	$\sum \lambda^*_j > 1$

The significance of Seiford and Zhu's (1999) approach lies in the fact that the possible alternate optimal λ^*_j obtained from the CCR model only affect the estimation of RTS for those DMUs that truly exhibit constant returns to scale, and have nothing to do with the RTS estimation on those DMUs that truly exhibit increasing returns to scale or decreasing returns to scale. That is, if a DMU exhibits increasing returns to scale (or decreasing returns to scale), then $\sum \lambda^*_j$ must be less (or greater) than one, no matter whether there exist alternate optima of λ^*_j, because these DMUs do not lie in the MPSS region. This finding is also true for the u^*_0 obtained from the BCC multiplier models.

Thus, in empirical applications, we can explore RTS in two steps. First, select all the DMUs that have the same CCR and BCC efficiency scores regardless of the value of $\sum_{j=1}^{n} \lambda^*_j$ obtained from model (2.5). These DMUs are constant returns to scale. Next, use the value of $\sum_{j=1}^{n} \lambda^*_j$ (in any CCR model outcome) to determine the RTS for the remaining DMUs. We observe that in this process we can safely ignore possible multiple optimal solutions of λ_j.

Part of the material in this chapter is adapted from Banker *et al.* (2004)[37] with permission from Elsevier Science.

5.15 RELATED DEA-SOLVER MODELS FOR CHAPTER 5

BCC-I, -O As explained in Chapter 4, these models solve the BCC model in the input-orientation and in the output-orientation, respectively. They also identify the returns-to-scale characteristics of DMUs which are recorded in the worksheet "RTS." For inefficient DMUs, we identify returns to scale with the projected point on the efficient frontier. BCC-I uses the projection formula (4.17)-(4.18), while BCC-O uses the BCC version of the formula (3.74)-(3.75).

IRS-I, -O These codes solve the increasing (or non-decreasing) returns-to-scale model introduced in Section 5.7, i.e. the case $\sum_{j=1}^{n} \lambda_j \geq 1$ in input and output orientations.

DRS-I, -O These codes solve the decreasing (or non-increasing) returns-to-scale model in Section 5.7, i.e. the case $0 \leq \sum_{j=1}^{n} \lambda_j \leq 1$.

GRS-I, -O These correspond to the generalized returns-to-scale model, which has the constraint $L \leq \sum_{j=1}^{n} \lambda_j \leq U$. The values of $L(\leq 1)$ and $U(\geq 1)$ must be supplied through the Message-Box on the display by request.

5.16 PROBLEM SUPPLEMENT FOR CHAPTER 5

Problem 5.1

There is an ambiguity associated with characterizing B in Figure 5.3 as a point of constant returns-to-scale in that movement to the left produces a point on the efficient frontier where returns to scale is increasing. Can you supply a rationale for identifying B as a point of constant returns-to-scale?
Suggested Answer : Movement to the left from B would result in an output decrease that would be more than proportionate to the associated reduction in input so, generally speaking, one would need some justification for such a move. That is, the orientation would generally be toward increases in output so a movement which would reduce output more than proportionately to the associated input reduction would need justification. Similarly at C the orientation would favor movements to the left rather than the right.

Problem 5.2

Redraw Figure 5.3 to show what would happen to CCR- and BCC-efficiency if a new point, F, were added that is CCR-efficient. Discuss and analyze.
Suggested Answer : If the point F is on the broken line connecting B and C the MPSS region common to both BCC- and CCR-efficiency would be extended. An analytical characterization is as follows: Because F is CCR-efficient it would have an optimum with $\theta_o^* = 1$ and zero slacks for $\lambda_F^* = 1$ — the case of constant returns-to-scale in Theorem 5.3. This solution also satisfies the BCC model since $\sum_{j=1}^{n} \lambda_j^* = 1$. Because the CCR model is less constrained than the BCC model, we will *always* have optimal solutions with $\theta_o^* \leq \theta_B^*$. In this case,

however, $\theta_o^* = \theta_B^*$. The solution must therefore be optimal for the BCC as well as the CCR model.

This is the crux of Theorem 5.5. As can be seen from Figure 5.3 and the numerical examples in Section 4, one can have $\theta_B^* = 1$ for the BCC model and $\theta_o^* < 1$ for the CCR model with zero slacks. This means that it is possible to have points which are BCC but not CCR efficient. As the theorem notes, it is not possible to have the reverse situation so a point which is CCR-efficient must also be BCC-efficient.

Problem 5.3

Solve the hospital example in Table 1.5 in Chapter 1 using both (input-oriented) BCC-I and (output-oriented) BCC-O models in the DEA-Solver. Compare the returns-to-scale characteristics of hospitals that are BCC inefficient and are projected to the efficient frontiers by both models.

Suggested Answer : After computations you will find the following information by opening Worksheets "Score" and "RTS."

(1) **Input-oriented Case:** The BCC-score, reference set and returns to scale (RTS) are exhibited in the left part of Table 5.6 below. Among the BCC efficient DMUs, A, B and D belong to CRS, while G, J, K and L to DRS. The returns-to-scale characteristics of BCC inefficient DMUs can be identified by referring to Theorem 5.8. For example, the reference set of DMU H consists of A, D and K of which A and D belong to CRS and K to DRS. Hence the projected activity of H displays DRS, since its reference set is a mixture of CRS and DRS. Similarly, DMUs F and I belong to DRS. However, the reference set of DMUs C and E belongs to CRS. So, we applied the procedure (i) on Page 133 to their projected activities and found that they belong to CRS.

(2) **Output-oriented Case:** The right half of Table 5.6 shows the results of the output-oriented BCC analysis. Since the output-oriented projection is different from the input-oriented case, BCC inefficient DMUs may have a different reference set from the one in the input orientation. For example, DMU C's reference set is composed of B, D and G in this case. Hence the output-oriented projection of C belongs to DRS, since the reference set is a mixture of CRS and DRS.

Notes

1. Throughout this chapter, we assume that the data set (X, Y) is semipositive.

2. R.D. Banker and R.M. Thrall (1992), "Estimating Most Productive Scale Size Using Data Envelopment Analysis," *European Journal of Operational Research* 62, pp.74-84.

3. R.D. Banker, I. Bardhan and W.W. Cooper (1996), "A Note on Returns to Scale in DEA," *European Journal of Operational Research* 88, pp.583-585.

4. R.D. Banker, H. Chang and W.W. Cooper (1996), "Equivalence and Implementation of Alternative Methods for Determining Returns to Scale in Data Envelopment Analysis," *European Journal of Operational Research* 89, pp.473-481.

Table 5.6. Results of Input-oriented/Output-oriented BCC Cases

	Input-orientation			Output-orientation		
DMU	Score	Ref Set	RTS	Score	Ref Set	RTS
A	1		CRS	1		CRS
B	1		CRS	1		CRS
C	0.8958	*B D*	CRS	0.9245	*B D G*	DRS
D	1		CRS	1		CRS
E	0.8818	*A B*	CRS	0.7666	*A B D L*	DRS
F	0.9389	*D J*	DRS	0.9548	*D J*	DRS
G	1		DRS	1		DRS
H	0.7988	*A D K*	DRS	0.8259	*A D K L*	DRS
I	0.9893	*A B L*	DRS	0.9899	*A B L*	DRS
J	1		DRS	1		DRS
K	1		DRS	1		DRS
L	1		DRS	1		DRS

5. R.D. Banker and R. Morey (1986), "Efficiency Analysis for Exogenously Fixed Inputs and Outputs" *Operations Research* 34, pp.513-521.

6. T. Ahn, A. Charnes and W.W. Cooper (1989), "A Note on the Efficiency Characterizations Obtained in Different DEA Models," *Socio-Economic Planning Sciences* 22, pp.253-257.

7. See R.D. Banker (1984), "Estimating Most Productive Scale Size Using Data Envelopment Analysis," *European Journal of Operational Research* 17, pp.35-44. See also R.D. Banker, A. Charnes and W.W. Cooper (1984), "Some Models for Estimating Technical and Scale Inefficiencies in Data Envelopment Analysis," *Management Science* 30, pp.1078-1092. A brief review of the history of this and related concepts is to be found in F.R. Førsund — "On the Calculation of the Scale Elasticity in DEA Models," *Journal of Productivity Analysis* 7, 1996, pp.283-302 — who traces the idea of a technically optimal scale to R. Frisch's classical volume, *Theory of Production* (Dordrecht: D.Reidel, 1965). See also L.M. Seiford and J. Zhu (1999) "An Investigation of Returns to Scale in Data Envelopment Analysis," *Omega* 27, pp.1-11.

8. This transformation is done explicitly in the Appendix to W.W. Cooper, R.G. Thompson and R.M. Thrall (1996), "Extensions and New Developments in DEA," *Annals of Operations Research* 66, pp.3-45.

9. K. Tone (1996), "A Simple Characterization of Returns to Scale in DEA," *Journal of the Operations Research Society of Japan* 39, pp.604-613.

10. R.M. Thrall (1996a), "Duality, Classification and Slacks in DEA," *Annals of Operations Research* 66, pp.109-138.

11. W.W. Cooper, K.S. Park and J.T. Pastor (1999), "RAM: A Range Adjusted Measure of Efficiency," *Journal of Productivity Analysis* 11, pp.5-42.

12. R.M. Thrall (1996b), "The Lack of Invariance of Optimal Dual Solutions under Translation Invariance," *Annals of Operations Research* 66, pp.109-138.

13. R.D. Banker, A. Charnes and W.W. Cooper (1984), "Some Models for Estimating Technical and Scale Efficiencies in Data Envelopment Analysis," *Management Science* 30, pp.1078-1092.

14. F.R. Førsund (1996), "On the Calculation of Scale Elasticities in DEA Models," *Journal of Productivity Analysis* 7, pp.283-302.

15. R.D. Banker and R.M. Thrall (1992), "Estimation of Returns to Scale Using Data Envelopment Analysis," *European Journal of Operational Research* 62, pp.74-84.

16. R. Färe, S. Grosskopf and C.A.K. Lovell (1985), *The Measurement of Efficiency of Production,* Boston: Kluwer Nijhoff. Publishing Co., and *Production Frontiers,* Cambridge University Press.

17. H. Fukuyama (2000), "Returns to Scale and Scale Elasticity in Data Envelopment Analysis," *European Journal of Operational Research* 125, pp.93-112.

18. J.C. Panzar and R.D. Willig (1977), "Economies of Scale in Multi-Output Production," *Quarterly Journal of Economics* XLI, pp.481-493.

19. W.J. Baumol, J.C. Panzar and R.D. Willig (1982), *Contestable Markets,* New York: Harcourt Brace Jovanovich.

20. A. Charnes, W.W. Cooper, L.M. Seiford and J. Stutz (1982), "A Multiplicative Model for Efficiency Analysis," *Socio-Economic Planning Sciences* 16, pp.213-224.

21. R.D. Banker, A. Charnes, W.W. Cooper and A. Schinnar (1981), "A Bi-Extremal Principle for Frontier Estimation and Efficiency Evaluation," *Management Science* 27, pp.1370-1382.

22. A. Charnes, W.W. Cooper, L.M. Seiford and J. Stutz (1983), "Invariant Multiplicative Efficiency and Piecewise Cobb-Douglas Envelopments," *Operations Research Letters* 2, pp.101-103.

23. R.D. Banker and A. Maindiratta (1986), "Piecewise Loglinear Estimation of Efficient Production Surfaces," *Management Science* 32, pp.126-135.

24. R.D. Banker, W.W. Cooper, L.M. Seiford, R.M. Thrall and J. Zhu (2003), "Returns to Scale In Different DEA Models," *European Journal of Operational Research* (to appear).

25. W.W. Cooper, R.G. Thompson and R.M. Thrall (1996), "Extensions and New Developments in DEA, Appendix," *The Annals of Operations Research* 66, pp.3-45.

26. H. Varian (1984), *Microeconomic Analysis,* New York: W.W. Norton.

27. R.D. Banker and R. Morey (1986), "Efficiency Analysis for Exogenously Fixed Inputs and Outputs," *Operations Research* 34, pp.513-521.

28. K. Tone (2001), "On Returns to Scale under Weight Restrictions in Data Envelopment Analysis," *Journal of Productivity Analysis* 16, pp.31-47.

29. T. Sueyoshi (1999), "DEA Duality on Returns to Scale (RTS) in Production and Cost Analyses: An Occurrence of Multiple Solutions and Differences between Production-based and Cost-based RTS Estimates," *Management Science* 45, 1593-1608.

30. K. Tone and B.K. Sahoo (2005), "Cost-Elasticity: A Re-Examination in DEA," *Annals of Operations Research* (forthcoming).

31. S. Ray (2005), "Are Some Indian Banks too Large? An Examination of Size Efficiency in Indian Banking," *Journal of Productivity Analysis* (forthcoming).

32. A. Maindiratta (1990), "Scale and Size Efficiencies of Decision Making Units in Data Envelopment Analysis," *Journal of Econometrics* 46, pp.39-56.

33. P. Bogetoft and D. Wang (2005), "Estimating the Potential Gains from Mergers," *Journal of Productivity Analysis* 23, pp.145-171.

34. J. Zhu and Z. Shen (1995), "A Discussion of Testing DMU's Returns to Scale," *European Journal of Operational Research* 81, pp.590-596.

35. L.M. Seiford and J. Zhu (1999), "An Investigation of Returns to Scale under Data Envelopment Analysis," *Omega* 27, pp.1-11.

36. J. Zhu (2002), *Quantitative Models for Performance Evaluation and Benchmarking: Data Envelopment Analysis with Spreadsheets and DEA Excel Solver,* Kluwer Academic Publishers, Boston.

37. R.D. Banker, W.W. Cooper, L.M. Seiford, R.M. Thrall and J. Zhu (2004), "Returns to scale in different DEA models," *European Journal of Operational Research* (2004), 154, pp.345-362.

6 MODELS WITH RESTRICTED MULTIPLIERS

6.1 INTRODUCTION

We have been dealing with several DEA models corresponding to different production possibility sets, which originate from technical aspects of organizational activities. In this manner we have minimized the need for "a priori" knowledge or the need for recourse to assumptions which are "outside the data." There are situations, however, where additional information is available or where one is willing to make assumptions that lead to the imposition of conditions other than nonnegativity on the components of the multiplier vectors $\boldsymbol{v}$ and $\boldsymbol{u}$. The beginning two sections of this chapter will treat these topics in terms of "assurance region" and "cone-ratio" approaches. In particular we will show how these assurance region and cone-ratio approaches can be incorporated in DEA models and how the results can be interpreted. This will be followed by applications designed to illustrate uses of these concepts.

For a preview we might note that Thompson, Singleton, Thrall and Smith (1986) developed the "assurance region" approach to help in choosing a "best site" for the location of a high-energy physics laboratory when other approaches proved to be deficient in evaluating outputs like "contributions to fundamental knowledge."[1] Charnes, Cooper, Huang and Sun (1990)[2] developed the approach that they called "cone-ratio envelopment" in order to evaluate bank performances when unknown allowances for risk and similar factors needed to be taken into account.

6.2 ASSURANCE REGION METHOD

In the optimal weight (v_i^*, u_j^*) of DEA models for inefficient DMUs, we may see many zeros — showing that the DMU has a weakness in the corresponding items compared with other (efficient) DMUs. See Definition 2.1 and Problem 3.3 for troubles that can be associated with these efficiency characterizations. Large differences in weights from item to item may also be a concern. It was concerns like these that led to the development of the assurance region approach which imposes constraints on the relative magnitude of the weights for special items. For example, we may add a constraint on the ratio of weights for Input 1 and Input 2 as follows:

$$L_{1,2} \leq \frac{v_2}{v_1} \leq U_{1,2}, \tag{6.1}$$

where $L_{1,2}$ and $U_{1,2}$ are lower and upper bounds that the ratio v_2/v_1 may assume. The name *assurance region* (AR) comes from this constraint, which limits the region of weights to some special area. Generally, the DEA efficiency score in the corresponding envelopment model is worsened by additions of these constraints and a DMU previously characterized as efficient may subsequently be found to be inefficient after such constraints have been imposed.

We might note that the ratio of the multipliers is likely to coincide with the upper or lower bound in an optimal solution. Hence some care needs to be exercised in choosing these bounds so recourse to auxiliary information such as prices, unit costs, etc, is often used. Indeed, as we will later see, this approach provides a generalization of "allocative" and "price" efficiency approaches which require exact knowledge of prices and costs.

6.2.1 *Formula for the Assurance Region Method*

The assurance region, method is formulated for a DEA model by adding constraints like (6.1) for pairs of items if needed. For example, in the CCR model, the constraints (2.8)–(2.11) in Chapter 2 are augmented by the linear inequalities,

$$L_{1,2}\ v_1 \leq v_2 \leq U_{1,2}\ v_1. \tag{6.2}$$

Some authors like Roll, Cook and Golany[3] have treated the bounds as absolute — viz., $L_i \leq v_i \leq U_i$ — rather than as ratios like in (6.1). This usage is readily accommodated by using the concept of a "numeraire"[4] defined as a unit of account in terms of which all commodities (and services) are stated.

More generally, we can think of an "input numeraire" and an "output numeraire." We can then constrain all of the values of the input (output) variables in the following manner.

$$v_1\ l_{1,i} \leq v_i \leq v_1\ u_{1,i} \quad (i = 2, \ldots, m) \tag{6.3}$$

$$u_1\ L_{1,r} \leq u_r \leq u_1\ U_{1,r}. \quad (r = 2, \ldots, s) \tag{6.4}$$

Much in the manner in which money serves as a unit of account we can set $v_1 = u_1 = 1$ to obtain $(m + s - 2)$ constraints in absolute form. If some are not required, we can delete them from the constraints.

Thus, the CCR-AR model is:

$$
\begin{array}{llll}
(AR_o) & \max\limits_{\boldsymbol{v},\boldsymbol{u}} & \boldsymbol{u}\boldsymbol{y}_o & (6.5)\\
& \text{subject to} & \boldsymbol{v}\boldsymbol{x}_o = 1 & (6.6)\\
& & -\boldsymbol{v}X + \boldsymbol{u}Y \leq \mathbf{0} & (6.7)\\
& & \boldsymbol{v}P \leq \mathbf{0} & (6.8)\\
& & \boldsymbol{u}Q \leq \mathbf{0} & (6.9)\\
& & \boldsymbol{v} \geq \mathbf{0},\ \boldsymbol{u} \geq \mathbf{0}, & (6.10)
\end{array}
$$

where

$$
P = \begin{pmatrix}
l_{12} & -u_{12} & l_{13} & -u_{13} & \dots & \dots & \dots & \dots \\
-1 & 1 & 0 & 0 & \dots & \dots & \dots & \dots \\
0 & 0 & -1 & 1 & \dots & \dots & \dots & \dots \\
\dots & \dots & \dots & \dots & \dots & \dots & \dots & \dots \\
\dots & \dots & \dots & \dots & \dots & \dots & \dots & \dots
\end{pmatrix}
$$

and

$$
Q = \begin{pmatrix}
L_{12} & -U_{12} & L_{13} & -U_{13} & \dots & \dots & \dots & \dots \\
-1 & 1 & 0 & 0 & \dots & \dots & \dots & \dots \\
0 & 0 & -1 & 1 & \dots & \dots & \dots & \dots \\
\dots & \dots & \dots & \dots & \dots & \dots & \dots & \dots \\
\dots & \dots & \dots & \dots & \dots & \dots & \dots & \dots
\end{pmatrix}.
$$

However, notice that choice of the numeraire (v_1) in (6.1) is arbitrary so we could use constraints such as

$$l_{1,2}\ v_1 \leq v_2 \leq u_{1,2}\ v_1 \quad \text{and} \quad l_{2,3}\ v_2 \leq v_3 \leq u_{2,3}\ v_2$$

as well, with resulting in modifications of the matrices P and/or Q.

The computations should generally be carried out on the dual side since the envelopment model is usually easier to solve and interpret.

$$
\begin{array}{llll}
(DAR_o) & \min\limits_{\theta,\boldsymbol{\lambda},\boldsymbol{\pi},\boldsymbol{\tau}} & \theta & (6.11)\\
& \text{subject to} & \theta\boldsymbol{x}_o - X\boldsymbol{\lambda} + P\boldsymbol{\pi} \geq \mathbf{0} & (6.12)\\
& & Y\boldsymbol{\lambda} + Q\boldsymbol{\tau} \geq \boldsymbol{y}_o & (6.13)\\
& & \boldsymbol{\lambda} \geq \mathbf{0},\ \boldsymbol{\pi} \geq \mathbf{0},\ \boldsymbol{\tau} \geq \mathbf{0}. & (6.14)
\end{array}
$$

We now turn to relations between (AR_o) and (DAR_o), assuming that both problems have a finite positive optimum. (Refer to Problem 6.5.) Let an optimal (and max-slack[5]) solution of (DAR_o) be $(\theta^*, \boldsymbol{\lambda}^*, \boldsymbol{\pi}^*, \boldsymbol{\tau}^*, \boldsymbol{s}^{-*}, \boldsymbol{s}^{+*})$, where slacks $\boldsymbol{s}^{-*}$ and $\boldsymbol{s}^{+*}$ are defined in vector-matrix notation as:

$$\boldsymbol{s}^{-*} = \theta^*\boldsymbol{x}_o - X\boldsymbol{\lambda}^* + P\boldsymbol{\pi}^* \quad \text{and} \tag{6.15}$$

$$\boldsymbol{s}^{+*} = -\boldsymbol{y}_o + Y\boldsymbol{\lambda}^* + Q\boldsymbol{\tau}^*. \tag{6.16}$$

Based on the solution, we define "AR-efficiency" as:

Definition 6.1 (AR-Efficiency) *The DMU associated with* $(\boldsymbol{x}_o, \boldsymbol{y}_o)$ *is* AR-efficient, *if and only if it satisfies*

$$\theta^* = 1, \quad \boldsymbol{s}^{-*} = \boldsymbol{0} \quad \text{and} \quad \boldsymbol{s}^{+*} = \boldsymbol{0}.$$

An improvement of an AR-inefficient $(\boldsymbol{x}_o, \boldsymbol{y}_o)$ obtained by using (DAR_o) can be represented:

$$\widehat{\boldsymbol{x}}_o = \theta^* \boldsymbol{x}_o - \boldsymbol{s}^{-*} + P\boldsymbol{\pi}^* \quad (= X\boldsymbol{\lambda}^*) \tag{6.17}$$

$$\widehat{\boldsymbol{y}}_o = \boldsymbol{y}_o + \boldsymbol{s}^{+*} - Q\boldsymbol{\tau}^* \quad (= Y\boldsymbol{\lambda}^*). \tag{6.18}$$

These projection formulas now replace (3.28) and (3.29) to allow for the extra variables $\boldsymbol{\pi}^*$ and $\boldsymbol{\tau}^*$ which appear in the envelopment model (DAR_o) as a result of the bounds placed on the variables in the multiplier model.

We now assert:

Theorem 6.1 *The activity* $(\widehat{\boldsymbol{x}}_o, \widehat{\boldsymbol{y}}_o)$ *is AR-efficient.*

Proof. Since $(\widehat{\boldsymbol{x}}_o, \widehat{\boldsymbol{y}}_o) = (X\boldsymbol{\lambda}^*, Y\boldsymbol{\lambda}^*) \in P$ (the production possibility set), we can evaluate its AR-efficiency by solving the program:

$$\begin{aligned} (DAR_e) \quad \min \quad & \theta_e \\ \text{subject to} \quad & \theta_e \widehat{\boldsymbol{x}}_o - X\boldsymbol{\lambda}_e + P\boldsymbol{\pi}_e - \boldsymbol{s}_e^- = \boldsymbol{0} \\ & Y\boldsymbol{\lambda}_e + Q\boldsymbol{\tau}_e - \boldsymbol{s}_e^+ = \widehat{\boldsymbol{y}}_o \\ & \boldsymbol{\lambda}_e \geq \boldsymbol{0}, \boldsymbol{\pi}_e \geq \boldsymbol{0}, \boldsymbol{\tau}_e \geq \boldsymbol{0}, \boldsymbol{s}_e^- \geq \boldsymbol{0}, \boldsymbol{s}_e^+ \geq \boldsymbol{0}. \end{aligned} \tag{6.19}$$

(DAR_e) has a feasible solution $(\theta_e = 1, \boldsymbol{\lambda}_e = \boldsymbol{\lambda}^*, \boldsymbol{\pi}_e = \boldsymbol{0}, \boldsymbol{\tau}_e = \boldsymbol{0}, \boldsymbol{s}_e^- = \boldsymbol{0}, \boldsymbol{s}_e^+ = \boldsymbol{0})$ so $0 \leq \theta_e^* \leq 1$. The dual to (DAR_e) is:

$$\begin{aligned} (AR_e) \quad \max_{\boldsymbol{v}_e, \boldsymbol{u}_e} \quad & \boldsymbol{u}_e \widehat{\boldsymbol{y}}_o \\ \text{subject to} \quad & \boldsymbol{v}_e \widehat{\boldsymbol{x}}_o = 1 \\ & -\boldsymbol{v}_e X + \boldsymbol{u}_e Y \leq \boldsymbol{0} \\ & \boldsymbol{v}_e P \leq \boldsymbol{0} \\ & \boldsymbol{u}_e Q \leq \boldsymbol{0} \\ & \boldsymbol{v}_e \geq \boldsymbol{0}, \quad \boldsymbol{u}_e \geq \boldsymbol{0}. \end{aligned} \tag{6.20}$$

Let $\boldsymbol{v}_e^* = \boldsymbol{v}^*/\theta^*$ and $\boldsymbol{u}_e^* = \boldsymbol{u}^*/\theta^*$, where $(\boldsymbol{v}^*, \boldsymbol{u}^*)$ is an optimal solution of (AR_o). Then, using these relations together with the complementary slackness conditions

$$\boldsymbol{v}^* \boldsymbol{s}^{-*} = \boldsymbol{v}^* P\boldsymbol{\pi}^* = \boldsymbol{u}^* \boldsymbol{s}^{+*} = \boldsymbol{u}^* Q\boldsymbol{\tau}^* = 0, \tag{6.21}$$

we obtain

$$\boldsymbol{v}_e^* \widehat{\boldsymbol{x}}_o = \frac{\boldsymbol{v}^*}{\theta^*}(\theta^* \boldsymbol{x}_o - \boldsymbol{s}^{-*} + P\boldsymbol{\pi}^*) = \boldsymbol{v}^* \boldsymbol{x}_o = 1 \tag{6.22}$$

and

$$\boldsymbol{u}_e^* \widehat{\boldsymbol{y}}_o = \frac{\boldsymbol{u}^*}{\theta^*}(\boldsymbol{y}_o + \boldsymbol{s}^{+*} - Q\boldsymbol{\tau}^*) = \frac{\boldsymbol{u}^* \boldsymbol{y}_o}{\theta^*} = 1. \tag{6.23}$$

Hence, the solutions $(\theta_e^* = 1, \boldsymbol{\lambda}_e^* = \boldsymbol{\lambda}^*, \boldsymbol{\pi}_e^* = \mathbf{0}, \boldsymbol{\tau}_e^* = \mathbf{0}, \boldsymbol{s}_e^{-*} = \mathbf{0}, \boldsymbol{s}_e^{+*} = \mathbf{0})$ and $(\boldsymbol{v}_e^*, \boldsymbol{u}_e^*)$ are optimal to (DAR_e) and (AR_e), respectively.

Moreover, we can demonstrate, in a way similar to the proof of Theorem 3.2, that every optimal solution of (DAR_e) satisfies $\boldsymbol{s}_e^{-*} = \mathbf{0}$ and $\boldsymbol{s}_e^{+*} = \mathbf{0}$. Thus, the activity $(\widehat{\boldsymbol{x}}_o, \widehat{\boldsymbol{y}}_o)$ is AR-efficient. □

We also need to note that the projection formulas (6.17, 6.18) do not always yield $\widehat{\boldsymbol{x}}_o \leq \boldsymbol{x}_o$ and $\widehat{\boldsymbol{y}}_o \geq \boldsymbol{y}_o$, contrary to (3.28) and (3.29). Also, the *output-oriented* assurance region model can be defined analogously to the output-oriented CCR model.

6.2.2 General Hospital Example

Tables 6.1 and 6.2 show data of 14 general hospitals with two inputs (Doctors and Nurses) and two outputs (Outpatients and Inpatients) and the optimal weights obtained by the CCR model where a number such as .332E-03 means $.332 \times 10^{-3}$. There are many zeros in the weights of inefficient DMUs, especially in the doctor and outpatient items. This means that the inefficient hospitals are very likely to have a surplus in doctors and shortage of outpatients, compared with the hospitals in the reference set.

Table 6.1. Data for 14 Hospitals

	Input		Output	
Hos.	Doctors	Nurses	Outpatients	Inpatients
H1	3008	20980	97775	101225
H2	3985	25643	135871	130580
H3	4324	26978	133655	168473
H4	3534	25361	46243	100407
H5	8836	40796	176661	215616
H6	5376	37562	182576	217615
H7	4982	33088	98880	167278
H8	4775	39122	136701	193393
H9	8046	42958	225138	256575
H10	8554	48955	257370	312877
H11	6147	45514	165274	227099
H12	8366	55140	203989	321623
H13	13479	68037	174270	341743
H14	21808	78302	322990	487539

To approach this problem by the assurance region method, we impose constraints

$$0.2 \leq \frac{v_2}{v_1} \leq 5, \qquad 0.2 \leq \frac{u_2}{u_1} \leq 5. \tag{6.24}$$

Table 6.2. Efficiency and Weight of 14 Hospitals by CCR Model

Hospital	CCR Eff.	Doctor	Nurse	Outpatient	Inpatient
		v_1^*	v_2^*	u_1^*	u_2^*
H1	0.955	.332E-03	0	.959E-05	.167E-06
H2	1	.242E-03	.140E-05	.714E-05	.225E-06
H3	1	.104E-03	.204E-04	.339E-05	.325E-05
H4	0.702	.282E-03	.128E-06	0	.699E-05
H5	0.827	0	.245E-04	0	.384E-05
H6	1	.784E-04	.154E-04	.256E-05	.245E-05
H7	0.844	.133E-03	.102E-04	0	.505E-05
H8	1	.209E-03	.458E-07	.111E-07	.516E-05
H9	0.995	0	.233E-04	.426E-05	.136E-06
H10	1	.319E-04	.149E-04	.215E-05	.143E-05
H11	0.913	.162E-03	.739E-07	0	.402E-05
H12	0.969	.793E-04	.611E-05	0	.301E-05
H13	0.786	0	.147E-04	0	.230E-05
H14	0.974	0	.128E-04	0	.200E-05

The results are shown in Table 6.3.[6] All the weights are now non zero, and in

Table 6.3. Efficiency and Weight of 14 Hospitals with Assurance Region Method

Hos.	AR Eff.	Doctor	Nurse		Outpa.	Inpa.	
		v_1^*	v_2^*	v_2^*/v_1^*	u_1^*	u_2^*	u_2^*/u_1^*
H1	0.926	.139E-03	.278E-04	0.2	.728E-05	.211E-05	0.29
H2	1	.110E-03	.219E-04	0.2	.575E-05	.167E-05	0.29
H3	1	.103E-03	.206E-04	0.2	.102E-05	.512E-05	5.0
H4	0.634	.116E-03	.232E-04	0.2	.116E-05	.579E-05	5.0
H5	0.820	.470E-05	.235E-04	5.0	.653E-06	.327E-05	5.0
H6	1	.776E-04	.155E-04	0.2	.280E-05	.224E-05	0.8
H7	0.803	.862E-04	.172E-04	0.2	.859E-06	.429E-05	5.0
H8	0.872	.794E-04	.159E-04	0.2	.790E-06	.395E-05	5.0
H9	0.982	.449E-05	.224E-04	5.0	.355E-05	.711E-06	5.0
H10	1	.519E-04	.114E-04	0.2	.284E-05	.863E-06	0.30
H11	0.849	.656E-04	.131E-04	0.2	.653E-06	.327E-05	5.0
H12	0.930	.516E-04	.103E-04	0.2	.513E-06	.257E-05	5.0
H13	0.740	.283E-05	.141E-04	5.0	.393E-06	.197E-05	5.0
H14	0.929	.242E-05	.121E-04	5.0	.336E-06	.168E-05	5.0

many instances the lower or upper bounds are achieved, as might be expected with linear (inequality) constraints. As might also be expected with extra variables in (6.12)-(6.14) CCR efficiency is generally reduced in value. For instance, H8 becomes inefficient in Table 6.3 whereas it was efficient in Table 6.2. This change in status is traceable to the change in weights resulting from the assurance region constraints. Thus H8 was efficient in Table 6.2 where it has comparatively large weights for doctors compared with nurses and a low weight to outpatients compared with inpatients. These weight unbalances allowed H8 to be efficient. But, the constraints of the assurance region prohibited such extreme weighting divergences and this resulted in a decline in efficiency. To portray the situation graphically Figure 6.1 provides a comparison of CCR models with and without the assurance region constraints.

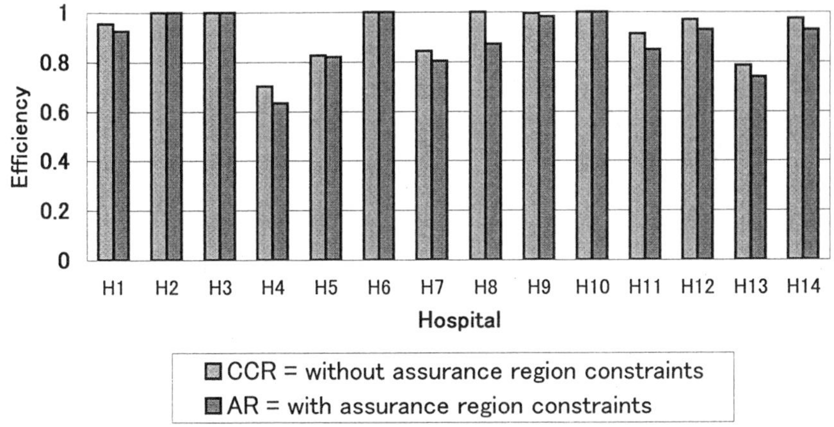

Figure 6.1. Efficiency Scores with and without Assurance Region Constraints

6.2.3 Change of Efficient Frontier by Assurance Region Method

To portray the influence of the assurance region method on the efficient frontier in the $(\boldsymbol{v},\ \boldsymbol{u})$-space we return to the case of Example 2.2 (two inputs and one output) represented by Figure 2.3 in Chapter 2. Now assume that we impose a constraint on the weights v_1 and v_2,

$$0.5 \leq \frac{v_2}{v_1} \leq 2. \tag{6.25}$$

That is, we require

$$0.5(v_1/u) \leq (v_2/u), \quad (v_2/u) \leq 2(v_1/u). \tag{6.26}$$

The feasible region in the $(\boldsymbol{v},\ \boldsymbol{u})$-space originally lay on or above the boundary indicated by the solid line in Figure 2.3 which changes to the area P in Figure

6.2 because of these constraints. The boundary of this new area P is composed of a part of line segments D and E, and two extremal rays corresponding to the constraints, and it is seen that the line segment $C : 8(v_1/u) + v_2/u = 1$ no longer contributes to P. In the original CCR model, C was efficient and, in fact, the values of v_1/u and v_2/u that make C efficient lay on the line segment $\overline{P_1P_2}$ in Figure 2.3. Now, however, this line segment is shut out from the region P by the assurance region constraints. Hence, we move the line segment C in parallel to touch P (at P_5) and it is given by the dotted line

$$8(v_1/u) + (v_2/u) = 1/0.8. \tag{6.27}$$

Thus the efficiency of DMU C drops to 0.8. (Refer to DMU A in Figure 2.4.)

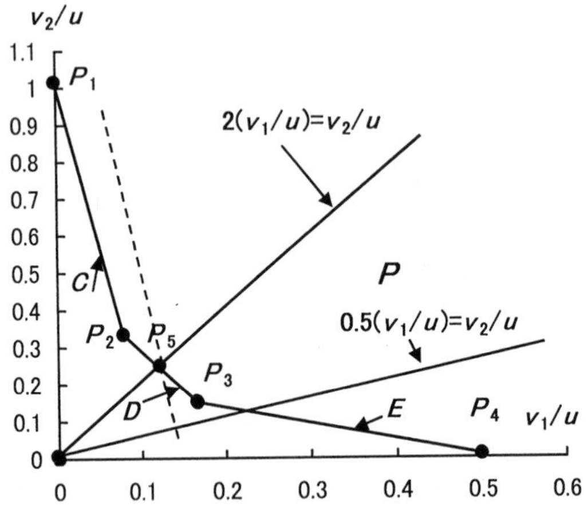

Figure 6.2. Assurance Region

6.2.4 *On Determining the Lower and Upper Bounds*

In this section we assume that all input and output items are technological (not monetary). Since, as was mentioned in the introductory chapter, the virtual inputs and outputs for the activity $(\boldsymbol{x}_k, \boldsymbol{y}_k)$ are expressed as:

$$C_k = v_1 x_{1k} + \cdots + v_m x_{mk} \quad \text{and} \quad P_k = u_1 y_{1k} + \cdots + u_s y_{sk},$$

we can interpret v_i (u_j) as the unit cost (price) of the input (output) item x_{ik} (y_{jk}). This interpretation corresponds to the primal-dual LP relation of the variables ($\boldsymbol{v}$ vs. $\boldsymbol{s}^-$) and ($\boldsymbol{u}$ vs. $\boldsymbol{s}^+$) in which the optimal virtual multiplier v_i^* (u_j^*) is related to the imputed (shadow) price of the i (j)-th input (output) item. See the discussion of "complementary slackness" immediately following (AR_e) in (6.20), above.

Let c_{ik} be the actual unit cost of the input item x_{ik}. In view of this interpretation, it would be reasonable to set the lower and upper bounds for v_j/v_i

by the following scheme:

$$l_{ij} = \min \frac{c_{jk}}{c_{ik}} \quad (k = 1, \ldots, n) \quad \text{and} \quad u_{ij} = \max \frac{c_{jk}}{c_{ik}} \quad (k = 1, \ldots, n). \tag{6.28}$$

We can similarly put bounds on output multipliers by using the actual output prices. Returning to (6.1) we can now see more clearly some of the advantages of the assurance region approaches by virtue of the following considerations. In many cases, actual prices (or unit costs) are not known exactly. Nevertheless it may be possible to put bounds on their values as in (6.1). In other cases a variety of different levels of prices (or unit costs) may have occurred in which case bounds can be again used with the resulting solution of the AR problem selecting a value. If these values are not satisfactory new bounds may be used, and so on.

6.3 ANOTHER ASSURANCE REGION MODEL

As a generalization of the assurance region model, we introduce weight restrictions to inputs or outputs in a manner as described by the following example (see Allen *et al.*(1997) [7]) :

$$0.1 \leq \frac{v_1x_1}{v_1x_1 + v_2x_2 + v_3x_3} \leq 0.3 \tag{6.29}$$

$$0.05 \leq \frac{v_2x_2}{v_1x_1 + v_2x_2 + v_3x_3} \leq 0.1 \tag{6.30}$$

$$0.45 \leq \frac{v_3x_3}{v_1x_1 + v_2x_2 + v_3x_3} \leq 0.7 \tag{6.31}$$

$$0.4 \leq \frac{u_1y_1}{u_1y_1 + u_2y_2} \leq 0.8 \tag{6.32}$$

In this example, we have three inputs x_1, x_2 and x_3 with respective weights v_1, v_2 and v_3, and two outputs y_1 and y_2 with respective weights u_1 and u_2. The ratio of the virtual weight of input 1 (v_1x_1) against the total virtual weight ($v_1x_1 + v_2x_2 + v_3x_3$) must be between the lower bound 0.1 and the upper bound 0.3. Similarly, $v_2x_2/(v_1x_1 + v_2x_2 + v_3x_3)$ is restricted to the range (0.05, 0.1). We can impose the same weight restriction on outputs as well. In this example we have no explicit restriction imposed on u_2y_2.

We call this model the *Assurance Region Global Model.*

Notice that (1) the sum of the lower bounds of inputs (outputs) must be less than or equal to unity and (2) the sum of the upper bounds must be greater than or equal to unity in the case all inputs (outputs) are restricted. We also recall the difference existing between this AR Global type model and the AR model. The (traditional) AR model imposes restriction on the virtual input multipliers, for instance,

$$L_2 \leq \frac{v_2}{v_1} \leq U_2 \tag{6.33}$$

$$L_3 \leq \frac{v_3}{v_1} \leq U_3. \tag{6.34}$$

This reduces to the following bounds.

$$\frac{x_1}{x_1 + U_2x_2 + U_3x_3} \le \frac{v_1x_1}{v_1x_1 + v_2x_2 + v_3x_3} \le \frac{x_1}{x_1 + L_2x_2 + L_3x_3}. \tag{6.35}$$

Thus, the relative weight of input 1 varies depending on the input values of the concerned DMU. The AR-Global model, however, restricts the relative weight to the designated (constant) bounds that are common to all DMUs. This scheme is intuitively more appealing than the traditional AR model.

6.4 CONE-RATIO METHOD

We now turn to the cone-ratio method which provides another approach. More general than the assurance region method, the cone-ratio approach lends itself to a variety of additional uses. However, we start by relating the cone-ratio to the assurance region approach.

6.4.1 Polyhedral Convex Cone as an Admissible Region of Weights

First we will observe the $\boldsymbol{v}$-space. Let us restrict the feasible region of the weight $\boldsymbol{v}$ to be in the polyhedral convex cone spanned by the k admissible nonnegative direction vectors $(\boldsymbol{a}_j)$ $(j = 1, \dots, k)$. Thus, a feasible $\boldsymbol{v}$ can be expressed as

$$\boldsymbol{v} = \sum_{j=1}^{k} \alpha_j \boldsymbol{a}_j \quad \text{with} \quad \alpha_j \ge 0 \ (\forall j) \tag{6.36}$$

$$= A^T \boldsymbol{\alpha}, \tag{6.37}$$

where $A^T = (\boldsymbol{a}_1, \dots, \boldsymbol{a}_k) \in R^{m \times k}$ and $\boldsymbol{\alpha}^T = (\alpha_1, \dots, \alpha_k)$. Let the polyhedral convex cone thus defined be V:

$$V = A^T \boldsymbol{\alpha}.$$

The two dimensional case is depicted in Figure 6.3.

Similarly, we restrict the feasible region of the output weight $\boldsymbol{u}$ in the polyhedral convex cone U spanned by the l admissible nonnegative direction vectors $(\boldsymbol{b}_j)$ $(j = 1, \dots, l)$ to define U,

$$U = \sum_{j=1}^{l} \beta_j \boldsymbol{b}_j \quad \text{with} \quad \beta_j \ge 0 \ (\forall j) \tag{6.38}$$

$$= B^T \boldsymbol{\beta}, \tag{6.39}$$

where $B^T = (\boldsymbol{b}_1, \dots, \boldsymbol{b}_l) \in R^{m \times l}$ and $\boldsymbol{\beta}^T = (\beta_1, \dots, \beta_l)$.

(Note 1.) We can demonstrate that the assurance region method is a special case of the cone-ratio method. In fact, if we impose a constraint such as

$$l_{1,2} \le \frac{v_2}{v_1} \le u_{1,2}$$

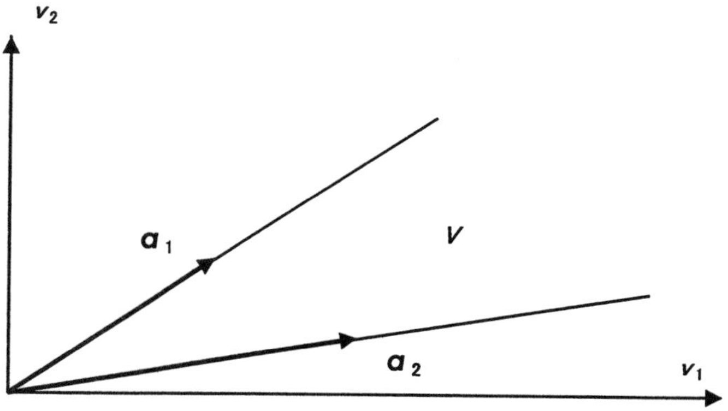

Figure 6.3. Convex Cone generated by Two Vectors

in the assurance region method, it corresponds to the selection of two admissible vectors in the cone-ratio method:

$$\boldsymbol{a}_1^T = (1, l_{1,2}, 0, 0, \ldots, 0) \quad \text{and} \quad \boldsymbol{a}_2^T = (1, u_{1,2}, 0, 0, \ldots, 0),$$

where only two elements of the directional vectors are allowed to be positive. However, the cone-ratio method is more general in dealing with directions with all elements being possibly nonnegative.

6.4.2 Formula for Cone-Ratio Method

Given the polyhedral convex cones V and U generated by A and B, we have the following CCR model:

$$(CRP_o) \quad \max_{\boldsymbol{v},\boldsymbol{u}} \quad \boldsymbol{u}\boldsymbol{y}_o \tag{6.40}$$

$$\text{subject to} \quad \boldsymbol{v}\boldsymbol{x}_o = 1 \tag{6.41}$$

$$-\boldsymbol{v}X + \boldsymbol{u}Y \leq \mathbf{0} \tag{6.42}$$

$$\boldsymbol{v} \in V \tag{6.43}$$

$$\boldsymbol{u} \in U, \tag{6.44}$$

with (CRP_o) coinciding with the ordinary CCR problem as represented in (3.2)-(3.7) when $V = R_m^+$ and $U = R_s^+$, the positive orthants in m and s space, respectively.

Using (6.37) and (6.39), we transform (CRP_o) into the following problem in row vectors $\boldsymbol{\alpha}$ and $\boldsymbol{\beta}$,

$$(CRP_o') \quad \max_{\boldsymbol{\alpha},\boldsymbol{\beta}} \quad \boldsymbol{\beta}(B\boldsymbol{y}_o) \tag{6.45}$$

$$\text{subject to} \quad \boldsymbol{\alpha}(A\boldsymbol{x}_o) = 1 \tag{6.46}$$

$$-\boldsymbol{\alpha}(AX) + \boldsymbol{\beta}(BY) \leq \mathbf{0} \tag{6.47}$$

$$\boldsymbol{\alpha} \geq \mathbf{0} \tag{6.48}$$

$$\boldsymbol{\beta} \geq \mathbf{0}. \tag{6.49}$$

The dual of (CRP'_o) may be expressed in a real variable θ and a vector variable $\boldsymbol{\lambda} = (\lambda_1, \ldots, \lambda_n)^T$ as follows:

$$(CRD'_o) \quad \min_{\theta,\boldsymbol{\lambda}} \quad \theta \tag{6.50}$$

$$\text{subject to} \quad \theta(A\boldsymbol{x}_o) - (AX)\boldsymbol{\lambda} \geq \mathbf{0} \tag{6.51}$$

$$(B\boldsymbol{y}_o) - (BY)\boldsymbol{\lambda} \leq \mathbf{0} \tag{6.52}$$

$$\boldsymbol{\lambda} \geq \mathbf{0}. \tag{6.53}$$

From these expressions, it is clear that the polyhedral cone-ratio method also coincides with a CCR envelopment model evaluating the same DMUs but with the transformed positive data $\bar{X}$ and $\bar{Y}$:

$$\bar{X} = AX \in R^{k\times n} \quad \text{and} \quad \bar{Y} = BY \in R^{l\times n}. \tag{6.54}$$

6.4.3 *A Cone-Ratio Example*

We will solve the same general hospital case as in Section 6.2.2. This has two input items (doctors and nurses) and two output items (outpatients and inpatients) covering 14 hospitals. The data set exhibited under (X^T, Y^T) in the first 4 columns of Table 6.4 has been transferred directly from Table 6.1.

Now let us choose two admissible input weight vectors, $\boldsymbol{a}_1$ and $\boldsymbol{a}_2$, and one

Table 6.4. Efficiency of 14 Hospitals by CR (Cone-Ratio) and CCR Models

Hos.	Doctor	Nurse	Outpa.*	Inpa.**	Cone-Ratio			CR	CCR
	X^T		Y^T		$\bar{X}^T$		$\bar{Y}^T$	Eff.	Eff.
H1	3008	20980	97775	101225	13498	44968	401450	0.821	0.95
H2	3985	25643	135871	130580	16806.5	55271	527611	0.867	1
H3	4324	26978	133655	168473	17813	58280	639074	0.991	1
H4	3534	25361	46243	100407	16214.5	54256	347464	0.592	0.70
H5	8836	40796	176661	215616	29234	90428	823509	0.811	0.82
H6	5376	37562	182576	217615	24157	80500	835421	0.955	1
H7	4982	33088	98880	167278	21526	71158	600714	0.771	0.84
H8	4775	39122	136701	193393	24336	83019	716880	0.814	1
H9	8046	42958	225138	256575	29525	93962	994863	0.943	0.99
H10	8554	48955	257370	312877	33031.5	106464	1196001	1	1
H11	6147	45514	165274	227099	28904	97175	846571	0.809	0.91
H12	8366	55140	203989	321623	35936	118646	1168858	0.898	0.96
H13	13479	68037	174270	341743	47497.5	149553	1199499	0.714	0.78
H14	21808	78302	322990	487539	60959	178412	1785607	0.891	0.97

* Outpa.=Outpatients
** Inpa.=Inpatients

weight for output, $\boldsymbol{b}_1$, as follows:

$$\begin{aligned}\boldsymbol{a}_1^T &= (1, 0.5)\\ \boldsymbol{a}_2^T &= (1,\ \ 2)\\ \boldsymbol{b}_1^T &= (1,\ \ 3).\end{aligned}$$

Thus, we have

$$A = \begin{pmatrix} 1 & 0.5 \\ 1 & 2 \end{pmatrix} \qquad B = \begin{pmatrix} 1 & 3 \end{pmatrix}.$$

The transformed data set ($\bar{X} = AX$ and $\bar{Y} = BY$) is also exhibited in Table 6.4 under the columns headed Cone-Ratio. We illustrate for H1,

$$\begin{pmatrix} 1 & 0.5 \\ 1 & 2 \end{pmatrix} \begin{pmatrix} 3008 \\ 20980 \end{pmatrix} = \begin{pmatrix} 13498 \\ 44968 \end{pmatrix} = \bar{X}$$

$$\begin{pmatrix} 1 & 3 \end{pmatrix} \begin{pmatrix} 97775 \\ 101225 \end{pmatrix} = \begin{pmatrix} 401450 \end{pmatrix} = \bar{Y}.$$

Notice that the output data are compressed to one item.

By using these $\bar{X}$ and $\bar{Y}$ values, we can solve the cone-ratio model in its CCR form obtained by inserting $\bar{X}$ and $\bar{Y}$ in (6.51)-(6.53). The results are shown in the CR column in Table 6.4, along with the original CCR efficiency obtained from column 2 in Table 6.2. As can be seen only one hospital (H10) is cone-ratio efficient, while five hospitals were CCR efficient.

6.4.4 *How to Choose Admissible Directions*

There are several ways to choose ($\boldsymbol{a}_j$) and ($\boldsymbol{b}_j$), from which the following pair are the most commonly used.

1. Use of the knowledge of experts for admissible ratios of weights.

2. First, solve the original CCR model and then select preferable DMUs among the efficient ones. Use the set of the optimal weights $\boldsymbol{v}^*$ and $\boldsymbol{u}^*$ for the preferable DMUs as the admissible directions.

In addition we can combine these two methods, as we now show, to obtain a third approach.

6.5 AN APPLICATION OF THE CONE-RATIO MODEL

We now provide an example of this third approach. The emphasis is on data adjustments as an alternative to the use of explicitly formulated constraints as in the assurance region approach. The objective is to deal with situations that are too complex to be dealt with by the latter method.

The problems we discuss are taken from Brockett *et al.* (1997),[8] a study which is directed to providing an example for the Texas Banking Commission to consider for use in improving its monitoring activities. In particular it is intended to provide an alternative to what is called the "risk adjusted capital" approach.

This concept, "risk adjusted capital," stems from an agreement called the "Basel Agreement" for dealing with adjusting the net worth requirements of individual banks in order to allow for risk differences arising from the various components of each bank's portfolio of loans and investments. This agreement was negotiated by the central banks of various countries with representatives meeting in Basel, Switzerland, in 1988. Subsequently adopted by both state and Federal government regulatory agencies in the U.S., this approach takes a form in which fixed percentages (the same for every bank) are applied to each class of items in a bank's portfolio of loans and investments in order to obtain "jackup factors" to estimate the increase in capital (net worth) required to cover the risk associated with each class of items, such as loans, mortgages, stocks, bonds, etc.[9] This is the origin of the term "risk adjusted capital."

This is a very rigid approach. Hence the objective of the study we are examining was to explore DEA as a possible alternative (or supplement). However, the example is only illustrative and was designed to provide the Texas Banking Department with a "feel" for what DEA could do. Focusing on banks in Texas, the period 1984-1985 was selected for the data because it represented a period of transition. The years prior to 1984 were good ones for Texas banks, after which an increasing number of bank failures then began to appear. The cut-off in 1985 was used because this trend was only beginning to become apparent, at least mildly, in this year. See Table 6.5 and note that, by 1988, the increasing number of bank failures in Texas exceeded the number of failures in all of the other 49 states.[10] The approach used by the Texas Banking Department failed to catch these developments, even when augmented by the risk adjusted capital approach, so this was considered to be a fair test for DEA.

Because this was an initiatory study, it was decided to confine the number of variables to the 4 inputs and the 4 outputs listed in Table 6.6. Reasons for the choice of these items are set forth in Brockett *et al.*(1997). We therefore here only break out some of the underlying components as shown in Table 6.6. Note that both balance sheet (stock) and income statement (flow) items are included whereas the Basel agreement is directed only to balance sheet items for their pertinence to risks and risk coverages. "Provision for Loan Losses" in Table 6.6 represents a flow item in the form of a charge against current income whereas "Allowance for Loan Losses" is a balance sheet item in the form of a "valuation reserve" which represents the cumulated amount of such "Provisions," minus the loan losses charged to this reserve.[11]

These input and output variables were selected in consultation with regulatory officials, and were intended to test the ability of DEA to allow for "risk" and "risk coverage" in its evaluations of efficient and inefficient performances. The inability of these officials to set bounds on specific variables made it impossible to use "assurance region" (or like) approaches. A strategy based on the cone-ratio envelopment approaches was therefore utilized which made it possible to use the expertise of these officials to identify a collection of "excellent

Table 6.5. Number of Bank Failures (through 10-31-88)

Year	Texas		U.S.	
	State banks	National banks	Total Texas	Nationwide
1980	0	0	0	10
1981	0	0	0	10
1982	4	3	7	42
1983	1	2	3	48
1984	2	4	6	79
1985	6	7	12	120
1986	14	12	26	145
1987	*24	28	52	188
1988	39	62	101	**175

* Includes 2 private uninsured banks.
** Includes 40 banks (including 9 State-chartered banks) which were closed, with some reopening under a different aegis as a result of FDIC assistance.
Source: *Audit of the Examination and Enforcement Functions of Texas Department of Banking,* A report to the Legislative Audit Committee (Austin, Texas, Office of the State Auditor, January 1989).

banks" — which were not necessarily confined to banks in Texas. The idea was to use information derived from the performance of these excellent banks to reflect the risk present in their asset-liability (balance sheet) and operating (income statement) structures. It was believed that this could help to identify problems in Texas banks that fell short of excellence in their performances.

This objective was to be accomplished by using the multiplier values of these excellent banks in a cone-ratio envelopment to adjust the data of each of the individual banks included in this study. However, the following modification to this approach was also used. The "excellent banks" selected by the Banking Department representatives were submitted to DEA efficiency evaluation (using the CCR model) along with the other banks. Dual (multiplier) values for excellent banks found to be inefficient were not used to effect the desired data adjustments because it was deemed desirable to use only banks which were efficient as determined by DEA and excellent as determined by expert evaluations in order to reflect managerial performances as well as risks. See the discussion of "Effectiveness" vs. "Efficiency" following (3.96) in Chapter 3. See also R.M. Nun (1989),[12] a report which reflects a finding of the U.S. Comptroller of the Currency to the effect that bank management incompetence was a major factor in the failures that resulted in a crisis in U.S. banking in the late 1980s.

The five excellent banks as listed in Table 6.7, are (1) Bankers Trust Co. (2) Citibank N.A. (3) Morgan Guaranty Trust Co NY (4) Wachovia Bank and Trust Co. N.A. — as shown in rows 1-4 — and (5) First Interstate Bank of Nevada N.A., listed in row 21. However, 2 of these excellent banks were found

Table 6.6. Inputs and Outputs

INPUTS
Interest Expense
Interest expenses on deposit
Expense for federal funds purchased and repurchased in domestic offices
Non-Interest Expense
Salaries and employees benefits
Occupancy expense, furniture, and equipment
Provision for Loan Loses
Total Deposits
Sum of interest bearing and noninterest bearing deposits
OUTPUTS
Interest Income
Interest and fees on loans
Income on federal funds sold and repurchases in domestic offices
Total Non-Interest Income
Allowances for Loan Loses
Total Loans
Loans, net of unearned income

Source: Adjusted data from FDIC call reports are used in this study.

to be inefficient in each year. See the discussion of Table 6.7 below. Therefore the number of excellent banks used to supply multiplier values was reduced from 5 to the 3 following ones in each year. For 1984: 1 (=Bankers Trust), 4 (=Wachovia) and 21 (=First Interstate of Nevada). For 1985: 1 (=Bankers Trust), 3 (=Morgan guaranty) and 21 (=First Interstate of Nevada).

Thus for each year we have a matrix of the form

$$D = \begin{pmatrix} v_{11}^* & \dots & v_{41}^* & & & \\ & \dots & & & O & \\ v_{13}^* & \dots & v_{43}^* & & & \\ & & & u_{11}^* & \dots & u_{41}^* \\ & O & & & \dots & \\ & & & u_{13}^* & \dots & u_{43}^* \end{pmatrix} \tag{6.55}$$

with the stars designating averages of the optimal multipliers represented by u^*, to be entered as elements of the B matrix of (6.51)-(6.52), and the v^* to be entered as elements of the A matrix so that

$$D\begin{pmatrix} X \\ Y \end{pmatrix} = \begin{pmatrix} AX \\ BY \end{pmatrix} = \begin{pmatrix} \bar{X} \\ \bar{Y} \end{pmatrix}, \tag{6.56}$$

as given in (6.54).

Table 6.7. CCR and Cone-Ratio Efficiency Scores (1984, 1985)*

		1984 Efficiency Scores		1985 Efficiency Scores	
No.	Bank title	CCR	Cone-R.	CCR	Cone-R.
1	BANKERS TRUST CO	1.0000	1.0000	1.0000	1.0000
2	CITIBANK NA	1.0000	1.0000	1.0000	0.9575
3	MORG. GUARANTY TR CO NY	0.9757	0.9985	1.0000	1.0000
4	WACHOVIA BK and TR CO NA	1.0000	1.0000	1.0000	0.9505
5	INTERFIRST BK AUSTIN NA	1.0000	0.9987	1.0000	0.7949
6	TX COMM BK AUSTIN NA	1.0000	0.9865	1.0000	0.9915
7	FIRST CITY BK OF DALLAS	1.0000	0.9997	1.0000	0.9947
8	INTERFIRST BK DALLAS NA	1.0000	0.9623	1.0000	0.9925
9	MBANK DALLAS NA	1.0000	0.9512	1.0000	0.8375
10	REPUBLIC BK DALLAS NA	0.9687	0.9289	1.0000	0.8260
11	INTERFIRST BK FT WOR. NA	1.0000	0.9650	0.9464	0.9017
12	TX AMER. BK FT WOR. NA	1.0000	1.0000	1.0000	0.9213
13	ALLIED BK OF TX	0.8373	0.8556	0.8158	0.6632
14	CAPITAL BK NA	1.0000	0.9943	1.0000	0.7864
15	FIR CITY NAT BK HOUSTON	1.0000	0.9632	1.0000	0.9812
16	INTERFIR BK HOUSTON NA	1.0000	0.9880	1.0000	0.9812
17	REPUBLIC BK HOUSTON NA	0.7794	0.7208	0.8585	0.7254
18	TX COMMERCE BK NA	1.0000	0.9260	1.0000	0.9034
19	FROST NA BK SAN ANTONIO	1.0000	0.9000	0.9568	0.7411
20	NB OF COMM SAN ANTONIO	0.9044	0.9513	0.9621	0.9794
21	FIRST INTRST BK NEVADA NA	1.0000	1.0000	1.0000	1.0000

* Source: P.L. Brockett, A. Charnes, W.W. Cooper and Z.M. Huang (1997). See Note 8 at the end this chapter.

Using these results the following (ordinary) CCR model was achieved for use on the thus transformed data

$$\begin{aligned} \min \quad & \theta \\ \text{subject to} \quad & \theta \bar{\boldsymbol{x}}_o \geq \bar{X}\boldsymbol{\lambda} \\ & \bar{\boldsymbol{y}}_o \leq \bar{Y}\boldsymbol{\lambda} \\ & \boldsymbol{\lambda} \geq \mathbf{0}. \end{aligned} \tag{6.57}$$

The results from the transformed data are shown in the second of the two columns exhibited in Table 6.7. The first of the two columns portrays results from a use of the CCR model on the untransformed (original) data so the comparisons can easily be made. As can be seen, some of the excellent banks have their efficiency scores changed with Morgan Guaranty improving its efficiency score in 1985 while Citibank and Wachovia in 1985 leave the status of being fully efficient. More importantly, many of the Texas banks lose the status of full efficiency which was accorded to them in 1984 when the original data were transformed by the cone-ratio envelopment used for 1984. More importantly, all of the Texas banks lose their efficiency status in 1985. In addition not a single Texas bank improved its efficiency score — although improvement is possible with these transformed data as is seen in the case of Morgan Guaranty. Moreover, every one of the Texas banks worsened its score when going from 1984 to 1985. Taken together this suggests that a worsening banking situation began to develop in 1984, as judged by the excellent-efficient performances present in the transformed data and the situation deteriorated further in 1985.

6.6 NEGATIVE SLACK VALUES AND THEIR USES

Table 6.8, below, allows us to delve deeper into these results in order to better understand what is happening with these transformations. This table provides details underlying the 1985 evaluation of Interfirst Bank of Fort Worth — the bank exhibited on line 11 in Table 6.7 — in the following manner. Column 1 shows the actual (non-transformed) data for this bank and column 2 shows the performance that was required to achieve efficiency under the heading "Value if Efficient."

As can be seen, all of the outputs were satisfactory but the only satisfactory performance among the 4 inputs appears for "deposits" in the final line. Both "interest expenses" and "non interest expenses" should have been reduced by the amounts shown in the column headed "Potential Improvement."

Special interest attaches to the Provision for Loan Loses an input which is used to cover potential risks due to this bank's lending activities. As can be seen, the value for this item is negative under "Potential Improvement." Hence the amount of this *input* should have been *increased* by \$4.7 (million).

The work of this study was suspended when part of the study task force was invited by the Texas State Auditor's Office to make a similar study for use in an audit of the State's Insurance Department. Hence the future extensions needed to perfect this DEA model for use in banking were never undertaken. Here we only note that this could include added constraints connecting balance sheet items such as Allowances for Loan Losses to corresponding income statement items like "Provision for Loan Losses." Nevertheless, the potential for future improvement is present so we study this topic a bit further as follows.

Consider the following reformulation of the constraints of (6.57),

$$\begin{aligned} \boldsymbol{s}^{+} &= (BY)\boldsymbol{\lambda} - B\boldsymbol{y}_o \\ \boldsymbol{s}^{-} &= -(AX)\boldsymbol{\lambda} + \theta(A\boldsymbol{x}_o) \\ \boldsymbol{0} &\leq \boldsymbol{\lambda},\ \boldsymbol{s}^{+},\ \boldsymbol{s}^{-}. \end{aligned} \tag{6.58}$$

This is in the same form as (6.51)-(6.53) with the inclusion of nonnegativity for the slacks.

Table 6.8. Printout for Cone-Ratio CCR Model - Interstate Bank of Fort Worth, 1985.

DECISION MAKING UNIT: 11 1985 INTERFIRST OF FORT WORTH

EFFICIENCY: 0.9017

IN MILLION DOLLARS	ACTUAL	VALUE IF EFFICIENT	POTENTIAL IMPROVEMENT	SUM OF IMPROVEMENTS
OUTPUTS				
INTINCOME	111.61	111.61	0.00	
NONINTINC	20.38	20.38	0.00	
ALLOWANCE	16.30	16.30	0.00	
NETLOANS	970.43	970.43	0.00	0.00
INPUTS				
INTEXPENS	91.51	79.51	12.00	
NONINTEXP	26.61	11.49	15.12	
PROVISION	7.00	11.70	-4.70	
DEPOSITS	1227.41	1227.41	0.00	22.42
				$22.42MILLIONS

We now write the following version for the CCR constraints to be used on the original data as follows,

$$\begin{aligned} \tilde{\boldsymbol{s}}^+ &= Y\boldsymbol{\lambda} - \boldsymbol{y}_o \\ \tilde{\boldsymbol{s}}^- &= -X\boldsymbol{\lambda} + \theta\boldsymbol{x}_o, \end{aligned} \tag{6.59}$$

where $\mathbf{0} \le \boldsymbol{\lambda}$ but the slacks are not otherwise constrained.

To show why these nonnegativity requirements are not imposed on the slacks in (6.59), we simplify matters by assuming that the A and B matrices have inverses.

Applying these inverses, A^{-1} and B^{-1}, to (6.58) produces

$$\begin{aligned} B^{-1}\boldsymbol{s}^+ &= Y\boldsymbol{\lambda} - \boldsymbol{y}_o \\ A^{-1}\boldsymbol{s}^- &= -X\boldsymbol{\lambda} + \theta\boldsymbol{x}_o \\ \mathbf{0} &\le \boldsymbol{\lambda}, \boldsymbol{s}^+, \boldsymbol{s}^-. \end{aligned} \tag{6.60}$$

Setting $\tilde{\boldsymbol{s}}^+ = B^{-1}\boldsymbol{s}^+$, $\tilde{\boldsymbol{s}}^- = A^{-1}\boldsymbol{s}^-$ we have a solution to (6.59). Moreover, if this solution is optimal for (6.58) then it is optimal for (6.59) since otherwise we could reverse this procedure and show that it is not optimal for (6.58).

Thus we have a solution of (6.60), in terms of original data, derived from (6.58) in terms of transformed data. However, we cannot guarantee that the slacks in (6.58) will satisfy nonnegativity. If this is wanted one needs to join the constraints in (6.58) and (6.59) to form the following augmented problem,

$$\boldsymbol{s}^+ = (BY)\boldsymbol{\lambda} - B\boldsymbol{y}_o \tag{6.61}$$

$$
\begin{aligned}
s^- &= -(AX)\boldsymbol{\lambda} + \theta(A\boldsymbol{x}_o) \\
\widetilde{s}^+ &= Y\boldsymbol{\lambda} - \boldsymbol{y}_o \\
\widetilde{s}^- &= -X\boldsymbol{\lambda} + \theta\boldsymbol{x}_o \\
0 &\leq \boldsymbol{\lambda}, \boldsymbol{s}^+, \boldsymbol{s}^-, \widetilde{\boldsymbol{s}}+, \widetilde{\boldsymbol{s}}^-.
\end{aligned}
$$

In this example we have assumed the existence of the ordinary inverse A^{-1}, B^{-1}. However, this could be extended to more general inverses[13] to show how the above derivation, from transformed to original data, could be made without this assumption. Finally, it is also possible to modify the problem (6.61) so that nonnegativity is imposed on only some of the variables in $\widetilde{\boldsymbol{s}}^+, \widetilde{\boldsymbol{s}}^-$. This, in fact, was what was done in this study and the slack variables for "Provision" was the only one not subject to a nonnegativity constraint. See Problem 6.2.

6.7 A SITE EVALUATION STUDY FOR RELOCATING JAPANESE GOVERNMENT AGENCIES OUT OF TOKYO

Takamura and Tone[14] have developed a consensus-making method for a national project in Japan. This section outlines the project in a manner that puts emphasis on the methodological aspects.

6.7.1 Background

Tokyo, with its population of more than 10 million persons is one of the largest megalopolises in the world. It serves as the center of economics and politics in nearly monopoly fashion in Japan. However, the efficiency of its functions as a city, e.g., culture, residence, transportation, health care, and anti-natural disaster measures, have dropped dramatically during the last two decades. In an effort to solve such urban problems, the Diet of Japan has decided to devolve some governmental functions from Tokyo, and a bill to transfer the Diet, Governmental Agencies and the Supreme Court to a new capital was approved by the Diet in 1992. The aim of this separation is to create a Washington D.C. in Japan, and to allow Tokyo to maintain itself in a manner similar to New York. This is a huge long-range national project with an initial budget more than 12 trillion yen (about 10 billion U.S. dollars).

A council for supervising this transfer to a new capital has been organized which consists of 19 "wise men" appointed by the Prime Minister. By the end of 1998, the Council selected 10 candidate sites for the new capital. They are Miyagi (A), Fukushima (B), Tochigi (C), Tochigi/Fukushima (D), Ibaraki (E), Shizuoka/Aichi (F), Gifu/Aichi (G), Mie (H), Mie/Kio (I) and Kio (J). Under the Council, a committee for evaluating these sites was organized in 1998 consisting of 6 experts in decision making, economics, law, civil engineering, environmental issues and assessment. One of the objects of this Committee is to survey potential methods for consensus formation among the Council members. The following method was proposed by Tone, as a Committee member, for this purpose.

6.7.2 The Main Criteria and their Hierarchy Structure

The Council selected the following three categories of criteria as crucial for evaluating the sites for the location of the new capital:

Influence on the future of the country:
Subjects related to the long-range strategic aspects of the relocation as they affect the entire country.

Conditions for establishing the capital city functions:
Terms related to the capital city functions.

Conditions for suitability of the new location:
Terms related to the suitability and feasibility of the relocation.

These three factors are further divided into eighteen concrete and detailed criteria (C1-C18) via two intermediate criteria (B1 and B2) as follows:

1. Influence on the future of the country:
   - (B1) Direction for reorganizing the national structure.
     This criterion is divided into:
     - (C1) Direction of reform of national land structure.
     - (C2) Rectification of the excessive concentration of activities in Tokyo.
   - (C3) Direction of culture formation.
   - (C4) Ease of correspondence with a new information network.
   - (C5) Speedy response in a large-scale disaster.
2. Conditions for establishing the capital city functions:
   - (C6) Ease of access to foreign countries.
   - (C7) Ease of access to Tokyo.
   - (C8) Ease of access to the entire country.
   - (C9) Appeal of the landscape.
   - (C10) Safety in the event of an earthquake disaster.
   - (C11) Safety in the event of a volcanic eruption.
3. Conditions for suitability of the new location:
   - (C12) Feasibility of the smooth acquisition of land.
   - (C13) Suitability of topographical features.
   - (C14) Safety against flood and sediment disasters.
   - (C15) Stability of water supply.
   - (C16) Suitability in terms of relations with existing cities.

- (B2) Environmental issues
  This criterion is divided into:
  - (C17) Harmony with the natural environment.
  - (C18) Possibility of lessening the environmental load.

Figure 6.4 summarizes the hierarchical structure expressed above.

6.7.3 Scores of the 10 Sites with respect to the 18 Criteria

Table 6.9 exhibits the score matrix $S = (S_{ij})$ $(i = 1, \ldots, 18 : j = 1, \ldots, 10)$ of the 10 candidate sites (A-J) with respect to the 18 criteria (C1-C18). These values were obtained from expert teams consisting of five specialists, on average. For example, (C4) ("Ease of correspondence with a new information system") was investigated by six members headed by Professor T. Ishii of Keio University, (C6), (C7) and (C8) ("Access to other areas") are evaluated by five members headed by Professor S. Morichi of Tokyo Institute of Technology. (C10) and (C11) ("Earthquake and volcano") are reviewed by ten members headed by Professor M. Fuchigami of University of Tokyo. The scores were measured by an absolute value ranging from 5 (the best) to 1 (the worst).

Table 6.9. Scores (S_{ij}) of 10 Sites (A-J) with respect to 18 Criteria (C1-C18)

	A	B	C	D	E	F	G	H	I	J	Avg	SD
C1	2.8	2.6	3.0	2.8	2.4	2.8	2.9	2.4	2.4	2.4	2.7	0.23
C2	3.0	3.0	3.0	3.0	2.5	3.5	3.5	4.0	4.0	4.0	3.4	0.53
C3	2.9	2.9	3.4	3.3	2.4	2.8	3.0	2.4	2.3	2.4	2.8	0.38
C4	3.0	2.3	2.3	2.3	3.0	4.3	4.3	2.3	2.3	2.3	2.8	0.82
C5	4.0	4.0	4.0	4.0	4.0	2.0	3.5	3.0	3.0	3.0	3.5	0.69
C6	2.2	2.1	2.5	2.5	3.7	4.2	4.4	5.0	3.8	3.8	3.4	1.02
C7	3.3	3.5	5.0	5.0	4.2	3.5	2.3	2.1	2.0	2.0	3.3	1.18
C8	3.4	3.3	3.9	3.9	3.6	5.0	4.7	4.3	4.3	4.3	4.1	0.55
C9	2.5	2.5	5.0	5.0	2.0	4.0	3.0	3.5	3.0	3.0	3.4	1.03
C10	4.0	5.0	4.0	4.0	5.0	1.0	2.0	2.0	2.0	2.0	3.1	1.45
C11	2.6	3.2	1.3	2.6	4.4	4.2	4.7	5.0	5.0	5.0	3.8	1.30
C12	3.6	2.8	3.6	3.2	3.4	2.0	4.1	1.6	2.6	3.3	3.0	0.76
C13	2.3	3.0	4.7	4.0	4.1	3.2	2.3	4.0	3.3	3.1	3.4	0.78
C14	3.7	4.2	3.7	4.1	3.0	2.9	4.5	3.0	3.6	3.6	3.6	0.55
C15	3.5	2.5	2.5	2.5	1.5	3.0	3.0	3.5	2.2	1.5	2.6	0.71
C16	4.2	3.3	3.0	3.2	2.4	4.1	3.5	3.5	3.5	3.5	3.4	0.52
C17	2.9	3.3	3.6	4.1	2.7	2.5	1.9	2.4	2.4	2.1	2.8	0.68
C18	2.6	3.8	3.0	3.6	3.4	4.2	3.0	2.9	2.6	2.3	3.1	0.59
Average	3.1	3.2	3.4	3.5	3.2	3.3	3.4	3.2	3.0	3.0	3.2	0.17

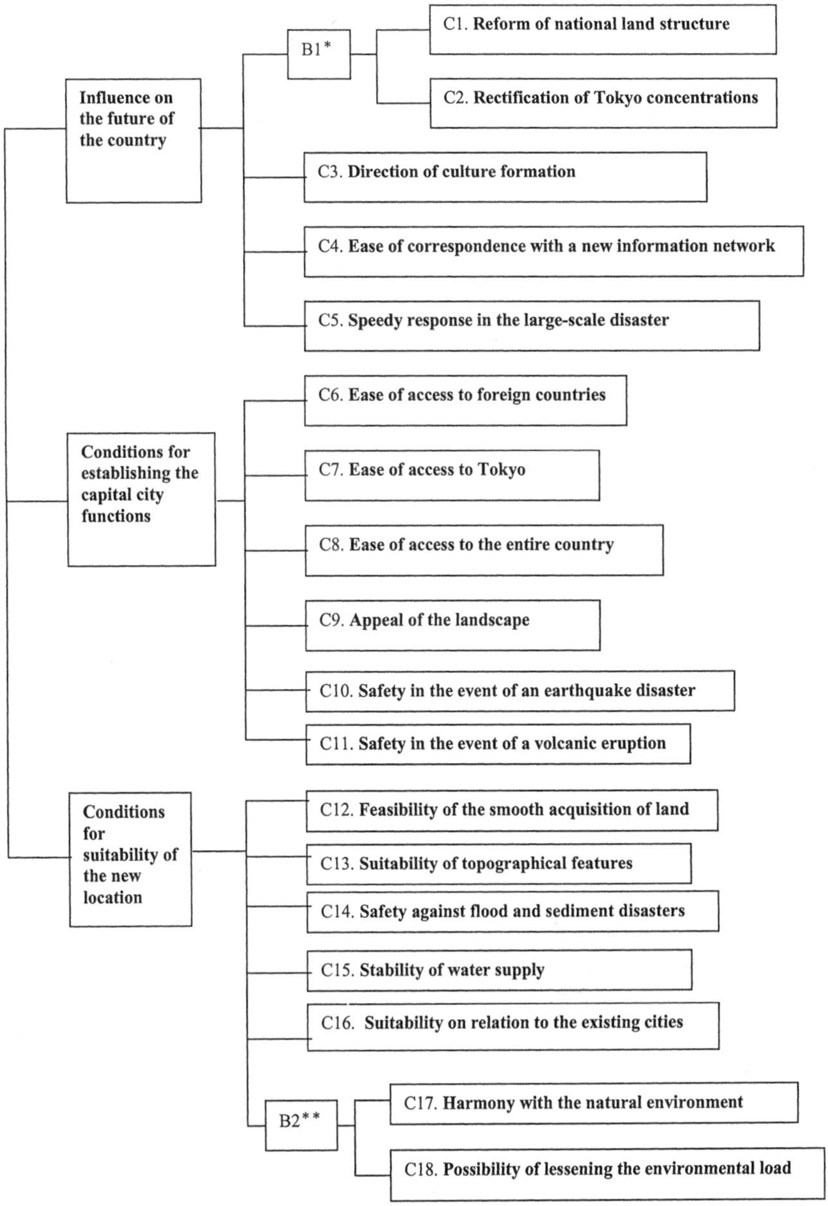

* B1=Direction for reorganizing the national structure
** B2=Environmental issues

Figure 6.4. The Hierarchical Structure for the Capital Relocation Problem

6.7.4 Weights of the 18 Criteria by the 18 Council Members (Evaluators)

We employed a multi-stage use of an AHP[15]-like method for 18 of the 19 Council members (excluding the Chairperson). This went as follows. At the first stage, the council member assigned their weights on criteria either by subjective judgments (not by AHP) or by AHP. In the latter case, we allowed incomplete paired comparisons so that the council member could skip the comparison when he or she possessed little or no confidence in comparing the criteria. At the end of the first stage, we gathered 18 sets of weights on 18 criteria. We then went on to the second stage, where we first showed the distribution of weight scores to the Council. Each member thus knew where he or she was located in the distribution. Taking account of this distribution, each member had the chance to alter his/her decision on criteria weights. Note that this is a form of the Delphi method. [16] We continued this process until convergence was obtained. Actually, this process converged at the third stage. The resulting statistics for this evaluations are exhibited in Table 6.10. Each member had 100 points allotted for the evaluation to be divided and assigned to the 18 criteria according to his/her individual judgments. Observe that criteria C5 (Speedy response to

Table 6.10. Statistics of Weights assigned the 18 Criteria (C1-C18) by 18 Council Members

	Average	Median	Max	Min
C1	5.64822	5.45588	10.00000	1.24549
C2	4.72971	4.72222	10.00000	1.00000
C3	5.14061	5.27778	8.00000	1.55039
C4	6.79305	5.97666	16.42193	2.00000
C5	9.33093	8.49438	16.47018	5.00000
C6	5.6461	5.68774	9.00000	1.70505
C7	6.11132	5.68774	13.0372	2.81531
C8	6.13195	5.87875	10.00000	2.52025
C9	4.41615	3.89019	9.66938	1.16727
C10	8.46151	7.54204	19.92784	3.00000
C11	5.85123	5.44406	10.31643	2.11388
C12	6.97198	7.44607	15.50388	1.37530
C13	4.57878	4.34375	7.75194	1.94630
C14	4.01225	4.59094	5.75751	1.55039
C15	4.45201	4.57263	8.00000	1.00000
C16	4.27136	4.16636	7.17284	1.37530
C17	4.17737	3.62176	7.97114	2.00000
C18	3.27548	2.89578	6.42064	1.09375

a large-scale disaster), C10 (Safety in the event of an earthquake disaster), C12 (Feasibility of the smooth acquisition of land) and C4 (Ease of correspondence with a new information network) have high scores, on average. However, a large variation in evaluations exists, as can be seen from the "Max" and "Min" columns. This reflects the existence of significant disagreement among the Council members. We note that the importance of the criterion B1 "Direction for reorganizing the national structure" is 10.378532 (the sum of C1 and C2) on average and that of B2 "Environmental issues" is 7.45285 (the sum of C17 and C18). Also the three top-criteria have the following weights: (Influence on the future of the country) = 31.64252, (Conditions for establishing the capital city functions) = 36.61826 and (Conditions for suitability of the new location) = 31.73923. Eventually, the top three criteria were found to have approximately equal weights.

6.7.5 Decision Analyses using Averages and Medians

By using the averages and medians of the weights in Table 6.10, we obtained the corresponding scores for each site, as exhibited in Table 6.11. This table suggests Sites C (Tochigi), D (Tochigi/Fukushima) and G (Gifu/Aichi) as promising candidates.

Table 6.11. Averages and Medians of Scores of the 10 Sites

	A	B	C	D	E	F	G	H	I	J
Average	321	325	344	351	333	316	340	310	299	298
Median	308	310	328	334	317	302	328	297	287	287

6.7.6 Decision Analyses using the Assurance Region Model

However, using the average suggests that only one "virtual" evaluator was "representative" of all members judgments. Thus, the variety of opinions across evaluators is not taken into account. Given the degree of scatter exhibited in Table 6.10, the use of such an "average" or "median" of weights must be employed cautiously from a consensus-making point of view.

Another way to look at the above approach is that the weights are common to all sites. We may call this a "fixed weight" approach, as contrasted with the following "variable weight" structure.

Given the score matrix $S = (S_{ij})$, we evaluate the total score of site $j = j_o$ using a weighted sum of S_{ij_o} as

$$\theta_{j_o} = \sum_{i=1}^{18} u_i S_{ij_o}, \tag{6.62}$$

with a nonnegative weight set (u_i). We assume that the weights can vary from site to site in accordance with the principle we choose for characterizing the sites.

Furthermore, the weights should reflect all evaluators preferences regarding the criteria. This can be represented by a version of the assurance region (AR) model proposed by Thompson *et al.* (1986). For every pair (i, j) of criteria, the ratio u_i/u_j must be bounded by L_{ij} and U_{ij} as follows,

$$L_{ij} \leq u_i/u_j \leq U_{ij}, \tag{6.63}$$

where the bounds are calculated by using evaluator k's weights (W_{ki}) on criterion i as

$$L_{ij} = \min_{k=1,\ldots,18} \frac{W_{ki}}{W_{kj}}, \quad U_{ij} = \max_{k=1,\ldots,18} \frac{W_{ki}}{W_{kj}}. \tag{6.64}$$

We now turn to the evaluation of candidate sites by means of the assurance region (AR) model of DEA. For this purpose, we employ two extreme cases as presented below.

6.7.7 *Evaluation of "Positive" of Each Site*

In order to evaluate the positives of site j_o, we choose the weights (u_i) in (6.62) so that they maximize θ_{j_o} under the condition that the same weights are applied in evaluating all other sites, and that the objective site should be compared relative to them. This principle is in accordance with that of DEA and can be formulated as follows:

$$\max \quad \theta_{j_o} = \sum_{i=1}^{18} u_i S_{ij_o} \tag{6.65}$$

$$\text{subject to} \quad \sum_i u_i S_{ij} \leq 1 \quad (\forall j) \tag{6.66}$$

$$L_{ij} \leq u_i/u_j \leq U_{ij}(\forall (i,j))$$

$$u_i \geq 0, \quad (\forall i)$$

where L_{ij} and U_{ij} are given by (6.64). The optimal score $\theta^*_{j_o}$ indicates the relative distance from the efficient frontier. The lower the score , the weaker the "positive" of the site.

It should be noted that DEA is here directed toward "effectiveness" rather than "efficiency" since we do not deal with resource utilization, as required for evaluating efficiency. See Section 3.9 in Chapter 3. Our concern is with achieving the already stated (or prescribed) goals. The initial goals, stated broadly, are made sufficiently precise with accompanying criteria for evaluation so that (a) proposed actions can be evaluated more accurately, and (b) once the proposals are implemented, any accomplishments (or lack thereof) can be subsequently identified and evaluated.

6.7.8 *Evaluation of "Negative" of Each Site*

In the above evaluations, each site was compared with the best performers. We will call this evaluation scheme as "positives," in that we observe the candidate from the positive side. Turning to the opposite side, we would like to evaluate the candidate sites from the worst side. For this purpose, we try seek the "worst" weights in the sense that the objective function in (6.65) is minimized. Thus, this principle can be formulated as follow:

$$\min \quad \phi_{j_o} = \sum_i u_i S_{ij_o} \tag{6.67}$$

$$\text{subject to} \quad \sum_i u_i S_{ij} \geq 1 \quad (\forall j) \tag{6.68}$$

$$L_{ij} \leq u_i/u_j \leq U_{ij} \quad (\forall (i,j)) \tag{6.69}$$

$$u_i \geq 0. \quad (\forall i) \tag{6.70}$$

By dint of the reversed inequality in (6.68), the optimal ϕ_{j_o} satisfies $\phi^*_{j_o} \geq 1$. If $\phi^*_{j_o} = 1$, then the site belongs to the worst performers group; otherwise, if $\phi^*_{j_o} > 1$, it rates higher than the worst performers group. Each site is compared with these worst performers and is gauged by its efficiency "negatives" as the ratio of distances from the "worst" frontiers in the same way as in ordinary DEA. (Yamada *et al.* [17] named this worst side approach "Inverted DEA.")

In order to make straightforward comparisons of the "negatives" and "positives" case scores, we invert $\phi^*_{j_o}$ as follows

$$\tau^*_{j_o} = 1/\phi^*_{j_o} \tag{6.71}$$

and call it the "negatives" score.

6.7.9 *Uses of "Positive" and "Negative" Scores*

Since the number of candidate sites (10) is smaller than the number of criteria (18), we have a shortfall in the number of degrees of freedom for discriminating efficiency among the 10 candidates even if we employ the AR models.

First, we estimated the lower/upper bounds, L_{ij} and U_{ij}, respectively, on the ratio of criteria i and j by (6.64). If sufficient discrimination among the candidates both in "positives" and "negatives" cannot be observed, we apply the following deletion processes. We delete the ks giving the *min* and *max* ratios in (6.64), and estimate the *min* and *max* again using (6.64). Thus, we remove two extreme ratios at each trial. Our experiments showed that after six deletions, using the remaining six (=18-12) ratios, we reached a sufficient and stationary evaluation of sites, as depicted in Figure 6.5. From this figure, we observe that two Sites, D (positive=1, negative=0.894) and G (positive=1, negative=0.898) are excellent in both "positives" and "negatives." The Site C is somewhat behind D and G.

6.7.10 *Decision by the Council*

From the above analyses, the Council acknowledged Sites C, D and G as the most promising candidates for their selection problem. However, since Site C (Tochigi district) is a part of Site D (Tochigi/Fukushima district), the Council

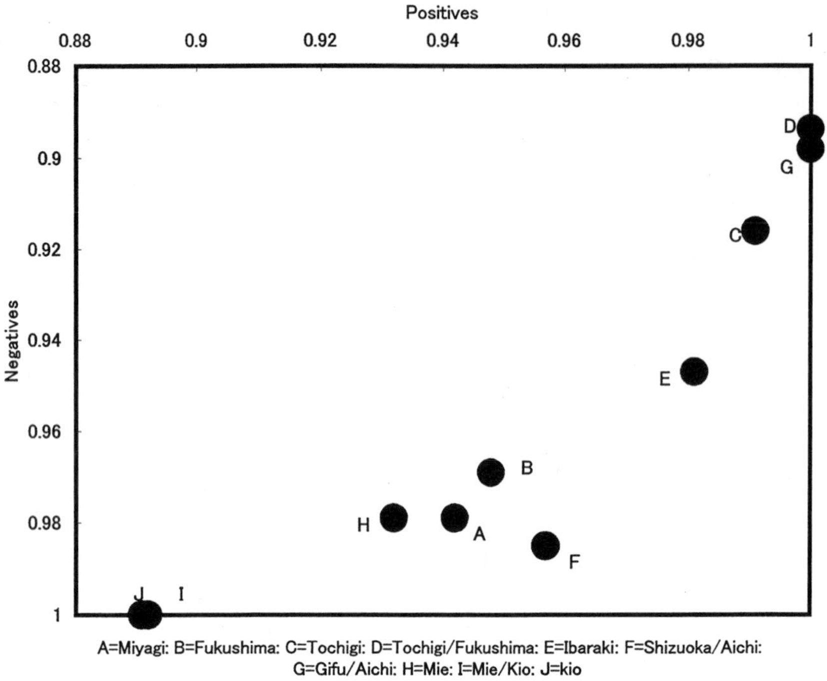

Figure 6.5. Positives and Negatives of the 10 Sites

decided to recommend Sites D and G as final. They also recommended Site I (Mie/Kio district) as the runner-up provided that this district would be powered by a high-speed transportation network. The report was delivered to the Prime Minister who transmitted it to the Diet where it still being considered with each of the three sites enthusiastically advertising their superiority.

6.7.11 Concluding Remarks

We have presented a report on the site selection process for relocating Government Agencies out of Tokyo, putting emphasis on the methodological aspects.

It is hoped that our methodologies will be utilized in a successful execution of this centennial project with satisfactory results.

We summarize key characteristics of the proposed methods below.

These methods assume the following two conditions:

1. Each site has been evaluated numerically with respect to the set of chosen criteria. These evaluations may be made objectively by using measurable yardsticks, or they may be done subjectively using expert knowledge.

2. Each evaluator can determine his/her own judgment on the relative importance of the criteria. For this purpose, AHP or direct subjective judgments may be utilized.

In cases where these conditions are satisfied, the proposed methods rank the candidate sites to identify a consensus within the evaluator group. Results obtained by the assurance region model have, in particular, several merits for both candidates and evaluators, as follows:

- **For Candidate Sites**: The results are acceptable to the candidate sites in the sense that the most preferable weights for the site are assigned within the allowable bounds of the evaluators. The optimal weights vary from site to site in that the best set of relative weights is assigned to the site. In a similar way, the relative weaknesses of each site can also be evaluated. These two measures are then utilized to characterize the candidate sites.

- **For Evaluators**: Each evaluator can be assured that his/her judgments on the criteria are taken into account and that the ratios of every pair of weights fall within his/her allowable range. Although in this case several evaluators ratios were excluded for discrimination purposes, this approach is more reasonable and acceptable than using the average (or median) weights of all evaluators, especially when there is a relatively high degree of scatter to consider.

The proposed approaches can also be applied to many other societal problems that involve group decision making.

6.8 SUMMARY OF CHAPTER 6

In this chapter, we introduced the assurance region and cone-ratio methods for combining subjective and expert evaluations with the more objective methods of DEA.

1. Usually expressed in the form of lower and upper bounds, the assurance region method puts constraints on the ratio of input (output) weights or multiplier values. This helps to get rid of zero weights which frequently appear in solutions to DEA models. The thus evaluated efficiency score generally drops from its initial (unconstrained) value. Careful choice of the lower and upper bounds is recommended.

2. Not covered in this chapter is the topic of "linked constraints" in which conditions on input and output multipliers are linked. See Problem 6.3.

3. The cone-ratio method confines the feasible region of virtual multipliers $\boldsymbol{v}$, $\boldsymbol{u}$, to a convex cone generated by admissible directions. Formulated as a "cone ratio envelopment" this method can be regarded as a generalization of the assurance region approach.

4. Example applications were used to illustrate uses of both of the "assurance region" and "cone ratio envelopment" approaches.

6.9 NOTES AND SELECTED BIBLIOGRAPHY

The assurance region (AR) method was developed by Thompson, Singleton, Thrall and Smith (1986).[18] They used DEA to analyze six Texas sites for location of a high energy physics lab (called the SSC = Supercolliding Super Conductor) which was directed to advancing fundamental knowledge in physics. Five of the six sites were DEA efficient. This was not satisfactory so they then used survey data and expert opinion to specify bounds for the virtual multipliers. The AR method identified only one efficient DMU for the location of SSC and this candidate site was selected by Texas and won in a national competition conducted by the U.S. Department of Energy in 1988 as the location for the SSC. Further development of the AR method can be found in Thompson, Langemeir, Lee, Lee and Thrall (1990)[19] and Roll and Golany (1993)[20]. See also Dyson and Thanassoulis (1988).[21] The use of DEA for site selection has been fully developed in Athanassopoulos and Storbeck (1995)[22] and Desai, Haynes and Storbeck (1994).[23]

The cone-ratio (CR) method was developed by Sun (1988)[24], Charnes, Cooper, Wei and Huang (1989)[25] and Charnes, Cooper, Huang and Sun (1990), [26] which they applied to large U.S. commercial banks. See also Brockett, Charnes, Cooper, Huang and Sun (1997)[27] which includes an Appendix on how to derive original (untransformed) data and solutions.

6.10 RELATED DEA-SOLVER MODELS FOR CHAPTER 6

AR-I-C (Input-oriented Assurance Region model under the Constant returns-to-scale assumption).

This code solves the AR problem expressed by (6.11)-(6.14). The main body of the data set, $(X,\ Y)$, for this model is put on an Excel worksheet in the same way as the CCR model. The assurance region (AR) is augmented by adding AR constraints such as (6.1) under the main data set separated by one blank row. See Figure B.2.1 in Section B.5.2 for a sample format and refer to the explanations above the figure. The main results will be obtained in the following worksheets.

- The efficiency score θ^* and the reference set (λ^*) are recorded in "Score."
- The optimal (dual) multipliers $\boldsymbol{v}^*, \boldsymbol{u}^*$ in (6.6) and the weighted data $\{x_{ij}v_i^*\}$, $\{y_{rj}u_r^*\}$ are displayed in "Weight" and "WeightedData." They satisfy the assurance region constraints.
- The projection onto efficient frontiers by (6.17) and (6.18) is stored in "Projection."
- Slacks $\boldsymbol{s}^{-*}$ and $\boldsymbol{s}^{+*}$ defined by (6.15) and (6.16) are included in "Slack."

AR-I-V (Input-oriented assurance region model under the variable returns-to-scale assumption).

This code solves the AR model under the added constraint $\sum_{j=1}^{n} \lambda_j = 1$. In addition to results similar to those in AR-I-C, this code identifies the returns-to-scale characteristics of each DMU under the AR constraints. This identification is based on Tone (2001).[28]

- The returns to scale of each DMU is recorded in the worksheet "RTS." This RTS is characterized under the AR environments. Thus it may differ from the one obtained by the BCC-I model, which has no constraints on the multipliers $\boldsymbol{v}$, $\boldsymbol{u}$, and hence the structure of the supporting hyperplanes changes from that of the BCC model. The RTS of inefficient DMUs is identified as that of the projected DMU on the efficient frontier in the input-oriented manner.

AR-I-GRS (Input-oriented assurance region model under the general returns-to-scale assumption).

This code solves the AR model under the added constraint $L \leq \sum_{j=1}^{n} \lambda_j \leq U$. The lower bound $L(\leq 1)$ and upper bound $U(\geq 1)$ must be supplied through keyboard. The defaults are $L = 0.8$ and $U = 1.2$.

AR-O-C (Output-oriented assurance region model under the constant returns-to-scale assumption).

This code is the output-oriented version of AR-I-C. We exhibit the optimal efficiency score θ^* as the inverse of the optimal output expansion rate so that it is less than or equal to one. This will make the comparisons with the input-oriented case straightforward.

AR-O-V(GRS) (Output-oriented assurance region model under the variable (general) returns-to-scale assumption).

The code AR-O-V also identifies the RTS of each DMU in the output-oriented manner.

ARG-I(O)-C(V or GRS) (Input (Output)-oriented Assurance Region Global models under the constant (variable or general) returns-to-scale assumption.)

This code solves the ARG problem expressed by (6.29)-(6.32). The main body of the data set, (X, Y), for this model is put on an Excel worksheet in the same way as the CCR model. The assurance region global (ARG) is augmented by adding ARG constraints under the main data set separated by one blank row. See Figure B.2.1 in Section B.5.2 for a sample format and refer to the explanations above the figure. The main results will be obtained in the same scheme as the AR models.

6.11 PROBLEM SUPPLEMENT FOR CHAPTER 6

Problem 6.1

In addition to "cone-ratio envelopment" can you identify other approaches to data transformations that are used to effect evaluations?

Suggested Response : One such possibility is the use of data from benchmark firms to obtain ratios that are then used to evaluate performances of other firms relative to these benchmarks. Other examples are (1) the use of index numbers of prices as divisors to obtain measures of "real" vs. "nominal" (dollar) performances of productivity and (2) the use of discount rates to replace future income streams with a single "present value."

Problem 6.2

Part(1) : Show that (6.59) can lead to trouble in the form of infinite solutions to (6.57) when the slacks are not constrained to be nonnegative.

Part(2) : Analyze the dual to this problem and use the analysis to show how such troubles can be avoided.

Part(3) : Show that such troubles do not occur when the transformed data are used as in (6.58).

Suggested Response :
Part(1): Consider the following CCR formulation in terms of the original data,

$$\begin{aligned}
\min \quad & \theta \\
\text{subject to} \quad & \theta x_{io} = \sum_{j=1}^{n} x_{ij}\lambda_j + \widetilde{s}_i^{\,-}, \quad i = 1, \ldots, m \\
& y_{ro} = \sum_{j=1}^{n} y_{rj}\lambda_j - \widetilde{s}_i^{\,+}, \quad r = 1, \ldots, s
\end{aligned}$$

where the $\lambda_j \geq 0 \; \forall j$ but the slacks not otherwise constrained. Setting all $\lambda_j = 0$ we can choose $-\widetilde{s}_r^{\,+} = y_{ro} \; \forall r$ and then let $\widetilde{s}_i^{\,-} = \theta x_{io}$ go to negative infinity in both $\widetilde{s}_i^{\,-}$ and θ.

Part(2): Consider the dual to this last problem in the form

$$\begin{aligned}
\max \quad & \sum_{r=1}^{s} u_r y_{ro} \\
\text{subject to} \quad & \sum_{r=1}^{s} u_r y_{rj} - \sum_{i=1}^{m} v_i x_{ij} \leq 0, \; j = 1, \ldots, n \\
& \sum_{i=1}^{m} v_i x_{io} = 1 \\
& u_r, v_i = 0 \; \forall i, r.
\end{aligned}$$

This has no solution because the condition $\sum_{i=1}^{m} v_i x_{io} = 1$ is not compatible with $v_i = 0, \; \forall i$. However, changing this to $\widetilde{s}_i^{-} \geq 0$ for some of the primal (= envelopment model) slacks removes this difficulty. Both problems will then have solutions so that, by the dual theorem of linear programming, they will have finite and equal optima. □

(Note, however, that it is only the $v_i^*, u_r^* > 0$ that can serve as multipliers. Hence only the slacks associated with these variables can directly contribute to the optimum objective. The other variables can contribute only indirectly.)

Part(3): Write the dual to (6.57) — and (6.58) — as

$$\begin{aligned}
\max \quad & \sum_{r=1}^{s} u_r \bar{y}_{ro} \\
\text{subject to} \quad & \sum_{r=1}^{s} u_r \bar{y}_{rj} - \sum_{i=1}^{m} v_i \bar{x}_{ij} \leq 0, \; j = 1, \ldots, n \\
& \sum_{i=1}^{m} v_i \bar{x}_{io} = 1 \\
& u_r, \; v_i \geq 0 \; \forall i, r.
\end{aligned}$$

Choosing $v_k = 1/\bar{x}_{ko}$ and all other variables zero satisfies all constraints. Hence we have exhibited a solution. Moreover, the constraints require

$$\max \sum_{r=1}^{s} u_r \bar{y}_{ro} \leq \sum_{i=1}^{m} v_i \bar{x}_{io} = 1$$

so the solutions are bounded. The dual theorem of linear programming asserts that if one member of a dual pair has a finite optimum then the other member will have a finite and equal optimum. Thus infeasible solution possibilities in the envelopment model and the non-existence of solutions in the dual (multiplier) model are both avoided. □

Problem 6.3

Background : The assurance region constraints in this chapter were restricted to separate conditions on the inputs and the outputs, respectively. The matrix D in (6.55) is then interpretable in terms of two convex cones, one in the input and the other in the output space. Extending this matrix to

$$\begin{pmatrix} A & O \\ O & B \\ F_1 & F_2 \end{pmatrix}$$

eliminates these characterizations by providing a linkage between the constraints associated with F_1 and F_2. A use of such "linkages" in the form of assurance region constraints may be needed to deal with conditions on profits, or other relations between inputs and outputs — as noted in R.G. Thompson, P.S. Dharmapala and R.M. Thrall, "Linked-cone DEA Profit Ratios and Technical Inefficiencies with Applications to Illinois Coal Mines," *International Journal of Production Economics,* 39, 1995, pp.99-115.

As these authors note in dealing with such "linkage constraints," especially in an assurance region approach to profit conditions, constraints like the following can have untoward effects in limiting profit evaluations,

$$\sum_{r=1}^{s} u_r y_{rj} - \sum_{i=1}^{m} v_i x_{ij} \le 0, \; j = 1, \ldots, n.$$

Assignment : Discuss the effects of such constraints on profits and suggest ways to handle the resulting problems.

Suggested Response : Such constraints are limitational because one wants results like

$$\sum_{r=1}^{s} u_r y_{rj} - \sum_{i=1}^{m} v_i x_{ij} > 0,$$

where the u_r and v_i are to be interpreted as prices and unit costs. Thompson, Dharmapala and Thrall suggest eliminating all such constraints, (and the data they contain,) in favor of formulations like the following.

$$\begin{aligned} \max \quad & \boldsymbol{u}\boldsymbol{y}_o \\ \text{subject to} \quad & \boldsymbol{v}\boldsymbol{x}_o = 1 \\ & \begin{pmatrix} A & O \\ O & B \\ F_1 & F_2 \end{pmatrix} \begin{pmatrix} \boldsymbol{u} \\ \boldsymbol{v} \end{pmatrix} \le \mathbf{0}, \end{aligned}$$

in which the condition $\boldsymbol{v}\boldsymbol{x}_o = 1$ is retained but the constraints employing the other data are discarded.

The idea is to use the above formulation as a means of determining whether profits are possible in the following manner.

Step 1 Determine $\max \boldsymbol{u}\boldsymbol{y}_o = MPR$.

Step 2 Determine $\min \boldsymbol{u}\boldsymbol{y}_o = mPR$.

Then use the following conditions.

- $MPR > 1$ indicates a positive profit potential.
- $MPR < 1$ only losses are possible.
- $mPR > 1$ positive profits are assured.
- $mPR < 1$ indicates a loss potential.

The following single input - single output example is provided by Thompson, Dharmapala and Thrall. Let $y_o = 1$ and $x_o = 1$ and require $3v \le u, u \le 5v$. The above formulation then becomes

$$\begin{aligned} \max \quad & u \\ \text{subject to} \quad & v = 1 \\ & 3v - u \le 0 \\ & -5v + u \le 0 \\ & u, v \ge 0. \end{aligned}$$

Step 1 gives $MPR = 5 > 1$. Reorienting the objective, as required in Step 2, gives $mPR = 3 > 1$ so this DMU$_o$ is assured of a profit.

Comment : Numerous DEA studies have shown that profitable firms, and even the most profitable firms, are not necessarily Pareto-Koopmans efficient, i.e., in the sense of Definition 3.3 it is possible to improve some inputs and outputs without worsening other inputs and outputs. Hence this approach to the use of "linked" assurance region constraints can help to identify where attention can be advantageously directed.

Problem 6.4

Prove that if the linear program (DAR_o) associated with the assurance region method in Section 6.2 is feasible with a positive optimum ($\theta^* > 0$), then the reference set defined by $E_o = \{j|\lambda_j^* > 0\}$ ($j \in \{1, \ldots, n\}$) is not empty.

Suggested Answer : An optimal solution $(\theta^*, \boldsymbol{\lambda}^*, \boldsymbol{\pi}^*, \boldsymbol{\tau}^*, \boldsymbol{s}^{-*}, \boldsymbol{s}^{+*})$ of (DAR_o) satisfies:

$$\theta^* \boldsymbol{x}_o = X\boldsymbol{\lambda}^* - P\boldsymbol{\pi}^* + \boldsymbol{s}^{-*}.$$

Let an optimal solution of (AR_o) be $(\boldsymbol{v}^*, \boldsymbol{u}^*)$. By multiplying the above equation by $\boldsymbol{v}^*$, we have:

$$\theta^* \boldsymbol{v}^* \boldsymbol{x}_o = \boldsymbol{v}^* X\boldsymbol{\lambda}^* - \boldsymbol{v}^* P\boldsymbol{\pi}^* + \boldsymbol{v}^* \boldsymbol{s}^{-*}.$$

From a constraint of (AR_o), it holds $\boldsymbol{v}^* \boldsymbol{x}_o = 1$ and the complementarity relations assert $\boldsymbol{v}^* P\boldsymbol{\pi}^* = 0$ and $\boldsymbol{v}^* \boldsymbol{s}^{-*} = 0$. Hence we have:

$$\theta^* = \boldsymbol{v}^* X\boldsymbol{\lambda}^* > 0.$$

Therefore $\boldsymbol{\lambda}^*$ must be semipositive, i.e., $\boldsymbol{\lambda}^* \geq \mathbf{0}$ and $\boldsymbol{\lambda}^* \neq \mathbf{0}$. Hence E_o is not empty. □

Notes

1. R.G. Thompson, F.D. Singleton, Jr., R.M. Thrall and B.A. Smith (1986), "Comparative Site Evaluations for Locating a High-Energy Physics Lab in Texas," *Interfaces* 16, pp. 35-49. See also R.G. Dyson and E.Thanassoulis (1988), "Reducing Weight Flexibility in Data Envelopment Analysis," *Journal of the Operational Research Society,* 39, pp.563-576. Finally, see Notes and Selected Bibliography in Section 6.9 of this chapter for references on uses of DEA for site selection.

2. A. Charnes, W.W. Cooper, Z.M. Huang and D.B. Sun (1990), "Polyhedral Cone-Ratio DEA Models with an Illustrative Application to Large Commercial Banks," *Journal of Econometrics* 46, pp.73-91. For a treatment that studies this approach as an alternative to the more rigid approach to risk evaluation under the "Basel Agreement" for controlling risks in bank portfolios see P.L. Brockett, A. Charnes, W.W. Cooper, Z.M. Huang and D.B. Sun, "Data Transformations in DEA Cone-Ratio Approaches for Monitoring Bank Performance," *European Journal of Operational Research,* 98, 1997, pp.250-268.

3. Y. Roll, W.D. Cook and B. Golany (1991), "Controlling Factor Weights in Data Envelopment Analysis," *IIE Transactions* 23, pp.2-9.

4. This usage is due to Vilfredo Pareto. See the references in Chapter 3, Section 3.10: "Notes and Selected Bibliography."

5. See Definition 3.1.

6. This problem can be solved by using "AR-I-C" (the input-oriented assurance region model under the constant returns-to-scale assumption) in DEA-Solver.

7. R. Allen, A. Athanassopoulos, R.G. Dyson and E. Thanassoulis (1997), "Weights restrictions and value judgements in data envelopment analysis," *Annals of Operations Research* 73, pp.13-34.

8. P.L. Brockett, A. Charnes, W.W. Cooper, Z. Huang and D.B. Sun (1997), "Data Transformations in DEA Cone Ratio Envelopment Approaches for Monitoring Bank Performance," *European Journal of Operational Research* 98, pp.250-268.

9. For detailed discussions see the references cited in Brockett *et al.*(1997).

10. In response to this trend, Barr, Seiford, and Siems with the Federal Reserve Bank of Dallas developed a bank failure prediction model based on DEA which outperforms all other failure prediction models in the banking literature. See R.S. Barr, L.M. Seiford and T.F. Siems (1994), "Forcasting Bank Failure: A Non-Parametric Frontier Estimation Approach," *Recherches Economiques de Louvain* 60, pp.417-429 and R.S. Barr, L.M. Seiford and T.F. Siems (1993), "An Envelopment-Analysis Approach to Measuring the Managerial Efficiency of Banks," *Annals of Operations Research*, 45, pp.1-19 for details.

11. See the definition and discussion of "Reserve (=Allowance) for Bad Debts" on page 433 in *Kohler's Dictionary for Accountants,* 6th Edition (Englewood Cliffs, N.J., Prentice-Hall Inc., 1983.)

12. R.M. Nun (1989), "Bank Failure: The Management Factor" (Austin, TX., Texas Department of Banking).

13. See A. Ben Israel and T.N. Greville, *Generalized Inverses* (New York, John Wiley & Sons Inc., 1974).

14. Y. Takamura and K. Tone (2003), "A Comparative Site Evaluation Study for Relocating Japanese Government Agencies out of Tokyo," *Socio-Economic Planning Sciences* 37, pp.85-102.

15. AHP (Analytic Hierarchy Process) was invented by T.L. Saaty (1980), *Analytic Hierarchy Process,* New York: McGraw-Hill. See also K. Tone (1989), "A Comparative Study on AHP and DEA," *International Journal of Policy and Information* 13, pp.57-63.

16. See J.A. Dewar and J.A. Friel (2001), "Delphi Method," in S.I. Gass and C.M. Harris, eds., *Encyclopedia of Operations Research and Management Science* (Norwell, Mass., Kluwer Academic Publishers) pp.208-209.

17. Y. Yamada, T. Matsui and M. Sugiyama (1994), "An Inefficiency Measurement Method for Management Systems," *Journal of the Operations Research Society of Japan* 37, 2, pp.158-167.

18. See the Note 1 reference.

19. R.G. Thompson, L.N Langemeir, C. Lee, E. Lee and R.M. Thrall (1990), "The Role of Multiplier Bounds in Efficiency Analysis with Application to Kansas Farming," *Journal of Econometrics* 46, pp.93-108.

20. Y. Roll and B. Golany (1993), "Alternate Methods of Treating Factor Weights in DEA," *OMEGA* 21, pp.99-109.

21. R.G. Dyson and E. Thanassoulis (1988), "Reducing Weight Flexibility in Data Envelopment Analysis," *Journal of the Operational Research Society* 39, pp.563-576.

22. A.D. Athanassopoulos and J.E. Storbeck (1995), "Non-Parametric Models for Spatial Efficiency," *The Journal of Productivity Analysis* 6, pp.225-245.

23. A. Desai, K. Haynes and J.E. Storbeck "A Spatial Efficiency Framework for the Support of Locational Decisions," in *Data Envelopment Analysis: Theory, Methodology, and Applications,* A. Charnes, W. W. Cooper, Arie Y. Lewin, and Lawrence M. Seiford (editors), Kluwer Academic Publishers, Boston, 1994.

24. D.B. Sun (1988), "Evaluation of Managerial Performance in Large Commercial Banks by Data Envelopment Analysis," Ph.D. dissertation, Graduate School of Business, University of Texas, Austin, TX.

25. A. Charnes, W.W. Cooper, Q.L. Wei and Z.M. Huang (1989), "Cone Ratio Data Envelopment Analysis and Multi-objective Programming," *International Journal of Systems Science* 20, pp.1099-1118.

26. See the Note 2 reference.

27. See the Note 8 reference.

28. K. Tone (2001), "On Returns to Scale under Weight Restrictions in Data Envelopment Analysis," *Journal of Productivity Analysis* 16, pp.31-47.

7 DISCRETIONARY, NON-DISCRETIONARY AND CATEGORICAL VARIABLES

7.1 INTRODUCTION

Section 3.9 in Chapter 3 introduced the topic of non-discretionary variables and applied it to examples drawn from a study of Texas schools. In that study it was necessary to allow for variations in "minority," "economically disadvantaged" and "low English proficiency" students who had to be dealt with in different schools. These input variables were "non-discretionary." That is, they could not be varied at the discretion of individual school managers but nevertheless needed to be taken into account in arriving at relative efficiency evaluations.

The approach taken in Chapter 3, as adapted from Banker and Morey [1] (1986), took form in the following model:

$$\begin{aligned}
\min \quad & \theta - \varepsilon\left(\sum_{i \in D} s_i^- + \sum_{r=1}^{s} s_r^+\right) && (7.1)\\
\text{subject to} \quad & \theta x_{io} = \sum_{j=1}^{n} x_{ij}\lambda_j + s_i^-, \quad i \in D \\
& x_{io} = \sum_{j=1}^{n} x_{ij}\lambda_j + s_i^-, \quad i \in ND \\
& y_{ro} = \sum_{j=1}^{n} y_{rj}\lambda_j - s_r^+, \quad r = 1, \ldots, s.
\end{aligned}$$

where all variables (except θ) are constrained to be nonnegative. See (3.79)-(3.82).

Here the symbol D and ND refer to "Discretionary" and "Non-Discretionary," respectively. Focussed on the inputs, as in (7.1) above, such non-discretionary variables may also be extended to outputs.

This topic can also be treated in different ways. For instance, Charnes *et al.* (1987)[2] extended the Additive model in order to accommodate non-discretionary variables in the following form.

$$
\begin{aligned}
\max \quad & \sum_{i=1}^{m} s_i^- + \sum_{r=1}^{s} s_r^+ \qquad (7.2)\\
\text{subject to} \quad & \sum_{j=1}^{n} x_{ij}\lambda_j + s_i^- = x_{io}, \; i = 1, \ldots, m \\
& \sum_{j=1}^{n} y_{rj}\lambda_j - s_r^+ = y_{ro}, \; r = 1, \ldots, s \\
& s_i^- \leq \beta_i x_{io}, \; i = 1, \ldots, m \\
& s_r^+ \leq \gamma_r y_{ro}, \; r = 1, \ldots, s
\end{aligned}
$$

where the β_i, γ_r represent parameters (to be prescribed) and all variables are constrained to be nonnegative.

Assigning values from 0 to 1 accords different degrees of discretion to input i with $\beta_i = 0$ characterizing this input as completely non-discretionary and $\beta_i = 1$ changing the characterization to completely discretionary. Similarly setting $\gamma_r = 0$ consigns output r to a fixed (non-discretionary) value while allowing $\gamma_r \to \infty$, or, equivalently, removing this constraint on s_r^+ allows its value to vary in a freely discretionary manner.

As can be seen, the latter model can be used to eliminate the nonzero slack that can appear with the Banker-Morey approach associated with (7.1). Such nonzero slack is omitted from the *measure* of efficiency in the B-M approach, of course, but it can be informative to, say, higher levels of management as impounded in the constraints. For instance, the ability to handle additional "economically disadvantaged" students noted in Table 3.7 could be used in redistricting this school to improve the performances of *other* schools in which this constraint is "tight" — as exhibited by zero slack and a positive value for the dual (multiplier) variable associated with this constraint in other schools.

This is not the end of the line, of course, so additional approaches to the treatment of non-discretionary variables will be undertaken in this chapter. There are related topics that will also be treated. One such topic involves the use of categorical (classificatory) variables which assign DMUs to different classes such as "large," "small," "medium," etc.

Another such topic involves the introduction of constraints on the range of values allowable for the variables. The "assurance regions" and "cone-ratio envelopments" developed in Chapter 6 approached this topic in terms of the multiplier variables. Here the constraints will be imposed on variables in the envelopment models.

7.2 EXAMPLES

Examples from Figure 7.1 can help us understand the treatment in these two models. This figure geometrically portrays five DMUs which use two inputs

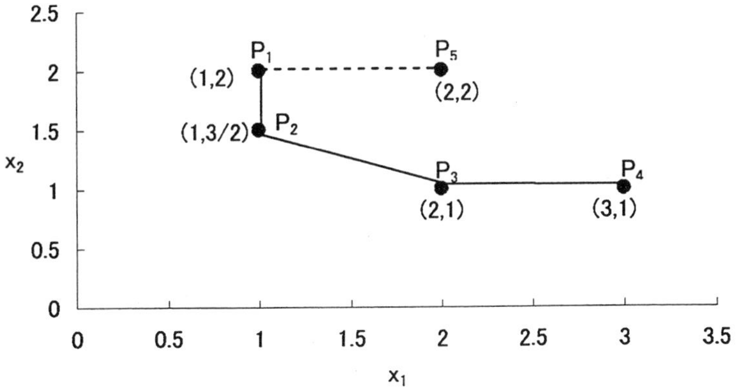

Figure 7.1. One Input Exogenously Fixed, or Non-Discretionary

to produce one unit of the same output. The points $P_1, \ldots, P_5$ represent the similarly numbered DMUs. The first parenthesized number is the value for the input amount x_1 and the second number is the value of the input amount x_2. We assume that the variable x_2 is non-discretionary. To evaluate P_5 via (7.1) we use

$$
\begin{aligned}
\min \quad & \theta - \varepsilon(s_1^- + s^+) \\
\text{subject to} \quad & 2\theta = \lambda_1 + \lambda_2 + 2\lambda_3 + 3\lambda_4 + 2\lambda_5 + s_1^- \\
& 2 = 2\lambda_1 + 3/2\lambda_2 + \lambda_3 + \lambda_4 + 2\lambda_5 + s_2^- \\
& 1 = \lambda_1 + \lambda_2 + \lambda_3 + \lambda_4 + \lambda_5 - s^+
\end{aligned}
$$

with all variables constrained to be nonnegative. Noting that s_2^- is not in the objective (where the efficiency evaluation measures appear) we find that the following solutions are alternative optima,

$$
\begin{aligned}
\theta^* &= 1/2, \quad \lambda_1^* = 1 \\
\theta^* &= 1/2, \quad \lambda_2^* = 1, \quad s_2^{-*} = 1/2
\end{aligned}
$$

where, in each case, all other variables are zero.

Recourse to (7.2) replaces the preceding example with

$$
\max \quad s_1^- + s_2^- + s^+
$$

$$
\begin{aligned}
\text{subject to} \quad & \lambda_1 + \lambda_2 + 2\lambda_3 + 3\lambda_4 + 2\lambda_5 + s_1^- = 2 \\
& 2\lambda_1 + 3/2\lambda_2 + \lambda_3 + \lambda_4 + 2\lambda_5 + s_2^- = 2 \\
& \lambda_1 + \lambda_2 + \lambda_3 + \lambda_4 + \lambda_5 - s^+ = 1 \\
& s_1^- \le 2\beta_1 \\
& s_2^- \le 2\beta_2 \\
& s^+ \le 1\gamma
\end{aligned}
$$

where, again, all variables are also nonnegative. The second input is the only non-discretionary variable so we set

$$\beta_2 = 0, \ \beta_1 = 1, \ \text{and} \ \gamma \to \infty.$$

This choice yields

$$\lambda_1^* = 1, \ s_1^{-*} = 1$$

and all other variables zero. Because s_2^- is restricted to a zero value in the constraint as well as in the objective, the preceding alternate optimum is eliminated and the solution is uniquely optimal.

When all variables are discretionary a DMU will be efficient if its associated vector of observations enters into an optimal basis with a positive coefficient. Such vectors are referred to as "active" members of the basis in order to distinguish them from other vectors which enter with zero coefficients because they are needed only to complete a basis and, hence, need not be efficient. See Problems 7.5 and 7.6, below.

As can be seen, P_1 has been elevated to the status of being efficient because it is an active member of the optimal basis used to evaluate P_5. The value of $s_2^{-*} = 1/2$ in the alternate optimum signals that this efficient status might change when the variable $x_2 = 2$ is changed from discretionary to non-discretionary. This information can be of interest to, say, higher levels of management, as was noted in our discussion of Table 3.5 in Chapter 3.

A common thread in the approaches used in (7.1) and (7.2) is the fact that nonzero slacks which appear only in the constraints do not enter into the efficiency evaluations. Hence for CCR (or BCC) efficiency, as represented in (7.1), Definition 3.2 is sharpened to the following

Definition 7.1 *Full (CCR or BCC) efficiency is achieved for DMU_o if and only if both of the following conditions are satisfied*

(i) $\theta^* = 1$

(ii) All slacks in the objective *are zero.*

For the extended additive model, as represented in (7.2), Definition 4.3 is modified to

Definition 7.2 *All slacks at zero in the objective is a necessary and sufficient condition for full efficiency with (7.2).*

The refinements in these definitions were not needed when all variable values could be altered at discretion of the managers. The consequences of the presence of non-discretionary variables can be examined a bit further by evaluating

P_1 with $x_2 = 2$ fixed at its observed value. Thus using (7.1) we obtain

$$\begin{aligned} \min \quad & \theta - \varepsilon(s_1^- + s^+) \\ \text{subject to} \quad & 1\theta = \lambda_1 + \lambda_2 + 2\lambda_3 + 3\lambda_4 + 2\lambda_5 + s_1^- \\ & 2 = 2\lambda_1 + 3/2\lambda_2 + 1\lambda_3 + 1\lambda_4 + 2\lambda_5 + s_2^- \\ & 1 = 1\lambda_1 + 1\lambda_2 + 1\lambda_3 + 1\lambda_4 + 1\lambda_5 - s^+ \end{aligned}$$

where all variables except θ are constrained to be nonnegative. This has alternate optima,

$$\begin{aligned} &\theta^* = 1, \quad \lambda_2^* = 1, \quad s_2^{-*} = 1/2 \\ &\theta^* = 1, \quad \lambda_1^* = 1 \end{aligned}$$

and all other variables zero. In either case P_1 is characterized as efficient because $\theta^* = 1$ and $s_2^{-*} = 1/2$ do not appear in the objective. See Definition 7.1.

7.3 NON-CONTROLLABLE, NON-DISCRETIONARY AND BOUNDED VARIABLE MODELS

Further discussion and development of the above examples are provided in the problems at the end of this chapter. Here we turn to an application which uses a different (but related) formulation and terminology as in the following formulations.

7.3.1 Non-controllable Variable (NCN) Model

$$(NCN) \quad \min_{\theta, \boldsymbol{\lambda}} \quad \theta \tag{7.3}$$

$$\begin{aligned} \text{subject to} \quad & \theta\, \boldsymbol{x}_o^C \geq X^C \boldsymbol{\lambda} \\ & \boldsymbol{y}_o^C \leq Y^C \boldsymbol{\lambda} \\ & \boldsymbol{x}_o^N = X^N \boldsymbol{\lambda} && (7.4) \\ & \boldsymbol{y}_o^N = Y^N \boldsymbol{\lambda} && (7.5) \\ & L \leq \boldsymbol{e\lambda} \leq U \\ & \boldsymbol{\lambda} \geq \mathbf{0}. \end{aligned}$$

Here we are using a matrix-vector formulation in which X^C, Y^C refer to matrices of "controllable" variables and $\boldsymbol{x}_o^C$, $\boldsymbol{y}_o^C$ refer to the corresponding vectors of observed values for the DMU_o being evaluated. The matrices X^N, Y^N, on the other hand, refer to data on the non-controllable variables that are to be evaluated relative to the vectors $\boldsymbol{x}_o^N$, $\boldsymbol{y}_o^N$ for this same DMU_o. Finally the last constraint imposes an upper bound, U, and a lower bound, L, on the variable choices with $\boldsymbol{e\lambda} = \sum_{j=1}^n \lambda_j$ in conformance with (5.37) in Chapter 5.

To simplify matters in the additions and extensions we wish to make, we have not included slacks in the objective of (7.3). For the non-controllable variables this conforms to the definition we have now provided. We have, however, also omitted the slacks for the non-controllable variables so this needs to be allowed for in the application we now introduce.

7.3.2 An Example of a Non-Controllable Variable

Table 7.1 shows the data for public libraries in the 23 Wards of the Tokyo Metropolitan Area in 1986. As the measurement items of efficiency we use the floor area (unit=$1000m^2$), the number of books (unit=1000), staffs and the populations of wards (unit=1000) as inputs and the number of registered residents (unit=1000) and borrowed books (unit=1000) as outputs. Using these data, we evaluated the relative efficiency using both the input-oriented CCR model and the CCR model with population as a non-controllable variable (NCN). Table 7.2 depicts the results.

Table 7.1. Data for Public Libraries in Tokyo

No.	Ward	INPUT				OUTPUT	
		Area ($1000m^2$)	Books (1000)	Staff	Populat. (1000)	Regist. (1000)	Borrow. (1000)
L1	Chiyoda	2.249	163.523	26	49.196	5.561	105.321
L2	Chūo	4.617	338.671	30	78.599	18.106	314.682
L3	Taito	3.873	281.655	51	176.381	16.498	542.349
L4	Arakawa	5.541	400.993	78	189.397	30.810	847.872
L5	Minato	11.381	363.116	69	192.235	57.279	758.704
L6	Bunkyo	10.086	541.658	114	194.091	66.137	1438.746
L7	Sumida	5.434	508.141	61	228.535	35.295	839.597
L8	Shibuya	7.524	338.804	74	238.691	33.188	540.821
L9	Meguro	5.077	511.467	84	267.385	65.391	1562.274
L10	Toshima	7.029	393.815	68	277.402	41.197	978.117
L11	Shinjuku	11.121	509.682	96	330.609	47.032	930.437
L12	Nakano	7.072	527.457	92	332.609	56.064	1345.185
L13	Shinagawa	9.348	601.594	127	356.504	69.536	1164.801
L14	Kita	7.781	528.799	96	365.844	37.467	1348.588
L15	Kōto	6.235	394.158	77	389.894	57.727	1100.779
L16	Katushika	10.593	515.624	101	417.513	46.160	1070.488
L17	Itabashi	10.866	566.708	118	503.914	102.967	1707.645
L18	Edogawa	6.500	467.617	74	517.318	47.236	1223.026
L19	Suginami	11.469	768.484	103	537.746	84.510	2299.694
L20	Nerima	10.868	669.996	107	590.601	69.576	1901.465
L21	Adachi	10.717	844.949	120	622.550	89.401	1909.698
L22	Ōta	19.716	1258.981	242	660.164	97.941	3055.193
L23	Setagaya	10.888	1148.863	202	808.369	191.166	4096.300

In evaluating the efficiency of a library, the population of the area is an important (input) factor. If we apply the input-oriented CCR model for this evaluation, the CCR-projection by (3.22) scales down all inputs by multiplying the optimal score $\theta^* (\leq 1)$ and further deletes slacks if any. "Population" is not an exception and hence the "CCR Projection" listed in Table 7.2 resulted in reductions of the population for inefficient libraries. In other words, the CCR score θ^* is evaluated under the assumption that it is possible to scale down radially (proportionally) all inputs so long as the reduced inputs remain in the production possibility set. However, "Population" is non-controllable and so we apply the model expressed by (7.4) for this NCN variable in order to evaluate the efficiency of libraries. As exhibited under CCR Score and NCN Score in Table 7.2, the latter differs from the CCR score and this shows the influence of this constraint (hence a modification of the production possibility set). Finally, as intended, the NCN-projection of population is the same as the original data listed under Data Population.

Table 7.2. Efficiency of Libraries by CCR and NCN

No	CCR Score	NCN Score	Data Population	CCR Projection	NCN Projection
L1	0.35	0.301	49.196	17.219	49.196
L2	0.792	0.643	78.599	62.237	78.599
L3	0.573	0.651	176.381	101.119	176.380
L4	0.719	0.618	189.397	136.116	189.397
L5	1	1	192.235	192.235	192.235
L6	1	1	194.091	194.091	194.091
L7	0.697	0.71	228.535	159.230	228.535
L8	0.58	0.7	238.691	138.520	238.691
L9	1	1	267.385	267.385	267.385
L10	0.705	0.793	277.402	195.313	277.402
L11	0.569	0.65	330.609	188.088	330.609
L12	0.758	0.773	332.609	252.235	332.609
L13	0.747	0.687	356.504	266.485	356.504
L14	0.722	0.8	365.844	263.972	365.844
L15	0.844	1	389.894	262.813	389.894
L16	0.582	0.766	417.513	211.251	417.513
L17	1	1	503.914	503.914	503.914
L18	0.787	1	517.318	258.248	517.318
L19	1	1	537.746	537.746	537.746
L20	0.849	0.958	590.601	399.189	590.601
L21	0.787	0.942	622.550	378.043	622.550
L22	0.785	0.701	660.164	518.192	660.164
L23	1	1	808.369	808.369	808.369

7.3.3 Non-discretionary Variable (NDSC) Model

The NCN model bases on the formulae (7.3). The non-controllable constraints (7.4) and (7.5) demand that the non-controllable input/output variables on the left should be expressed exactly as *equality* by a nonnegative combination of the corresponding non-controllable variables in the data set on the right. This reflects situations like the following. When comparing the performance of supermarkets using the population of the district as the non-controllable variable, the right side population should be compared with the same size one as expressed by the left side of the equation.

However, if other situations (constraints) are preferred, i.e., *greater than or equal* ($\geq$) constraints in (7.4) and *less than or equal* ($\leq$) constraints in (7.5), the NDSC (non-discretionary variable) model can be utilized. This is the model introduced in (7.1), i.e.,

$$
\begin{aligned}
\min \quad & \theta - \varepsilon\left(\sum_{i\in D} s_i^- + \sum_{r=1}^{s} s_r^+\right) \qquad (7.6)\\
\text{subject to} \quad & \theta x_{io} = \sum_{j=1}^{n} x_{ij}\lambda_j + s_i^-, \quad i \in D \\
& x_{io} = \sum_{j=1}^{n} x_{ij}\lambda_j + s_i^-, \quad i \in ND \\
& y_{ro} = \sum_{j=1}^{n} y_{rj}\lambda_j - s_r^+, \quad r = 1,\ldots,s.
\end{aligned}
$$

where all variables (except θ) are constrained to be nonnegative.

Although the non-discretionary (ND) variables do not enter the objective in (7.6) they, nevertheless, affect the efficiency score. Turning to the dual, as represented in (7.7), below, we can assign these effects a numerical value. This can be seen by writing the dual to (7.6) as follows.

$$
\begin{aligned}
& \max_{\boldsymbol{v},\boldsymbol{u}} \sum_{r=1}^{s} u_r y_{ro} - \sum_{i\in ND} v_i x_{io} \\
& \text{subject to} \\
& \sum_{r=1}^{s} u_r y_{rj} - \sum_{i\in ND} v_i x_{ij} - \sum_{i\in D} v_i x_{ij} \leq 0, \ j = 1,\ldots,n \\
& \sum_{i=1}^{m} v_i x_{io} = 1 \qquad (7.7)\\
& v_i \geq \varepsilon, \ i \in D \\
& v_i \geq 0, \ i \in ND \\
& u_r \geq \varepsilon. \ r = 1,\ldots,s
\end{aligned}
$$

Notice now that the objective of the dual in (7.7) has a property that is opposite to that of (7.6). Only the non-discretionary input multipliers enter into the objective of (7.7), i.e., the multiplier model, whereas only the discretionary (D) input variables enter into the objective of (7.6), i.e., the envelopment model. Thus, a value of the optimal multiplier v_i^* provides the *rate* at which the optimal value of (7.6) is reduced by increasing x_{io} for this member of the ND and $v_i^* x_{io}$ is the *amount* by which the optimal value of (7.6) is reduced by this member of the set ND when the corresponding constraint is critical in the envelopment model. On the other hand if this constraint is loose, so that positive slack is present for this constraint, the optimal value of the objective in (7.6) is not affected. Because this non-discretionary factor is present in more than needed amounts the optimal $v_i^* = 0$.

We illustrate the latter property by forming the dual to the preceding problem in Section 7.2 used to evaluate DMU P_5 in Figure 7.1,

$$\begin{aligned}
&\max u - 2v_2 \\
&\text{subject to} \\
&1 = 2v_1 \\
&0 \geq u - v_1 - 2v_2 \\
&0 \geq u - v_1 - \frac{3}{2}v_2 \\
&0 \geq u - 2v_1 - v_2 \\
&0 \geq u - 3v_1 - v_2 \\
&0 \geq u - 2v_1 - 2v_2 \\
&v_1,\ u \geq \varepsilon,\ v_2 \geq 0.
\end{aligned}$$

This has the following optimum solution,

$$u^* = v_1^* = 1/2$$

with all other variables zero, including $v_2^* = 0$. Hence the presence of the non-discretionary variable represented by x_{25} has no effect in the performance evaluation.

In the above example, no harm is experienced by $x_{25} = 2$ being in excess of what is needed since the slack variable makes it unnecessary to use all of what is available. However, there are cases in which an excessive amount of an input may cause a reduction in performance. This can occur with a discretionary variable as in the case of congestion where too much of a discretionary input is used in excessive amounts that cause a reduction in output. See Deng (2003). [3] It can also occur with non-discretionary variables. An excessive amount of rain that damages farm crops is an example. The latter kind of excess will show up in the objective of the multiplier model when the non-discretionary variables are evaluated explicitly while the former will show up in the dual (envelopment model). See Problem 7.2, below.

7.3.4 Bounded Variable (BND) Model

There are several extensions of the these models —NCN and NDSC— among which we will introduce models with upper/lower bounds constraints. For example, when we evaluate the efficiency of baseball stadiums and take the number of spectators as an output variable, this value cannot exceed the maximum capacity of the stadium for each DMU. Thus, the maximum capacity should be considered as an upper bound for the number of spectators.

In order to cope with these situations, we will relax the non-controllable constraints (7.4) and (7.5) into constraints with upper/lower bounds as follows:

$$\boldsymbol{l}_o^{N_x} \leq X^N \boldsymbol{\lambda} \leq \boldsymbol{u}_o^{N_x} \tag{7.8}$$

$$\boldsymbol{l}_o^{N_y} \leq Y^N \boldsymbol{\lambda} \leq \boldsymbol{u}_o^{N_y} \tag{7.9}$$

where $(\boldsymbol{l}_o^{N_x},\ \boldsymbol{u}_o^{N_x})$ and $(\boldsymbol{l}_o^{N_y},\ \boldsymbol{u}_o^{N_y})$ are vectors of the lower and upper bounds to the non-discretionary inputs and outputs of DMU$_o$, respectively. $\boldsymbol{x}_o^N$ and $\boldsymbol{y}_o^N$ are not explicitly included in the formulae, because it is assumed that they lie in between the two bounds. The extensions including bounded non-discretionary variables are as follows:

(a) Input Oriented Bounded Variable Model

$$\begin{aligned}
(BND_o)\quad \min\quad & \theta & (7.10)\\
\text{subject to}\quad & \theta\, \boldsymbol{x}_o^C \geq X^C \boldsymbol{\lambda} \\
& \boldsymbol{y}_o^C \leq Y^C \boldsymbol{\lambda} \\
& \boldsymbol{l}_o^{N_x} \leq X^N \boldsymbol{\lambda} \leq \boldsymbol{u}_o^{N_x} \\
& \boldsymbol{l}_o^{N_y} \leq Y^N \boldsymbol{\lambda} \leq \boldsymbol{u}_o^{N_y} \\
& L \leq \boldsymbol{e\lambda} \leq U \\
& \boldsymbol{\lambda} \geq \boldsymbol{0}.
\end{aligned}$$

(b) Output Oriented Bounded Variable Model

$$\begin{aligned}
(BNDO_o)\quad \max\quad & \eta & (7.11)\\
\text{subject to}\quad & \boldsymbol{x}_o^C \geq X^C \boldsymbol{\lambda} \\
& \eta \boldsymbol{y}_o^C \leq Y^C \boldsymbol{\lambda} \\
& \boldsymbol{l}_o^{N_x} \leq X^N \boldsymbol{\lambda} \leq \boldsymbol{u}_o^{N_x} \\
& \boldsymbol{l}_o^{N_y} \leq Y^N \boldsymbol{\lambda} \leq \boldsymbol{u}_o^{N_y} \\
& L \leq \boldsymbol{e\lambda} \leq U \\
& \boldsymbol{\lambda} \geq \boldsymbol{0}.
\end{aligned}$$

7.3.5 An Example of the Bounded Variable Model

As an example of the bounded variable model, we introduce the evaluation of the relative efficiency of the twelve Japanese professional baseball teams in the year 1993. We employ 2 inputs and 2 outputs for this evaluation as follows:

Input 1 : Average annual salary of managers, including those of coaching staff.

Input 2 : Average annual salary of players: the top ranked 9 fielders and 6 pitchers.

Output 1 : "Team power" as explained below.

Output 2 : "Attendance" as explained below.

"Team power" is measured as a function of the percentage of victories per year, the batting averages of fielders, the number of home runs, the number of stolen bases and the defense rating of pitchers. We utilized results from a "principal component" analysis to define the team power of each team. Actually, the contribution ratio of the first principal component was 53% and was judged to be interpretable as "Team power." The second component seems to represent "Mobile power" and its proportion is 23%. For this study we employed "Team power" as a representative index of the total power of the team. "Attendance" is measured as the ratio of the total annual attendance vs. the annual maximum capacity of the team's home stadium and expressed as a percentage. Hence this number cannot exceed 100%. Table 7.3 exhibits these data for 12 teams. Salaries of managers and players, as inputs, are in tens of thousands of Japanese yen (ten thousand Japanese yen $\approx$ 83 U.S. dollars in 1993).

Table 7.3. Data of 12 Japanese Baseball Teams in 1993

	INPUT		OUTPUT	
Team	Manager	Player	Team power	Attendance
Swallows	5250	6183	21.72	80.87
Dragons	2250	5733	21.02	93.76
Giants	6375	8502	18.44	100.00
Tigers	3125	4780	18.49	76.53
Bay Stars	3500	4042	18.05	79.12
Carp	3125	5623	18.55	51.95
Lions	5500	10180	21.25	56.37
Fighters	3625	5362	20.74	57.44
Blue Wave	2715	4405	20.39	58.78
Buffalos	3175	6193	20.59	53.00
Marines	2263	5013	17.78	43.47
Hawks	3875	3945	16.80	82.78
Average	3732	5830	19.49	69.51

The purpose of this study is to evaluate the relative efficiency of these 12 teams using salaries as inputs and seasonal records as outputs. Salary control, especially reduction in the average salary, is very difficult for the owner of a team, so we employed the output-oriented models.

Using an output-oriented CCR model we obtained the results exhibited in Table 7.4 under the heading CCR, where the "Projection" of "Attendance" to efficient frontiers is also recorded in the next to last column. Full efficiency was attained by the Dragons, Bay Stars, Blue Wave and Hawks within this model. These teams had good outputs relative to comparatively low salaries. In contrast, the well paid Giants and Lions ranked as the lowest two. As can be seen, the CCR-Projections of Swallows, Giants and Lions exceed 100%. This, however, is impossible because of the capacity limit. Thus, the efficiency scores obtained under this assumption are not always matched to their "Attendance" situation.

Table 7.4. Projection of Attendance by CCR and Bounded Models

Team	CCR Score	Bounded Score	Attendance	CCR Projection	Bounded Projection
Swallows	0.7744	0.7589	80.87	104.40	82.51
Dragons	1	1	93.76	93.76	93.76
Giants	0.6139	0.5088	100.00	162.90	100.00
Tigers	0.9204	0.8829	76.53	83.15	76.53
Bay Stars	1	1	79.12	79.12	79.12
Carp	0.7552	0.7552	51.95	79.61	79.61
Lions	0.4852	0.5474	56.37	146.25	100.00
Fighters	0.8356	0.8356	57.44	71.55	71.55
Blue Wave	1	1	58.78	58.78	58.78
Buffalos	0.7948	0.7948	53.00	91.27	91.27
Marines	0.9048	0.9048	43.47	78.01	78.01
Hawks	1	1	82.78	82.78	82.78

Now we treat the output "Attendance" as a bounded variable with an upper bound of 100% and use the status quo for every team as a lower bound. In this model, "Team power" is the only variable connected directly to the efficiency evaluation. "Attendance" is treated as a second output which constrains the feasible region and thus indirectly influences the objective value. "Team power" is connected to attendance, however, because when the team power increases then "Attendance" also grows. The results are exhibited in Table 7.4 under the heading "Bounded Projection." As can be seen, all projections listed under this heading remain within the prescribed bounds. The efficiency scores evaluated under these bounds as listed under "Bounded Score" are (a) changed only slightly from those in the CCR case, (b) none of the previously identified efficient teams lost this status, and (c) even some of the teams with Attendance reduced by the 100% bound increased their efficiency score. See the increase from 0.4852 to 0.5474 for the Lions.

Remark : It might be argued that the value under CCR projection provides information on where added capacity is needed. Information of this kind is also available from the dual (multiplier) values when the constraints associated with these bounds are critical. Notice, for instance, that these constraints are critical for the Giants and Lions, which are both at 100% under "bounded projection," but this is not true for the Swallows even though this team's CCR projection exceeded 100%.

7.4 DEA WITH CATEGORICAL DMUS

There are other managerial situations over which managers of particular organizations do not have total control. For instance, in evaluating the performance of a branch store of a supermarket, it is necessary to consider the sales environment of the store, including whether it has severe competition, is in a normal business situation, or in a relatively advantageous one. If we evaluate the efficiency of the above supermarkets as "scratch" players, the evaluation would be unfair to the stores in the highly competitive situation and would be too indulgent to the stores in the advantageous one. Hence we need to provide something in the way of a "handicap" when evaluating them.

A hierarchical category is suitable for handling such situations. As for the supermarket example, we classify stores facing severe competition as *category 1*, in a normal situation as *category 2* and in an advantageous one as *category 3*. Then we evaluate stores in category 1 only within the group, stores in category 2 with reference to stores in category 1 and 2 and stores in category 3 within all stores in the model. Thus, we can evaluate stores under operating handicaps which take into account their particular environments, and we also use this information to evaluate stores in higher categories.

There are two kinds of categorical variables. One is *non-controllable* by decision makers, as pointed out above. The other is under the control of decision makers. We will consider the non-controllable type first.

7.4.1 An Example of a Hierarchical Category

The public libraries in Tokyo (in Table 7.1, above) can be classified into three categories as follows. Category 1 consists of libraries in the business district of central Tokyo, category 2 in the shopping area around the business district, and category 3 in the residential area on the outskirts. (See the column headed by "Category" in Table 7.5.) Libraries in category 1 are in a severe situation for library utilization, since residents around them are few compared with libraries in the other two categories. Category 3 is in the most advantageous situation. So we evaluate the efficiency of libraries in category 1 only within the category while libraries in category 2 are evaluated with reference to categories 1 and 2 and libraries in category 3 are evaluated with reference to all libraries. We utilized the data in Table 7.1 excluding "population" from inputs. Thus three

Table 7.5. Categorization of Libraries

	Categorization			CCR	
Library	Category	Score	Reference	Score	Reference
L1	1	0.377	L6	0.226	L23
L2	1	0.879	L5 L6	0.638	L23
L3	1	0.936	L4 L6	0.54	L23
L4	1	1	L4	0.593	L23
L5	1	1	L5	0.911	L17 L23
L6	1	1	L6	0.745	L23
L7	2	0.743	L5 L9	0.65	L19 L23
L8	2	0.648	L5 L15	0.539	L17
L9	2	1	L9	0.907	L19 L23
L10	2	0.815	L9 L15	0.705	L19 L23
L11	2	0.646	L5 L9 L15	0.539	L17 L23
L12	2	0.835	L9	0.719	L19 L23
L13	2	0.794	L9 L15	0.657	L17 L23
L14	2	0.835	L9	0.715	L23
L15	2	1	L15	0.844	L17 L23
L16	2	0.687	L9 L15	0.582	L23
L17	3	1	L17	1	L17
L18	3	0.787	L19 L23	0.787	L19 L23
L19	3	1	L19	1	L19
L20	3	0.849	L19 L23	0.849	L19 L23
L21	3	0.787	L23	0.787	L23
L22	3	0.681	L23	0.681	L23
L23	3	1	L23	1	L23

inputs (area, books, staff) and two outputs (registered, borrowed) are evaluated under categorical conditions.

The results of computations based on this categorization are given in Table 7.5. It can be seen that the reference set for all DMUs in category 1 consists of only category 1 DMUs. In the reference set for DMUs in category 2, however, L5 is included from category 1, and L23 appears only in the reference set of category 3. This is in contrast to the CCR model applied to all libraries altogether in which case L17, L19 and L23 have full efficiency score and appear as reference DMUs used to evaluate other libraries. See the column headed "CCR" in Table 7.5. These three libraries belong to category 3, i.e. the area which is most favorable to library utilization.

7.4.2 Solution to the Categorical Model

The categorical, model can be incorporated into any DEA model, such as CCR, BCC, IRS, DRS and GRS which are discussed in Section 5.7 of Chapter 5. In designing LP computations for the categorical model, we should notice that the DMUs in the upper categories cannot be chosen as basic variables for DMUs in a lower category.

7.4.3 Extension of the Categorical Model

In the preceding subsections, we dealt with the non-controllable category case, in which decision makers cannot choose one level from a category at will. However, in some cases, the category is under the control of decision makers. For example, suppose that the DMUs are a set of shops with three levels of service, i.e., *poor, average* and *good.* A shop owner in the *poor* service category has the option to remain *poor* or upgrade to *average* or *good.* An *average* shop can move within the *average* level or further up to the *good* level.

Suppose that each shop (DMU) can be rated as having one of L different categorical classes, where the categories range from category 1 (representing the lowest service orientation) to category L (the highest). The problem here is then to find, for each DMU, the DEA projected point in the same or higher category levels.

In the algorithm below, we consider the case for DMU_o, which is currently at level l $(1 \le l \le L)$ and try to find the reference set and the DEA projected points on the frontier with levels in the same category or higher. As for the DEA model employed, we can choose any model, e.g. CCR, BCC, Additive, etc.

[**Algorithm**] (Controllable Categorical Level)

For $h = l, l+1, \ldots, L$, repeat the following steps:

Step 1.

Organize a set of DMUs composed of level
h or *higher* and the DMU_o to be evaluated.
Evaluate the efficiency of DMU_o with respect to this group
via the DEA model chosen. Go to Step 2.

Step 2.

(i) If DMU_o is found to be efficient, go to Step 3.
(ii) If DMU_o is inefficient, then record its reference set
and reference (projected) point on the frontier.
If $h = L$, go to Step 3. Otherwise, replace h by $h + 1$
and go back to Step 1.

Step 3.

Examine the reference set, reference point, and category level,
obtained from Step 2, and choose the most appropriate point
and category level for DMU_o.

One of the characteristics of this algorithm is that it allows DMUs at different levels to be used as a candidate reference point. For example, suppose a *poor* DMU_1 has its reference set composed of an *average* DMU_2 and a *good* DMU_3 with weights 0.75 and 0.25. A categorical service level of $0.75 \times$ *average* $+ \ 0.25 \times$ *good* is assumed for the projected point, i.e. it has a quality level close to *average* and slightly upgraded to *good.*

[Example]

Table 7.6 exhibits nine single input and single output DMUs, each having either a *poor, average* or *good* category level. Figure 7.2 shows them graphically, where the symbols ●, ▲ and ■ correspond to *poor, average* and *good*, respectively. We applied this algorithm using the BCC model and obtained the results shown in the far right column of Table 7.6, where the number in parenthesis designates the λ-value assigned to the referent DMU.

Table 7.6. Nine DMUs with Three Category Levels

DMU	Input	Output	Category	Reference set
A	3	1	*poor*	$A(1)$
B	7	7	*poor*	$D(.6),\ E(.4)$
C	12	6	*poor*	$D(.8),\ E(.2) \mid G(1)$
D	4	5	*average*	$D(1)$
E	6	10	*average*	$E(1)$
F	11	11	*average*	$E(.667),\ H(.333) \mid G(1)$
G	8	11	*good*	$G(1)$
H	9	13	*good*	$H(1)$
I	13	15	*good*	$I(1)$

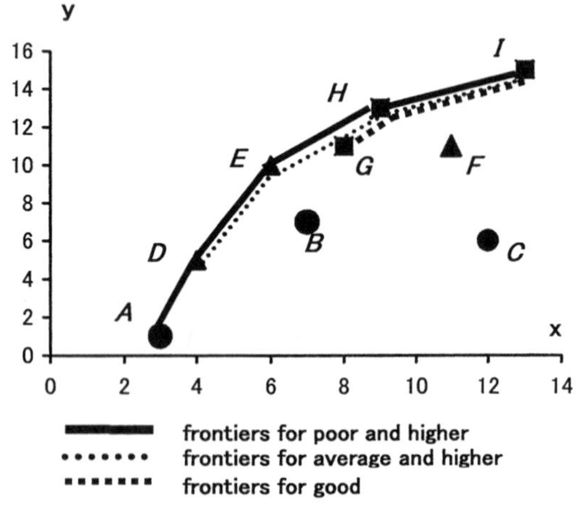

Figure 7.2. DMUs with Controllable Category Levels

DMU A (*poor*) is judged to be efficient, even compared with DMUs in the same or higher category levels and it is reasonable for A to remain in this category. DMU B (*poor*) is enveloped by D (*average*) with $\lambda_D = 0.6$ and E (*average*) with $\lambda_E = 0.4$ and is suitable for upgrading to *average*. DMU C (*poor*) has two sets of reference points, one composed of D (*average*) with $\lambda_D = 0.8$ and E (*average*) with $\lambda_E = 0.2$ and the other G (*good*) with $\lambda_G = 1$. So, two possibilities are available for DMU C to upgrade its level. Similarly, DMU F has two possibilities and this kind of information can be taken into account by higher management, for instance, in assigning bonuses, etc., based on actual or attained performance.

7.5 COMPARISONS OF EFFICIENCY BETWEEN DIFFERENT SYSTEMS

The DEA models assume that the production possibility set P is convex and, in fact, if two activities $(\boldsymbol{x}_1, \boldsymbol{y}_1)$ and $(\boldsymbol{x}_2, \boldsymbol{y}_2)$ belong to P, then every point on the line segment connecting these two points belongs to P.[4] However, there are situations where this assumption is not valid. For example, an activity $(\boldsymbol{x}_1, \boldsymbol{y}_1)$ uses one kind of instrument, while an activity $(\boldsymbol{x}_2, \boldsymbol{y}_2)$ adopts another, so we cannot reasonably assume any activity exists in between them.

7.5.1 Formulation

To see how this problem can be treated suppose the DMUs under consideration belong exclusively to one of two systems i.e. Systems A and B. (Although we deal with two systems, the discussions below can be easily extended to more general cases). We divide input X into X_A and X_B and output Y into Y_A and Y_B. The convexity assumption holds *within* the same system but does not hold *between* the two systems. The production possibility set $\{(\boldsymbol{x}, \boldsymbol{y})\}$ is then assumed to satisfy the following constraints:

$$
\begin{aligned}
&\boldsymbol{x} \geq X_A \boldsymbol{\lambda}_A + X_B \boldsymbol{\lambda}_B \\
&\boldsymbol{y} \leq Y_A \boldsymbol{\lambda}_A + Y_B \boldsymbol{\lambda}_B \\
&L z_A \leq \boldsymbol{e}\boldsymbol{\lambda}_A \leq U z_A \\
&L z_B \leq \boldsymbol{e}\boldsymbol{\lambda}_B \leq U z_B \\
&z_A + z_B = 1 \\
&\boldsymbol{\lambda}_A \geq \boldsymbol{0}, \ \boldsymbol{\lambda}_B \geq \boldsymbol{0} \\
&z_A, \ z_B = 0 \text{ or } 1.
\end{aligned}
\tag{7.12}
$$

That is, the problem is found to be an integer program.

In this situation, the efficiency of DMU $(\boldsymbol{x}_o, \boldsymbol{y}_o)$ can be evaluated by the following mixed integer LP problem with z_A, z_B as binary variables that assume only the values 0 and 1,

$$
\begin{aligned}
\min \quad & \theta \\
\text{subject to} \quad & \theta \boldsymbol{x}_o \geq X_A \boldsymbol{\lambda}_A + X_B \boldsymbol{\lambda}_B
\end{aligned}
\tag{7.13}
$$

$$\begin{aligned}
&\boldsymbol{y}_o \leq Y_A\boldsymbol{\lambda}_A + Y_B\boldsymbol{\lambda}_B \\
&Lz_A \leq \boldsymbol{e}\boldsymbol{\lambda}_A \leq Uz_A \\
&Lz_B \leq \boldsymbol{e}\boldsymbol{\lambda}_B \leq Uz_B \\
&z_A + z_B = 1 \\
&\boldsymbol{\lambda}_A \geq \boldsymbol{0},\ \boldsymbol{\lambda}_B \geq \boldsymbol{0} \\
&z_A,\ z_B = \{0, 1\}.
\end{aligned}$$

From the results secured, we can evaluate not only the efficiency of each DMU but we can also compare the two systems by observing the efficiency of DMUs in each system. For this purpose, the nonparametric rank-sum test statistics described in the next section will be useful.

7.5.2 Computation of Efficiency

As is shown below, we can solve (7.13) by enumeration rather than by using a mixed integer 0-1 program,

1. Set $z_A = 1,\ z_B = 0$ and solve the LP problem above. That is, we evaluate the efficiency of DMU $(\boldsymbol{x}_o,\ \boldsymbol{y}_o)$, based on System A. Let the optimal objective value be θ_A. If the corresponding LP problem is infeasible, we define $\theta_A = \infty$.

2. Set $z_A = 0,\ z_B = 1$ and solve the LP problem above. Let the optimal objective value be θ_B, which is infinity if the corresponding LP problem is infeasible.

3. We obtain the efficiency of DMU $(\boldsymbol{x}_o,\ \boldsymbol{y}_o)$ by

$$\theta_o^* = \min\{\theta_A,\ \theta_B\}.$$

7.5.3 Illustration of a One Input and Two Output Scenario

Figure 7.3 exhibits DMUs in two sales systems, A and B, with one input and two outputs. Output 1 is the number of customers per salesman and output 2 is profit per salesman. In this case, if we neglected the distinction between the systems, the efficient frontier would be the line segments connecting A1, A2, B7 and B10. For reasons described above, however, the efficient frontier is represented by the bold lines connecting A1, A2, A6, P, B7 and B10 which no longer span convex production possibility sets like those we have employed in many of our previous examples.

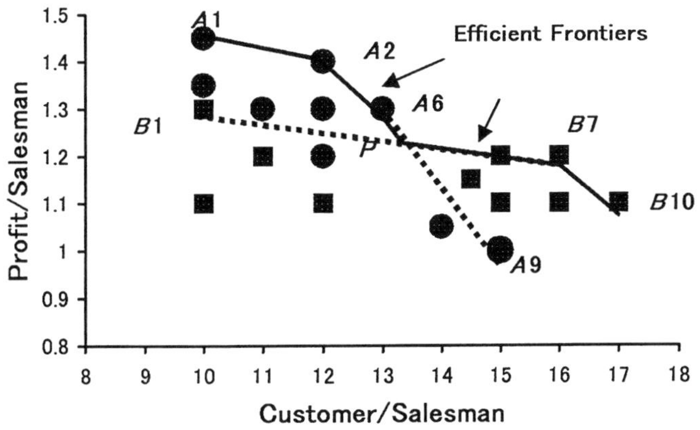

Figure 7.3. Comparisons between Stores using Two Systems

Table 7.7 shows the data set of this example under the heading "Data," while the efficiency of stores obtained by using the algorithm and the CCR model, i.e. $L = 0, U = \infty$, are listed under the heading "Comparisons" in the same table. Since the efficient frontiers of System B are spanned by line segments connecting B1, B7 and B10, it can be seen that DMUs in System A which are above these frontiers have the efficiency score θ_B greater than one as displayed in the table. Similarly, the efficient frontiers of System A are spanned by A1, A2, A6 and A9 and hence DMUs in B which are above the frontiers have the efficiency value θ_A greater then one. A comparison of the two systems is shown in Table 7.8 where we notice that it is possible that the reference set includes inefficient DMUs (e.g. B1).

7.6 RANK-SUM STATISTICS AND DEA

As an instance of DEA applications of these ideas, comparisons of efficiencies between two types of DMUs, e.g. department stores vs. supermarkets, or electric power industries in the U.S.A vs. those in Japan, have constituted important research subjects. It is often necessary to test statistically the difference between two groups in terms of efficiency. Do differences occur by chance or are they statistically significant? This section deals with such statistical issues. Since the theoretical distribution of the efficiency score in DEA is usually unknown, we are forced to deal with nonparametric statistics for which the distribution of the DEA scores are statistically independent. For this purpose, the rank-sum-test developed by Wilcoxon-Mann-Whitney may be used to identify whether the differences between two groups are significant.

Table 7.7. Comparisons of Stores in Two Systems

	Data			Comparisons			
Store	In.	Out1	Out2	θ_A	θ_B	θ^*	Ref. Set
A1	1	10	1.45	1	1.115	1	A1
A2	1	12	1.4	1	1.091	1	A2
A3	1	10	1.35	0.941	1.038	0.941	A1 A2
A4	1	12	1.3	0.926	1.011	0.926	A1 A2
A5	1	12	1.3	0.962	1.023	0.962	A2 A6
A6	1	13	1.3	1	1.034	1	A6
A7	1	12	1.2	0.923	0.955	0.923	A6
A8	1	14	1.05	0.969	0.875	0.875	B7
A9	1	15	1.0	1	0.893	0.893	B7 B10
B1	1	10	1.3	0.912	1	0.912	A1 A2
B2	1	11	1.2	0.885	0.943	0.885	A2 A6
B3	1	10	1.1	0.808	0.864	0.808	A2 A6
B4	1	12	1.1	0.892	0.886	0.886	B1 B7
B5	1	14.5	1.15	1.023	0.949	0.949	B1 B7
B6	1	15	1.2	1.062	0.989	0.989	B1 B7
B7	1	16	1.2	1.108	1	1	B7
B8	1	15	1.1	1.031	0.929	0.929	B7 B10
B9	1	16	1.1	1.077	0.964	0.964	B7 B10
B10	1	17	1.1	1.133	1	1	B10

Table 7.8. Comparisons of Two Systems

System	System A	System B
No. Eff. Stores	3	2
Average Eff.	0.947	0.932
Frequency of Reference to Other System	6	3

7.6.1 Rank-Sum-Test (Wilcoxon-Mann-Whitney)

This method is one of the nonparametric statistical tests based on the ranking of data. Given statistically independent data belonging to two groups, this test serves to test whether the hypothesis that the two groups belong to the same population or whether they differ significantly.

Let the data in two groups be represented by $A = \{a_1, a_2, \ldots, a_m\}$ and $B = \{b_1, b_2, \ldots, b_n\}$. We then merge A and B to arrive at a sequence C in which the data are arranged in descending order. An example is as follows,

$$A = \{0.42, 0.51, 0.45, 0.82, 0.9, 1, 0.72, 0.92, 0.65, 0.87\}\ (m = 10),$$

$$B = \{0.12, 0.32, 0.18, 0.42, 0.56, 0.9, 0.7, 0.22, 0.44, 0.02, 1\}\ (n = 11).$$

We then merge these sequences into a new sequence, C, with length $m + n = 10 + 11 = 21$:

$$C = \{\underline{1}, 1, \underline{0.92}, \underline{0.9}, 0.9, \underline{0.87}, \underline{0.82}, \underline{0.72}, 0.7, \underline{0.65}, 0.56, \underline{0.51}, \underline{0.45}, 0.44, \underline{0.42}, 0.42, 0.32, 0.22, 0.18, 0.12, 0.02\}$$

in which the underlined numbers belong to A.
Then we rank C from 1 to $N(= m + n)$. If there is a tie, we use the midrank for the tied observation.

The above example yields the ranking below,

$$R = \{\underline{1.5}, 1.5, \underline{3}, \underline{4.5}, 4.5, \underline{6}, \underline{7}, \underline{8}, 9, \underline{10}, 11, \underline{12}, \underline{13}, 14, \underline{15.5}, 15.5, 17, 18, 19, 20, 21\}.$$

For example, the top two numbers in the sequence C have the same value 1 so they are ranked as (1+2)/2=1.5 which is their midrank in R and the 4th and 5th in the sequence C have a tie, 0.9, so their midrank in R is (4+5)/2=4.5.

Next, we sum the ranking of the A data indicated by the underlined numbers,

$$S = 1.5 + 3 + 4.5 + 6 + 7 + 8 + 10 + 12 + 13 + 15.5 = 80.5.$$

Then statistic, S, follows an approximately normal distribution with mean $m(m+n+1)/2$ and variance $mn(m+n+1)/12$ for $m, n \geq 10$. By normalizing S, we have:

$$T = \frac{S - m(m+n+1)/2}{\sqrt{mn(m+n+1)/12}}. \tag{7.14}$$

T has an approximately standard normal distribution. Using T we can check the null hypothesis that the two groups have the same population at a level of significance α. We will reject the hypothesis if $T \leq -T_{\alpha/2}$ or $T \geq T_{\alpha/2}$, where $T_{\alpha/2}$ corresponds to the upper $\alpha/2$ percentile of the standard normal distribution. In the above example, we have $T = -2.0773$. If we choose $\alpha = 0.05$ (5%), then it holds that $T_{0.025} = 1.96$. Since $T = -2.0773 < -1.96 = -T_{0.025}$, we reject the null hypothesis at the significance level 5%. This test, which is attributed to Wilcoxon, is essentially equivalent to the Mann-Whitney test.

7.6.2 *Use of the Test for Comparing the DEA Scores of Two Groups*

A simple and straightforward application of the test mentioned above is to use it for identifying the difference between DEA scores in two groups of DMUs. Let the two groups be A and B. We pool them into $C(= A \cup B)$. Then, we employ an appropriate DEA model to measure the efficiency of each DMU in C. The rank-sum-test can be applied to check the hypothesis that groups A and B have the same population of efficiency scores.

7.6.3 *Use of the Test for Comparing the Efficient Frontiers of Two Groups*

Sometimes, it is required to compare the efficient frontiers of group A with those of group B and to see whether there is a significant frontier shift between the two.

For this purpose, we first run DEA separately for the two groups (once for A and once for B) and obtain the efficient DMUs in each group. Although we can compare the two groups using these efficient DMUs within each group, the number of DMUs for evaluation will decrease to a small portion of the total, and we will be confronted with a degrees-of-freedom problem for comparisons. To avoid this difficulty, we project inefficient DMUs in the group A (B) onto their respective efficient frontiers to obtain a new (combined) efficient frontier consisting of the same number of DMUs as in the original group. Then, we pool the two groups of efficient DMUs and run DEA. The rank-sum-test will be applied to the DEA results.

This method, which will eliminate inefficiency within each group makes it possible to compare the performance of two types of organization when both are operating efficiently.

7.6.4 *Bilateral Comparisons Using DEA*

The preceding two methods use a *mixed* evaluation of DMUs in two groups, by which it is meant that each DMU is evaluated on both *inter* and *within* group bases. Thus, the reference set of a DMU in A may consist of DMUs in A (*within*), or B (*inter*). In contrast to this, there is a view that each DMU in A (B) should be evaluated with respect to DMUs in the opposite group B (A). This *inter* comparison will result in a sharper discrimination between two groups. We can formulate this idea in the following way for each DMU $a \in A$.

$$\min \quad \theta \tag{7.15}$$

$$\text{subject to} \quad \sum_{j \in B} \boldsymbol{x}_j \lambda_j \leq \theta \boldsymbol{x}_a \tag{7.16}$$

$$\sum_{j \in B} \boldsymbol{y}_j \lambda_j \geq \boldsymbol{y}_a \tag{7.17}$$

$$\lambda_j \geq 0 \quad (\forall j \in B). \tag{7.18}$$

In the two input (x_1, x_2) and single unitized output ($y = 1$) case, Figures 7.4 and 7.5 illustrate a use of this program. In Figure 7.4, the DMU $a \in A$ is enveloped by DMUs in B and the optimal θ^* is given by

$$\theta^* = \frac{\text{OQ}}{\text{OP}} < 1.$$

In Figure 7.5, the DMU a cannot be enveloped by DMUs in B and a should be expanded radially to Q to be enveloped. The optimal θ^* is given by

$$\theta^* = \frac{\text{OQ}}{\text{OP}} > 1.$$

Thus, the optimal θ^* is no longer bounded by 1. The region of θ^* is therefore

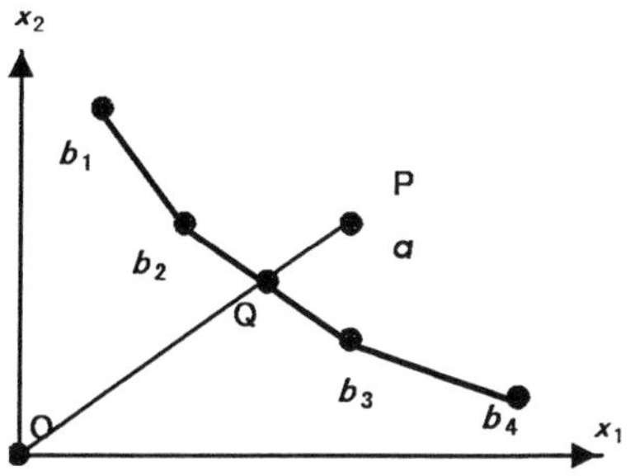

Figure 7.4. Bilateral Comparisons–Case 1

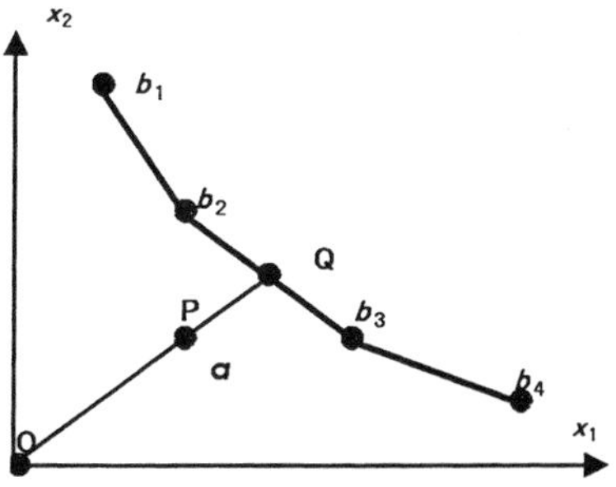

Figure 7.5. Bilateral Comparisons–Case 2

expanded upward and a sharper contrast between two groups can be expected than in the case of the *mixed* evaluation scheme. In order to make the LP feasible, we assume $X > O$ and $Y > O$. Bilateral comparisons of DMUs in B with respect to A follow in the same way. By virtue of its nonparametric character we can then apply the rank-sum-test to the thus obtained efficiency score θ^*, whether or not the groups A and B have the same distribution of efficiency values.

7.6.5 An Example of Bilateral Comparisons in DEA

We applied the bilateral comparisons to the library problem exhibited in Table 7.5 for testing the statistical difference in efficiency between the libraries in the business district (Category 1) and those in the residential area (Category 3). Thus Group A consists of six libraries L1-L6 and Group B consists of seven libraries L17-L23. We utilized numbers of books and staffs as inputs and numbers of registered residents and borrowed books as outputs (see Table 7.1). The results from these comparisons are exhibited in Table 7.9, where the efficiency scores against the counterpart libraries are listed along with the reference set and the ranking of the efficiency scores. The rank sum of Group A is $S = 63$ and the statistic calculated from (7.14) gives

$$T = \frac{63 - 6 \times (6+7+1)/2}{\sqrt{6 \times 7 \times (6+7+1)/12}} = 3.$$

Since, for $\alpha = 0.01$ (1%), the upper $\alpha/2$ percentile of the standard normal distribution is $T_{0.005} = 2.58$, we reject the null hypothesis that the efficiency scores of these two groups belong to the same distribution at the 1% significance level. That is, Group B outperforms Group A in efficiency.

Table 7.9. Example of Bilateral Comparisons

Group	DMU	Score	Reference set	Rank
		Against B		
A	L1	0.226006153	L23	13
A	L2	0.637737534	L23	10
A	L3	0.540054844	L23	12
A	L4	0.593020886	L23	11
A	L5	0.911284882	L17 L23	8
A	L6	0.744964289	L23	9
	Average	0.608844765	Rank Sum	63
		Against A		
B	L17	1.283765162	L5 L6	5
B	L18	1.309557862	L6	4
B	L19	1.769104754	L6	1
B	L20	1.408073355	L6	3
B	L21	1.26630526	L5 L6	6
B	L22	1.000331642	L6	7
B	L23	1.612422038	L5 L6	2
	Average	1.378508582	Rank Sum	28

7.6.6 Evaluating Efficiencies of Different Organization Forms

We turn to a study of the U.S. insurance industry and show how the approach described in Section 7.6.3 can be used to address a problem that has not been satisfactorily addressed by other methods. For this purpose we note that the U.S. insurance industry has long been dominated by two different forms of organization: "mutual companies" which are organized in the manner of co-operatives so that policyholders are also the owners of the company and (2) "stock companies" which are organized in the manner of private corporations with stockholders as owners and policy holders as customers. This has led to many studies directed to discovering whether one of these organization forms is superior (i.e., more efficient) than the other.

For instance, using data on various accounting ratios — such as rate of return on sales, etc., — a supposedly decisive study by Fama and Jensen (1983) concluded that each of these two forms was efficient — indeed, more efficient — for the niche it occupied.[5] The presumption was that organizations compete for survival. The fact that both forms continue to survive therefore documents the claimed result.

In a more recent study, however, Brockett *et al.* (1998)[6] argued that, first, Fama and Jensen confused observed performance (efficient or not) with the potential capabilities of these two types of organization. Second, Brockett *et al.* argue that a use of ratios used separately on the assumption of independence of the resulting tests did not provide a satisfactory basis for evaluating the performance of institutions, such as insurance companies, which must have multiple interacting objectives if they are to meet the needs of an industry charged with a "public interest."[7] Brockett *et al.* — turning to DEA — eliminated the need for *assuming* that technical efficiency is attained by every DMU — an assumption which underlies the "agency theory" approaches used by Fama and Jensen.

To satisfy the first and third of these study objectives, Brockett *et al.* followed the approaches described in the immediately preceding section of this chapter. Thus, in agreement with Section 7.6.3 they identified two efficient frontiers — one for mutual companies and another for stock companies. They then used the results from applying DEA to the data for each of these two groups to project each DMU onto its respective (mutual or stock) efficient frontiers. See the formulas in (3.28) and (3.29). This eliminated the need for *assuming* that efficiency had been attained and it also moved each DMU up to the boundary of the capabilities which the evidence showed was possible for its form of organization.

This done, Brockett *et al.* again utilized DEA to establish a new (combined) frontier against which each entity (=mutual or stock company) could be accorded a new (relative) efficiency rating. Note therefore that this rating showed how each entity could perform under the condition that it had already achieved the best performance that its form of organization allowed. Finally the Mann-Whitney test procedures were applied (a) to see if the rankings for two groups differed significantly and (b) to determine which group dominated when the hypothesis maintained under (a) was rejected.

This study covered other items such as the forms of sales and marketing systems used by these companies. However, we do not go into further detail and focus, instead, on the issue of relative efficiency of these two organization forms. Thus applying the approach we just described to 1989 data for n_1=1114 stock companies and n_2=410 mutual companies in the U.S. property-liability insurance industry,[8] Brockett *et al.* concluded that the stock form was uniformly more efficient (and significantly so) in all of the dimensions studied. *Inter alia*, they documented this finding by noting that stock companies could merge and acquire (or be acquired by) other companies in ways that were not available to mutual companies in order to obtain access to resources, talents and markets. Recent activities in "de-mutualization" by some of the largest U.S. mutual companies tends to support the study results, especially when it is noted that no movements have occurred in the opposite direction. Hence we conclude with Brockett *et al.* that this use of DEA uncovered latent possibilities that were not detected in the approaches used by others, as well as Fama and Jensen.

7.7 SUMMARY OF CHAPTER 7

In this chapter we expanded the ability of DEA to deal with variables that are not under managerial control but nevertheless affect performances in ways that need to be taken into account when effecting evaluations. Non-discretionary and categorical variables represent two of the ways in which conditions beyond managerial control can be taken into account in a DEA analysis. Uses of upper or lower bounds constitute yet another approach and, of course, these approaches can be combined in a variety of ways. Finally, uses of Wilcoxon-Mann-Whitney statistics were introduced for testing results in a nonparametric manner when ranking can be employed.

Illustrative examples were supplied along with algorithms that can be used either separately or with the computer code DEA-Solver. We also showed how to extend DEA in order to deal with production possibility sets (there may be more than one) that are not convex. Finally we provided examples to show how new results can be secured when DEA is applied to such sets to test the efficiency of organization forms (and other types of activities) in ways that were not otherwise available.

7.8 NOTES AND SELECTED BIBLIOGRAPHY

Banker and Morey (1986a) developed the first model for evaluating DEA efficiency with exogenously fixed, inputs and outputs, with other models following as in Charnes *et al.* (1987), Adolphson *et al.* (1991)[9] and Tone (1993).[10] (See also Problems 7.9 and 7.10.)

Models treating categorical inputs and outputs were also introduced by Banker and Morey (1986b).[11] They formulated the controllable categorical variable problem within the framework of a mixed-integer LP model under the BCC model. Kamakura (1988)[12] pointed out shortcomings of their formulation and presented another mixed-integer model. Rousseau and Semple (1993)[13] presented a new algorithm for this case, as an extension of Kamakura's work (1988). This method eliminated difficulties of computation that had accompanied earlier mixed integer models. It should be noted that the algorithms introduced in Sections 7.3 and 7.4 can be coupled with any DEA model, in contrast to earlier methods. (See Tone (1997).[14])

Tone (1995)[15] proposed the bounded variable models in Section 7.3. The materials in Section 7.5 (comparisons of efficiency between different systems) was introduced by Tone (1993).[16]

The use of DEA to effect bilateral comparisons were first explored in Charnes, Cooper and Rhodes (1981).[17] The rank-sum test in Section 7.5.3 was developed by Brockett and Golany (1996).[18] See also Brockett and Levine (1984)[19] for a detailed development.

7.9 RELATED DEA-SOLVER MODELS FOR CHAPTER 7

NCN-I-C(V),NCN-O-C(V) These codes solve the input (output)-oriented non-controllable variable model expressed by (7.3) under a constant (variable) returns-to-scale assumption. The sample data format can be found in Figure B.3 in Section B.5.3 of Appendix B, where non-controllable input or output are designated by (IN) or (ON)), respectively. The results of the computation are recorded in worksheets "Summary," "Score," ("Rank,") "Projection," "Graph1," and "Graph2."

NDSC-I(O)-C(V, or GRS) These codes solve the input (output)-oriented non-discretionary variable model under the constant (variable or general) returns-to-scale assumption that is described in (7.6). The input data format is the same with the NCN.

BND-I-C(V or GRS), BND-O-C(V or GRS) These codes solve the input (output)-oriented bounded variable model expressed by (7.10) or (7.11) under a constant (variable or general) returns-to-scale assumptions. See Figure B.4 in Section B.5.4 of Appendix B for reference, where (IB)Doc., (LB)Doc. and (UB)Doc. designate "Doc.(doctor)" as a bounded variable and its lower/upper bounds (LB)Doc. and (UB)Doc. Joint use of an input variable and a bounded variable is permitted. It is thus possible to register Doc. as (I) and at the same time as (IB). See Problems 7.9 and 7.10. The results of the computation are recorded in worksheets "Summary," "Score," "Rank," "Projection," "Graph1," and "Graph2."

CAT-I-C(V), CAT-O-C(V) These codes solve the input (output)-oriented hierarchical category model as introduced in Section 7.4.1. See Figure B.5 in Appendix B for an example. The main body of the data format is the same as for the CCR case, with an additional column showing the categorical level. Its heading is arbitrary, e.g. Category. The category number ranges 1, 2, ..., L, where DMUs in Level 1 are in the most disadvantageous condition and they will be compared only among themselves. Level 2 is in a better position than 1, and the DMUs in Level 2 will be compared with reference to DMUs in Levels 1 and 2 and so on. It is recommended that the level numbers be continuously assigned starting from 1. The results of the computation are recorded in worksheets "Summary," "Score," "Rank," "Projection," "Graph1," and "Graph2." Also, statistics by category class can be found in "Summary."

SYS-I-C(V), SYS-O-C(V) These codes deal with comparisons of efficiency between different systems, as introduced in Section 7.5. The data format is the same as for the SYS model, with the last column showing the system number each DMU belongs to. In contrast to the CAT model, this number is used only for discrimination purposes but it is recommended that the system numbers be continuously assigned starting from 1. The results of the computation are recorded in worksheets "Summary," "Score," "Rank," "Projection," "Graph1," and "Graph2." Also, statistics by system class can be found in "Summary."

Bilateral This code solves the bilateral comparisons model expressed by (7.15)-(7.18) in Section 7.6.4. It tests the null hypothesis that the efficiency of

DMUs in the concerned two groups belongs to the same statistical distribution using nonparametric rank-sum statistics. The data format is the same as for the CAT model, where the only level numbers allowed to be assigned are 1 or 2. The results of this test are shown in "Score," as well as in "Summary." This model has four variations as follows:

- Bilateral-CCR-I, Bilateral-BCC-I
- Bilateral-SBM-C, Bilateral-SBM-V

7.10 PROBLEM SUPPLEMENT FOR CHAPTER 7

Problem 7.1

Part (1): Formulate the dual to (7.1) and discuss its uses in evaluating performances.

Suggested Response: The dual to (7.1) can be represented as follows:

$$\begin{aligned}
&\max_{\mu,\nu} \sum_{r=1}^{s} \mu_r y_{ro} - \sum_{i \in ND} \nu_i x_{io} \\
&\text{subject to} \\
&\sum_{r=1}^{s} \mu_r y_{rj} - \sum_{i \in ND} \nu_i x_{ij} - \sum_{i \in D} \nu_i x_{ij} \leq 1 \quad (j = 1, \ldots, n) \\
&\sum_{i \in D} \nu_i x_{io} = 1 \\
&\nu_i \geq \varepsilon, i \in D; \ \nu_i \geq 0, i \in ND; \ \mu_r \geq \varepsilon, r = 1, \ldots, s.
\end{aligned}$$

Whereas (7.1) evaluates the discretionary variables in the set D in its objective, the above model evaluates the non-discretionary variables in the set ND in its objective. Between the two all possibilities for the evaluation are covered and are related to each other by the duality theorem of linear programming. See A.4 in Appendix A. Thus, if a ND constraint is critical it will have zero slack in the objective of (7.1) and $\nu_i^* > 0$ in the above dual (multiplier) model to evaluate the limitation it imposes on performance. If non-zero slack appears in (7.1) then the corresponding dual variable $\nu_i^* = 0$ in the multiplier model so this non-zero slack evaluates the complete discretion this constraint allows to management in the primal (=envelopment) model.

Part (2): Illustrate your interpretation by formulating and interpreting the dual to the problem that was used to evaluate P_5 immediately following Figure 7.1.

Suggested Response: Reference to the formulation of the dual in Problem 7.1 gives the following as the multiplier model that is dual to the model that was used to evaluate P_5 in the discussion following Figure 7.1,

$$\max \mu - 2\nu_2$$

$$
\begin{aligned}
&\text{subject to} \\
&1 = 2\nu_1 \\
&0 \geq \mu - \nu_1 - 2\nu_2 \\
&0 \geq \mu - \nu_1 - \frac{3}{2}\nu_2 \\
&0 \geq \mu - 2\nu_1 - \nu_2 \\
&0 \geq \mu - 3\nu_1 - \nu_2 \\
&0 \geq \mu - 2\nu_1 - 2\nu_2 \\
&\mu, \nu_1 \geq \varepsilon;\ \nu_2 \geq 0.
\end{aligned}
$$

The solution to this problem is $\mu^* = \nu_1^* = \frac{1}{2}$ with all other variables zero. This solution is confirmed by noting that $\mu^* = \theta^* = \frac{1}{2}$ as required by the duality theorem. Note, therefore, that $\nu_2^* = 0$ shows that this constraint did not enter into the evaluation of the performance of the DMU associated with P_5.

Problem 7.2

Part (1): Modify the problem following Figure 7.1 to reflect the fact too much rainfall damaged a farmer's crop. Then discuss how this affects the above dual (multiplier) model formulation and interpretation.

Suggested Response: This can be most easily expressed by replacing the inequality for the ND constraint (representing rainfall) by an equation in the primal problem used to evaluate P_5. This affects the dual, as formulated above, by allowing the ν_i associated with this constraint to assume negative as well as positive values and thus eliminates the condition $\nu_i \geq \varepsilon > 0$ for this variable.

Part (2): Apply the above modification to the preceding illustrative example.

Suggested Response: Replacing the inequality by an equation in the primal problem eliminates the alternate optimum possibility which had $s_2^{-*} > 0$ in the solutions to (7.1). The uniquely optimal solution to this primal is then $\theta^* = \frac{1}{2}$, $\lambda_1^* = 1$ and all other variables zero. This replacement by an equation in the primal results in a solution to the dual with $\nu_1^* = \frac{1}{2}$, $\nu_2^* = -\frac{1}{4}$ and all other variables zero to yield $-2\nu_2^* = \theta^* = \frac{1}{2}$, as required by the duality theorem.

Note that this value of $\nu_2^* = -\frac{1}{4}$ means that performance could have been improved at the rate $\nu_2^* = -\frac{1}{4}$ if $x_{2o} = 2$ had been lowered in the primal. For instance, if $x_{2o} = 2$ were replaced by $\widehat{x}_{2o} = 1$ the point would coincide with P_3, which is efficient, as shown in Figure 7.1, with the uniquely optimal solution $\theta^* = 1$, $\lambda_3^* = 1$ and all other variables zero. Hence the entire inefficiency is caused by the excess rainfall.

Remark: The usual techniques used in the linear programming literature for exploring such changes are not applicable in DEA because the changes occur on both sides of the inequalities. See A. Charnes, W.W. Cooper, A. Lewin and R. Morey (1985) "Sensitivity and Stability Analysis in DEA," Annals of Operations Research 2, pp.139-156.

Problem 7.3

Interpret (7.2) and discuss its use when (1) all β_i, $\gamma_r = 0$ and when (2) some or all of these parameters are assigned positive values.

Suggested Response : (1) If all β_i, $\gamma_r = 0$ then all inputs and outputs are non-discretionary. All slacks will be zero in the objective as well as the constraints. Hence no inefficiency is present and any solution is as good as any other solution because inefficiency must be associated with discretionary variables if it is to be correctable.
(2) If some or all parameters are assigned positive values then any nonzero slack which appears in the objective (a) incurs a penalty and (b) shows where corrections can be made. For instance, if all $\beta_i = 0$ and all $\gamma_r = \infty$ — so the latter collection of constraints can be omitted — then attention is focussed on the output improvements that are possible when all inputs are non-discretionary and have to be used in observed amounts.

Problem 7.4

Relate (7.3) to (7.2) and (7.1).

Suggested Response : If all β_i, $\gamma_r = 0$ in (7.2) then it can be simplified by omitting these constraints and replacing the corresponding inequality relations by equalities as in the sets indexed by N in (7.3). The remaining expressions in (7.3) can be brought into correspondence with (7.1) by (a) omitting the condition $L \leq e\boldsymbol{\lambda} \leq U$, or, equivalently, setting $L = 0$ and $U = \infty$ and (b) omitting the slacks in the objective of (7.1).

Problem 7.5

Background : Recall from Section 7.2 in this chapter that an "active" vector is one which enters into a solution with a positive coefficient. We can express this mathematically as follows,

$$\theta_o^* x_{io} = \sum_{j=1, j\neq k}^{n} x_{ij}\lambda_j^* + x_{ik}\lambda_k^* + s_i^{-*}, \quad i = 1, \ldots, m \tag{7.19}$$

$$y_{ro} = \sum_{j=1, j\neq k}^{n} y_{rj}\lambda_j^* + y_{rk}\lambda_k^* - s_r^{+*}, \quad r = 1, \ldots, s$$

with all variables nonnegative. Here we have simplified (7.1) by assuming that non-discretionary inputs or outputs are not to be considered. Then the vector with components (x_{ik}, y_{rk}) is "active" if $\lambda_k^* > 0$.

We are using "*" to represent an optimum value in the following theorem.

Theorem 7.1 *An input-output vector with components (x_{ik}, y_{rk}) will not enter actively into an optimum evaluation for any DMU_o unless the DMU_k with which it is associated is DEA-efficient.*

Assignment : Prove this theorem.

Suggested response : Although we have proved this theorem in the form of Theorem 3.3 using the multiplier model in Chapter 3, we will give another proof which uses the envelopment model and extends to BCC as well as CCR models. Assume that DMU_k is not efficient so that either or both of the conditions for efficiency as specified in Definition 3.2 must fail to be satisfied. Then consider, first, the case in which DMU_k's failure to be efficient is identified with $0 < \theta_k^* < 1$. The solution used to evaluate DMU_k can be written

$$\theta_k^* x_{ik} = \sum_{j=1}^{n} x_{ij} \widehat{\lambda}_j^* + \widehat{s}_i^{\,-*}, \quad i = 1, \ldots, m \tag{7.20}$$
$$y_{rk} = \sum_{j=1}^{n} y_{rj} \widehat{\lambda}_j^* - \widehat{s}_r^{\,+*}, \quad r = 1, \ldots, s$$

with all variables nonnegative and $0 < \theta_k^* < 1$.

Substitution in (7.19) gives

$$\theta_o^* x_{io} = \sum_{j=1, j\neq k}^{n} x_{ij}\lambda_j^* + s_i^{-*} + \frac{\lambda_k^*}{\theta_k^*}\left[\sum_{j=1}^{n} x_{ij}\widehat{\lambda}_j^* + \widehat{s}_i^{\,-*}\right], \quad i = 1, \ldots, m \tag{7.21}$$
$$y_{ro} = \sum_{j=1, j\neq k}^{n} y_{rj}\lambda_j^* - s_r^{+*} + \lambda_k^*\left[\sum_{j=1}^{n} y_{rj}\widehat{\lambda}_j^* - \widehat{s}_r^{\,+*}\right], \quad r = 1, \ldots, s$$

By hypothesis $0 < \theta_k^* < 1$. Also λ_k^* and its coefficients (in square brackets) are positive. Hence

$$\sum_{j=1}^{n} x_{ij}\lambda_j^* + s_i^{-*} = \theta_o^* x_{io} > \sum_{j=1, j\neq k}^{n} x_{ij}\lambda_j^* + s_i^{-*} + \lambda_k^*\left[\sum_{j=1}^{n} x_{ij}\widehat{\lambda}_j^* + \widehat{s}_i^{\,-*}\right], \tag{7.22}$$
$$i = 1, \ldots, m$$
$$y_{ro} = \sum_{j=1, j\neq k}^{n} y_{rj}\lambda_j^* - s_r^{+*} + \lambda_k^*\left[\sum_{j=1}^{n} y_{rj}\widehat{\lambda}_j^* - \widehat{s}_r^{\,+*}\right]. \quad r = 1, \ldots, s$$

Collecting terms on the right a solution is obtained which satisfies all constraints. Because strict inequality holds in all of the first m constraints, the value of θ_o^* can be reduced while continuing to satisfy all constraints. Hence the assumption $\lambda_k^* > 0$ for $0 < \theta_k^* < 1$ is not consistent with the optimality assumed for (7.19).

Now assume $\theta_k^* = 1$ but some slacks are not zero. In this case (7.21) becomes

$$\theta_o^* x_{io} = \sum_{j=1, j\neq k}^{n} x_{ij}\lambda_j^* + s_i^{-*} + \lambda_k^*\left[\sum_{j=1}^{n} x_{ij}\widehat{\lambda}_j^* + \widehat{s}_i^{\,-*}\right], \quad i = 1, \ldots, m \tag{7.23}$$

$$y_{ro} = \sum_{j=1, j\neq k}^{n} y_{rj}\lambda_j^* - s_r^{+*} + \lambda_k^* \left[\sum_{j=1}^{n} y_{rj}\widehat{\lambda}_j^* - \widehat{s}_r^{+*} \right]. \quad r = 1, \ldots, s$$

Now if any $\widehat{s}_i^{-*}$ or $\widehat{s}_r^{+*} \neq 0$ then $\lambda_k^* > 0$ would multiply these nonzero slacks to yield a new set with a greater sum than $\sum_{i=1}^m s_i^{-*} + \sum_{r=1}^s s_r^{+*}$. Therefore the latter sum could not be maximal as required for optimality in the objective for the second stage maximization of the CCR model.

Thus, if either $0 < \theta_k^* < 1$ or some slacks are not zero in (7.20) then $\lambda_k^* > 0$ cannot hold in (7.19). The same is true, *a fortiori*, if both $0 < \theta_k^* < 1$ and some slacks are not zero. The theorem is therefore proved. DMU_k cannot be active in the evaluation of any DMU_o. □

Corollary 7.1 *The above theorem is also true for the class of additive models.*

The proof of this corollary follows from the sum of slacks being maximal, as in case 2 of the above proof.

Corollary 7.2 *The above theorem, as proved for the CCR model, also holds for the BCC model.*

Proof : Non-negativity of the $\lambda_j^*, \widehat{\lambda}_j$ as required for the CCR model satisfies one part of the convexity condition prescribed for the BCC model. To show that the other condition is also satisfied when the BCC model is used instead of the CCR model, we collect the variables in (7.23) to define the following new set

$$\begin{aligned} \lambda_j^{new} &= \lambda_j^* + \lambda_k^* \widehat{\lambda}_j^*, \ j \neq k \\ \lambda_k^{new} &= \lambda_k^* \widehat{\lambda}_k^*. \end{aligned}$$

Therefore

$$\sum_{j\neq k} \lambda_j^{new} + \lambda_k^{new} = \sum_{j\neq k} \lambda_j^* + \lambda_k^* \left(\sum_{j\neq k} \widehat{\lambda}_j^* + \widehat{\lambda}_k^* \right).$$

We now show that this is equal to unity when

$$\sum_{j=1}^{n} \lambda_j^* = \sum_{j=1}^{n} \widehat{\lambda}_j^* = 1.$$

This result is immediate if λ_k^* is zero or unity. Therefore assume $0 < \lambda_k^* < 1$, in which case we have

$$\lambda_k^* \left(\sum_{j=1}^{n} \widehat{\lambda}_j^* \right) = \lambda_k^*$$

and

$$\sum_{j\neq k} \lambda_j^* = 1 - \lambda_k^*$$

so, summing, we have

$$\sum_{j \neq k} \lambda_j^* + \lambda_k^* \left(\sum_{j \neq k} \widehat{\lambda}_j^* + \widehat{\lambda}_k^* \right) = \sum_{j=1, j \neq k}^{n} \lambda_j^{new*} + \lambda_k^{new*} = 1,$$

as claimed. □

Problem 7.6

Using the point $P_1 = (1, 2)$ in Figure 7.1 and the model (7.2), show that the above theorem and corollaries need not hold when non-discretionary variables are involved.

Suggested Response : If $x_2 = 2$ is the only non-discretionary variable, the application of (7.2) to the data in Figure 7.1 yields

$$\begin{aligned}
\max \quad & s_1^- + s_2^- + s^+ \\
\text{subject to} \quad & \lambda_1 + \lambda_2 + 2\lambda_3 + 3\lambda_4 + 2\lambda_5 + s_1^- = 1 \\
& 2\lambda_1 + 3/2\lambda_2 + \lambda_3 + \lambda_4 + 2\lambda_5 + s_2^- = 2 \\
& \lambda_1 + \lambda_2 + \lambda_3 + \lambda_4 + \lambda_5 - s^+ = 1 \\
& s_1^- \leq 1 \\
& s_2^- \leq 2\beta_2
\end{aligned}$$

with all variables nonnegative and the condition $s^+ \leq \infty$ omitted. Setting $\beta_2 = 0$ yields $\lambda_1^* = 1$ and all other variables at zero so P_1 is a member of the active solution set and is also characterized as efficient with all slacks at zero in this optimum. Changing to $\beta_2 = 1/8$ yields the following optimum

$$\lambda_1^* = \lambda_2^* = 1/2, \; s_2^{-*} = 1/4$$

with all other variables at zero. Hence P_1 changes to inefficient status under definition 7.2. This change of status shows that the preceding theorem does not apply in this case because P_1 is needed in the expression of this optimum. In fact, changing from $\beta_2 = 1/8$ to $\beta_2 = 1/4$ gives an optimum with $\lambda_2^* = 1$, $s_2^{-*} = 1/2$ and all other variables zero. P_1 is no longer an active member of this optimum because it is no longer needed with this degree of discretion allowed.

Problem 7.7

Now use model (7.1), the Banker-Morey model, to evaluate P_1 in Figure 7.1. Then interpret and compare the results from this model with the last results secured from model (7.2), the Charnes-Cooper-Rousseau-Semple model, in the preceding question.

Suggested response : Applying (7.1) to the data in Figure 7.1 yields the following model

$$\min \quad \theta - \varepsilon(s_1^- + s^+)$$

$$\begin{aligned}\text{subject to}\quad & \theta = \lambda_1 + \lambda_2 + 2\lambda_3 + 3\lambda_4 + 2\lambda_5 + s_1^- \\ & 2 = 2\lambda_1 + 3/2\lambda_2 + \lambda_3 + \lambda_4 + 2\lambda_5 + s_2^- \\ & 1 = \lambda_1 + \lambda_2 + \lambda_3 + \lambda_4 + \lambda_5 - s^+ \\ & 0 \leq \lambda_1, \ldots, \lambda_5, s_1^-, s_2^-, s^+.\end{aligned}$$

An optimal solution to this problem is

$$\theta^* = 1,\ \lambda_2^* = 1,\ s_2^{-*} = 1/2$$

and all other variables are zero.

This solution is the same as the one previously secured with model (7.2) but its interpretation is different because no penalty is associated with s_2^{-*} as would be appropriate if this unutilized capacity was left idle because no alternative use for this resource was available. P_1 was characterized as inefficient with this same nonzero slack of $s_2^{-*} = 1/2$ because this value, 25% of this input, could have been used in a less valuable secondary usage. This secondary usage could have been accommodated, moreover, without worsening efficiency in the primary activity undertaken. Hence, in this case it is proper to penalize P_1 for its failure to make this nonzero slack available to this secondary use.

Problem 7.8

The Banker-Morey formulation of non-discretionary variables is described by (7.1) at the beginning of this chapter and the BND model in DEA-Solver can solve the bounded-variable model (7.10) in Section 7.3.

Assignment :
(1) How can you solve (7.1) by transforming it into (7.10)?
(2) Solve the example depicted in Figure 7.1 by the BND-I-C (input-oriented bounded variable model under constant returns-to-scale assumption) code.

Suggested Answer :
(1) Since the non-discretionary inputs in (7.1) are bounded above by the observed values, we can transform this problem into (7.10) by setting

$$\boldsymbol{l}_o^{N_x} = \boldsymbol{0},\ \boldsymbol{u}_o^{N_x} = \boldsymbol{x}_o^N,\ L = 0,\ U = \infty,\ Y^N = \emptyset\ (empty).$$

(2) The input data format for the BND model is explained in Appendix B (Section B.5.4). Following this, the input data for the problem described in Figure 7.1 must be prepared on an Excel Sheet as follows,

Problem 7.8	(I)x1	(IB)x2	(LB)x2	(UB)x2	(O)y
P1	1	2	0	2	1
P2	1	1.5	0	1.5	1
P3	2	1	0	1	1
P4	3	1	0	1	1
P5	2	2	0	2	1

Here "(IB)x2" designates x_2 as a bounded input variable, "(LB)x2" and "(UB)x2" denote its lower and upper bounds, respectively.

After running the BND-I-C code, we found the optimal solution as summarized below:

For P_1: $\theta^* = 1, s_2^{-*} = 1/2$ and is inefficient. Its reference is P_2.
For P_2: $\theta^* = 1$ and is efficient.
For P_3: $\theta^* = 1$ and is efficient.
For P_4: $\theta^* = 2/3, s_1^{-*} = 1$ and is inefficient. Its reference is P_3.
For P_5: $\theta^* = 1/2, s_1^{-*} = 1, s_2^{-*} = 1/2$ and is inefficient. Its reference is P_2.

Notice that the BND model maximizes the sum of slacks including non-discretionary variables and hence P_5's projection is P_2.

Problem 7.9

In Section 7.3.2 we introduced bounded variable models where it is implicitly assumed that the set of controllable variables (C) and that the set of bounded variables (N) are mutually exclusive (disjoint), i.e. $C \cap N = \emptyset$.

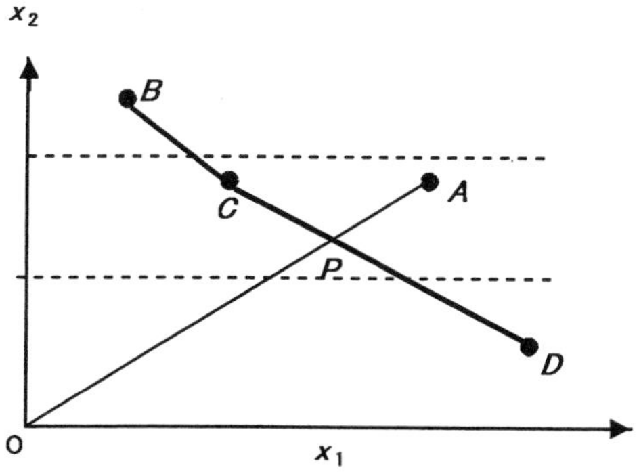

Figure 7.6. Case 1

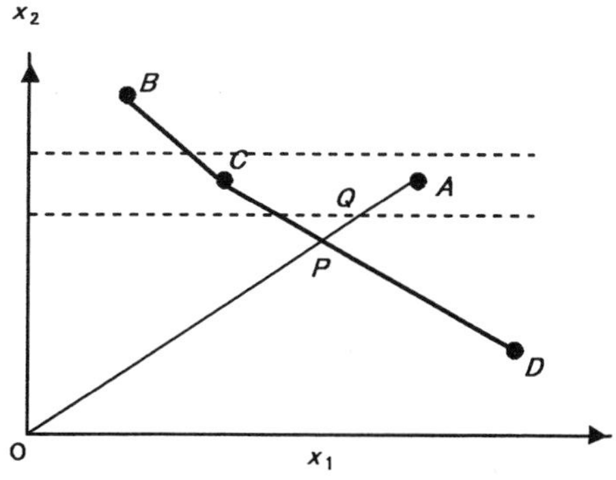

Figure 7.7. Case 2

Consider the case $C \cap N \neq \emptyset$, i.e. some (or all) bounded variables are directly related in the evaluation of the DMU efficiency scores, although they are still bounded.

Assignment : What is induced by this change of situation?

Suggested Response : We illustrate this in the case of two inputs (x_1, x_2) and one "unitized" output $(y = 1)$ where it is assumed that $C = \{x_1, x_2\}$ and $N = \{x_2\}$, i.e. x_2 is bounded but should nevertheless be accounted for directly in the efficiency evaluation. We observe two feasible cases as depicted in Figures 7.6 and 7.7. In both cases the efficient frontiers are spanned by bold line segments connecting B, C and D and we evaluate the efficiency of DMU A whose upper and lower bounds of x_2 are designated by broken lines.

Since we incorporated two inputs into evaluating the efficiency of A using an input-oriented model, the CCR projection is the point P on the frontier. In Figure 7.6, the x_2-value of P is in between the bounds and hence we have $\theta^*_{CCR} = \theta^*_{BND}$ for A. In case 2 in Figure 7.7, on the other hand, the projected point P is outside the bounds and A must therefore be projected to Q, a point which is not on the frontier defined by B, C and D, in order to satisfy the lower bound. Thus, in this case we have $\overline{OP}/\overline{OA} = \theta^*_{CCR} < \theta^*_{BND} = \overline{OQ}/\overline{OA}$ for A. However, these relations will differ by the selection of the sets C and N and hence this inequality need not always be true.

Problem 7.10

In succession to the preceding problem:
(i) Use the BND-I-C code in DEA-Solver for solving the following problem with two inputs (x_1, x_2) and single unitized output (y), where x_2 is discretionary but bounded and should be included in the evaluation of the efficiency score, i.e. the sets $C = \{x_1, x_2\}$ and $N = \{x_2\}$.

	INPUT		Lower bound	Upper bound	OUTPUT
	x_1	x_2	x_2	x_2	y
A	4	5	4.5	7	1
B	1	8	0	10	1
C	2	6	0	10	1
D	5	2	0	10	1

Thus, A's efficiency θ^* is evaluated by solving the following LP,

$$\begin{aligned} \min \quad & \theta \\ \text{subject to} \quad & 4\theta = 4\lambda_1 + \lambda_2 + 2\lambda_3 + 5\lambda_4 + s_1^- \quad (\text{for } x_1) \\ & 5\theta = 5\lambda_1 + 8\lambda_2 + 6\lambda_3 + 2\lambda_4 + s_2^- \quad (\text{for } x_2) \\ & 1 = \lambda_1 + \lambda_2 + \lambda_3 + \lambda_4 - s^+ \quad (\text{for } y) \\ & 4.5 \leq 5\lambda_1 + 8\lambda_2 + 6\lambda_3 + 2\lambda_4 \leq 7. \quad (\text{bounds of } x_2 \text{ for } A) \end{aligned} \tag{7.24}$$

(ii) Compare the results of (i) with the case where x_1 is the only controllable variable and x_2 is bounded, i.e. the sets $C = \{x_1\}$ and $N = \{x_2\}$.
In this case (7.24), which connects x_2 directly to the efficiency evaluation, is dropped from the above constraints for evaluating the efficiency of A.

Suggested Response :

(i) Prepare an Excel sheet as described below.

(1)	(I)x1	(I)x2	(IB)x2	(LB)x2	(UB)x2	(O)y
A	4	5	5	4.5	7	1
B	1	8	8	0	10	1
C	2	6	6	0	10	1
D	5	2	2	0	10	1

Here the variable x_2 appears as an input (I) and as a bounded (IB) variable. This kind of mixed (joint) use is permitted in the BND code.

The solution is summarized as follows,

The DMU A has $\theta^* = 0.9$ and its projection onto the efficient frontier is $x_1 = 3.125$, $x_2 = 4.5$ (the lower bound). All other DMUs are efficient. (This corresponds to Case 2 in Figure 7.7.)

(ii) In this case the Excel sheet is standard, like the one explained in Section B.5.4, and is as follows,

(2)	(I)x1	(IB)x2	(LB)x2	(UB)x2	(O)y
A	4	5	4.5	7	1
B	1	8	0	10	1
C	2	6	0	10	1
D	5	2	0	10	1

The results show that only B is efficient and all other DMUs are inefficient, i.e., $\theta_A^* = 0.375(\lambda_B^* = 1), \theta_C^* = 0.5(\lambda_B^* = 1), \theta_D^* = 0.2(\lambda_B^* = 1)$. Although the projections of C and D are directed to B, A is projected to $x_1 = 1.5, x_2 = 7, y = 1$. This is caused by the upper bound (7) on x_2 listed under (UB)x2 for A in the above table.

Figure 7.8 depicts the relationship between the "primary" input x_1 and output y. From this figure we can see that the efficient frontier is spanned by B

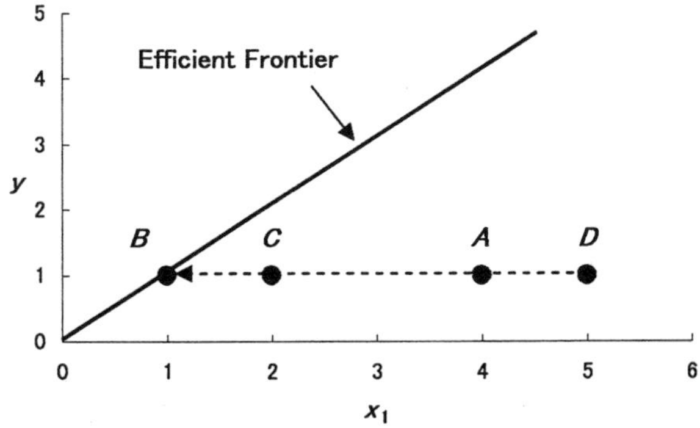

Figure 7.8. Controllable Input and Output

and all other DMUs are to be projected to B with coordinates ($x_1 = 1$, $x_2 = 8$) in order to become efficient. However, A cannot attain B since its upper bound of $x_2 = 7$ bars it from doing so.

As observed in this example, the choice of "primary" (discretionary) and "secondary" (bounded) variables affects the evaluation of efficiency substantially so care needs to be exercised in deciding which variables should be "primary."

Problem 7.11 (Categorical Levels Expanded)

In Section 7.4, we introduced the use of "categories" into efficiency evaluations. For the supermarket example discussed in the beginning of Section 7.4, we classified stores facing severe competition as *category* 1 (C_1), in a normal situation as *category* 2 (C_2) and in an advantageous situation as *category* 3 (C_3). Then we evaluated stores in C_1 only with respect to stores within the group. Stores in C_2 were then evaluated with reference to stores in C_1 and C_2. Finally, stores in C_3 were evaluated with all stores in the entire data set.

Generally, if DMUs in category C_i are used for evaluating DMUs in category C_j, we say C_i is at a *lower level* than C_j and denote this by using the symbol "$\succ$" as in

$$C_i \succ C_j.$$

The above example has level relations,

$$C_1 \succ C_2 \succ C_3.$$

This relation can be diagrammed by Figure 7.9.

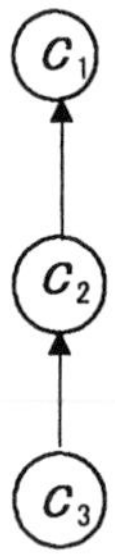

Figure 7.9. Categorization (1)

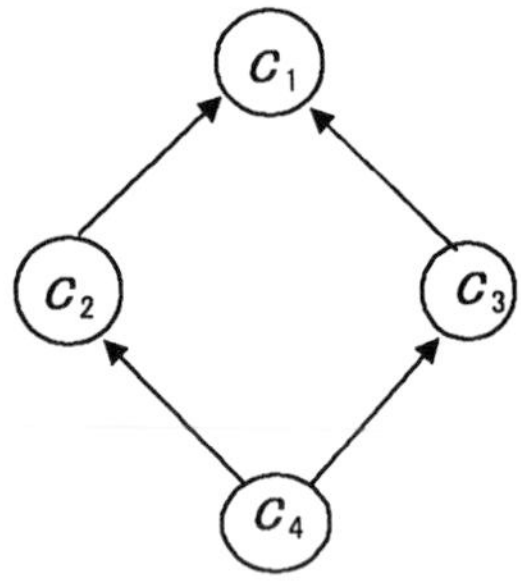

Figure 7.10. Categorization (2)

However, there are situations where the order of levels between categories cannot be readily identified. For example suppose we have four categories of book stores, i.e. stores with both magazine drive-through and free delivery service (C_4), stores with magazine drive-through but no free delivery service (C_2), stores with free delivery service but no magazine drive-through (C_3) and stores with neither magazine drive-through nor free delivery service (C_1). In this case the hierarchy level of C_2 and C_3 is not clearly identified. Hence the level relations can be exhibited by a diagram as Figure 7.10.

Assignment : Enumerate the reference group for each category group in Figure 7.10.

Suggested Response :

Category	Reference Category
C_1	C_1 C_2 C_3 C_4
C_2	C_2 C_4
C_3	C_3 C_4
C_4	C_4

By this is meant that C_1 can be used to evaluate all 4 reference categories. C_2 can be used to evaluate C_2 and C_4. C_3 can be used to evaluate C_3 and C_4. Finally, however, C_4 can be used to evaluate DMUs in only its own class. Thus we have

$$C_1 \succ C_2,\ C_1 \succ C_3,\ C_1 \succ C_4,\ C_2 \succ C_4,\ C_3 \succ C_4$$

which reduces to $C_1 \succ C_2 \succ C_4$ and $C_1 \succ C_3 \succ C_4$ but no further reduction is possible because no such relation holds for C_2 and C_3.

Problem 7.12 (Bilateral Comparisons)

Background : In Section 7.3.3 we introduced the data for the twelve Japanese professional baseball teams (see Table 7.3), with the first six teams belonging to "Central League" and the last six to "Pacific League." It is sometimes said that the Pacific League is not as attractive as the Central League among baseball fans.

Assignment : Use the Bilateral Comparisons model in DEA-Solver to verify this rumor by taking the average salary of "Manager" and "Player" as inputs and "Attendance" as output.

Suggested Answer : Prepare the following Excel sheet as the input data for bilateral comparisons.

Team	(I)Manager	(I)Player	(O)Attendance	League
Swallows	5250	6183	80.87	1
Dragons	2250	5733	93.76	1
Giants	6375	8502	100.00	1
Tigers	3125	4780	76.53	1
Bay Stars	3500	4042	79.12	1
Carp	3125	5623	51.95	1
Lions	5500	10180	56.37	2
Fighters	3625	5362	57.44	2
Blue Wave	2715	4405	58.78	2
Buffalos	3175	6193	53.00	2
Marines	2263	5013	43.47	2
Hawks	3875	3945	82.78	2

where column heading (I) refers to input and (O) refers to output while "League" 1 corresponds to Central and 2 to Pacific, as listed on the right. The "Bilateral" code in DEA-Solver gives the following results,

Rank	DMU	Score	League
1	Dragons	1.924754452	1
2	Tigers	1.133467367	1
3	Hawks	1.07198435	2
4	Bay Stars	1.05497859	1
5	Carp	0.767848928	1
6	Blue Wave	0.746425041	2
7	Giants	0.729163394	1
8	Swallows	0.71851069	1
9	Fighters	0.585915665	2
10	Marines	0.517530223	2
11	Buffalos	0.498371459	2
12	Lions	0.31897526	2

Thus, the rank sum of Central League is 27, while that of Pacific League is 51. The test statistic T of (7.14) is given by

$$T = \frac{27 - 6(6+6+1)/2}{\sqrt{6 \times 6(6+6+1)/12}} = -1.921537845.$$

The corresponding significance level is $\alpha = 0.05466$ so we reject the null hypothesis that the two leagues have the same distribution of efficiency scores — at a significance level of 5.466%.

On this basis we say that the Pacific League is less efficient than the Central League in terms of salaries (of managers and players) vs. attendance. This seems consistent with the assertion that the Pacific League is less attractive because this league uses more resources to service its attendance.

Notes

1. R.D. Banker and R. Morey (1986a), "Efficiency Analysis for Exogenously Fixed Inputs and Outputs," *Operations Research* 34(4), pp.513-521

2. A. Charnes, W.W. Cooper, J.J. Rousseau and J. Semple (1987), "Data Envelopment Analysis and Axiomatic Notions of Efficiency and Reference Sets," CCS Research Report 558 (Austin, Texas: University of Texas, Graduate School of Business, Center for Cybernetic Studies, Austin, Texas 78712).

3. H. Deng (2003), " Evaluating and Managing Congestion in Chinese Production," Ph.D. Thesis (Austin, TX: The Red McCombs School of Business in The University of Texas at Austin). Also available from University Micro Films, Inc., Ann Arbor, Mich.

4. Exceptions are the FDH and multiplicative models discussed in Chapter 4 and Chapter 5.

5. E.F. Fama and M.C. Jensen (1983a), "Separation of Ownership and Control," *Journal of Law and Economics* 26, pp.301-325. See also Fama and Jensen (1983b), "Agency Problems and Residual Claims," *Journal of Law and Economics* 26, pp.327-349.

6. P.L. Brockett, W.W. Cooper, J.J. Rousseau and Y. Wang (1998), "DEA Evaluations of the Efficiency of Organizational Forms and Distribution Systems in the U.S. Property and Liability Insurance Industry," *International Journal of System Science* 29, pp.1235-1247.

7. Solvency, defined as an ability to meet claims, for example, played a critically important role in Brockett *et al.* but not in the Fama-Jensen study. See P.L. Brockett, W.W. Cooper, L. Golden, J.J. Rousseau and Y. Wang "Evaluating Solvency and Efficiency Performances by U.S. Property-Liability Insurance Companies," *Journal of Risk and Insurance* (submitted).

8. The Mann-Whitney statistic does not require equal sample size. See E.L. Lehman (1979) *Nonparametrization: Statistical Methods Based on Ranks* (New York: Holden-Day).

9. D.L. Adolphson, G.C. Cornia, L.C. Walters, "A Unified Framework for Classifying DEA Models," *Operational Research '90,* edited by H.E. Bradley, Pergamon Press, pp.647-657 (1991).

10. K. Tone (1993), "On DEA Models," (in Japanese) *Communications of the Operations Research Society of Japan* 38, pp.34-40.

11. R.D. Banker and R.C. Morey (1986b), "The Use of Categorical Variables in Data Envelopment Analysis," *Management Science* 32(12), pp.1613-1627.

12. W.A. Kamakura (1988), "A Note on The Use of Categorical Variables in Data Envelopment Analysis," *Management Science* 34(10), pp.1273-1276.

13. J.J. Rousseau and J. Semple (1993), "Categorical Outputs in Data Envelopment Analysis," *Management Science* 39(3), pp.384-386.

14. K. Tone (1997), "DEA with Controllable Category Levels," in *Proceedings of the 1997 Spring National Conference of the Operations Research Society of Japan* pp.126-127.

15. K. Tone (1995), "DEA Models Revisited," *Communications of the Operations Research Society of Japan* 40, pp.681-685 (in Japanese).

16. K. Tone (1993), *Data Envelopment Analysis* (in Japanese) (Tokyo:JUSE Press, Ltd.).

17. A. Charnes, W.W. Cooper and E. Rhodes (1981), "Evaluating Program and Managerial Efficiency: An Application of Data Envelopment Analysis to Program Follow Through," *Management Science* 27, pp.668-697.

18. P.L. Brockett and B. Golany (1996), "Using Rank Statistics for Determining Programmatic Efficiency Differences in Data Envelopment Analysis," *Management Science* 42, pp.466-472.

19. P.L. Brockett and A. Levine (1984) *Statistics and Probability and Their Applications* (Fort Worth, TX. CBS College Publishing).

8 ALLOCATION MODELS

8.1 INTRODUCTION

The preceding chapters focused on the technical-physical aspects of production for use in situations where unit price and unit cost information are not available, or where their uses are limited because of variability in the prices and costs that might need to be considered. This chapter turns to the topic of "allocative efficiency" in order to show how DEA can be used to identify types of inefficiency which can emerge for treatment when information on prices and costs are known exactly. Technology and cost are the wheels that drive modern enterprises; some enterprises have advantages in terms of technology and others in cost. Hence, the management is eager to know how and to what extent their resources are being effectively and efficiently utilized, compared to other similar enterprises in the same or a similar field.

Regarding this subject, there are two different situations: one with common unit prices and costs for all DMUs and the other with different prices and costs from DMU to DMU. Section 2 of this chapter deals with the former case. However, the common price and cost assumption is not always valid in actual business and it is demonstrated that efficiency measures based on this assumption can be misleading. So we introduce a new cost-efficiency related model along with new revenue and profit efficiency models in Section 3. Section 4 develops a new formula for decomposition of the observed actual cost based on the new cost efficiency model. Using this formula, we can decompose actual cost as the sum of the minimum cost and the losses due to technical, price and allocative inefficiencies.

8.2 OVERALL EFFICIENCY WITH COMMON PRICES AND COSTS

8.2.1 *Cost Efficiency*

Figure 8.1 introduces concepts dealing with "allocative efficiency" that can be traced back to M.J. Farrell (1957) and G. Debreu (1951)[1] who originated many of the ideas underlying DEA. Färe, Grosskopf and Lovell (1985)[2] developed linear programming formulations of these concepts.

The solid lines in this figure are segments of an isoquant that represents all possible combinations of the input amounts (x_1, x_2) that are needed to produce the same amount of a single output. P is a point in the interior of the production possibility set representing the activity of a DMU which produces this same amount of output but with greater amounts of both inputs.[3]

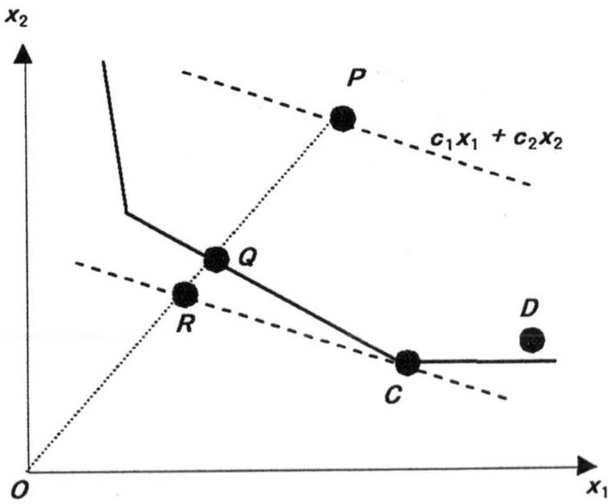

Figure 8.1. Technical, Allocative and Overall Efficiency

To evaluate the performance of P we can use the customary Farrell measure of radial efficiency. Reverting to the notation of Section 1.4, Chapter 1, we can represent this measure in ratio form as

$$0 \leq \frac{d(O,Q)}{d(O,P)} \leq 1,$$

and interpret this as the distance from O to Q relative to the distance from O to P. The result is the measure of technical efficiency that we have customarily represented as θ_o^*.

The components of this ratio lie on the dotted line from the origin through Q to P. To bring price-cost and, hence, "allocative efficiency" considerations into the picture we turn to the broken line passing through P for which the budget (or cost) line is associated with $c_1x_1 + c_2x_2 = k_1$. However, this cost can be reduced by moving this line in parallel fashion until it intersects the isoquant at C. The coordinates of C then give $c_1x_1^* + c_2x_2^* = k_0$ where $k_0 < k_1$ shows the amount by which total cost can be reduced. Further parallel movement in a downward direction is associated with reduced output so the position of the broken line passing through C is minimal at the prescribed output level. This optimal point C is obtained as the optimal solution $\boldsymbol{x}^*$ of the following LP (Farrell (1957)):

$$
\begin{aligned}
\text{[Cost]} \quad & \boldsymbol{c}\boldsymbol{x}^* = \min_{\boldsymbol{x},\boldsymbol{\lambda}} \boldsymbol{c}\boldsymbol{x} \\
\text{subject to} \quad & \boldsymbol{x} \geq X\boldsymbol{\lambda} \\
& \boldsymbol{y}_o \leq Y\boldsymbol{\lambda} \\
& \boldsymbol{\lambda} \geq \boldsymbol{0},
\end{aligned}
\tag{8.1}
$$

where $\boldsymbol{c} = (c_1, \ldots, c_m)$ is the common unit input-price or unit-cost vector.

Now we note that we can similarly determine the relative distances of R and Q to obtain the following ratio,

$$0 \leq \frac{d(O,R)}{d(O,Q)} \leq 1.$$

Farrell refers to this as a measure of "price efficiency" but the more commonly used term is "allocative efficiency." In either case it provides a measure of the extent to which the technically efficient point, Q, falls short of achieving minimal cost because of failure to make the substitutions (or reallocations) involved in moving from Q to C along the efficiency frontier.

There is one further measure that is commonly referred to as "overall efficiency," or "cost efficiency." We can represent this by means of the following ratio,

$$0 \leq \frac{d(O,R)}{d(O,P)} = \frac{\boldsymbol{c}\boldsymbol{x}^*}{\boldsymbol{c}\boldsymbol{x}_o} \leq 1. \tag{8.2}$$

This is a measure of the extent to which the originally observed values at P, represented in the denominator, have fallen short of achieving the minimum cost represented in the numerator.

To put this in a way that relates all three of these efficiency concepts to each other, note that

$$\frac{d(O,R)}{d(O,Q)} \cdot \frac{d(O,Q)}{d(O,P)} = \frac{d(O,R)}{d(O,P)}. \tag{8.3}$$

In sum, "overall (cost) efficiency" (on the right) is equal to the product of "allocative" times "technical efficiency," (on the left).

Furthermore, the technical efficiency can be decomposed into the pure technical efficiency and the scale efficiency as defined in Section 4.5 of Chapter 4. Thus, we have the following decomposition:

$$\text{Overall Eff.} = \text{Allocative Eff.} \times \text{Pure Technical Eff.} \times \text{Scale Eff.}. \tag{8.4}$$

Or,

$$\text{OE} = \text{AE} \times \text{TE} = \text{AE} \times \text{PTE} \times \text{SE}. \tag{8.5}$$

8.2.2 *Revenue Efficiency*

Given the common unit price vector $\boldsymbol{p} = (p_1, \ldots, p_s)$ for the output $\boldsymbol{y}$, we evaluate the revenue efficiency of DMU$_o$ as follows:

$$\begin{aligned} &[\text{Revenue}] \quad && \boldsymbol{p}\boldsymbol{y}^* = \max_{\boldsymbol{y},\boldsymbol{\lambda}} \boldsymbol{p}\boldsymbol{y} \\ &\text{subject to} && \boldsymbol{x}_o \geq X\boldsymbol{\lambda} \\ &&& \boldsymbol{y} \leq Y\boldsymbol{\lambda} \\ &&& L \leq \boldsymbol{e}\boldsymbol{\lambda} \leq U \\ &&& \boldsymbol{\lambda} \geq \boldsymbol{0}. \end{aligned} \tag{8.6}$$

We inserted an additional constraint on scale ($L \leq \boldsymbol{e}\boldsymbol{\lambda} \leq U$) in order to cope with various returns-to-scale assumptions.

This model allows substitutions in outputs. Let the optimal solution be $(\boldsymbol{y}^*, \boldsymbol{\lambda}^*)$. Then, the *revenue efficiency* is defined in ratio form as:

$$E_R \text{ (Revenue Efficiency)} = \frac{\boldsymbol{p}\boldsymbol{y}_o}{\boldsymbol{p}\boldsymbol{y}^*}. \tag{8.7}$$

We have $0 \leq E_R \leq 1$ and DMU$(\boldsymbol{x}_o, \boldsymbol{y}_o)$ is *revenue efficient* if and only if $E_R = 1$.

8.2.3 *Profit Efficiency*

To express the profit of DMU$_o$ we use the common unit price vector $\boldsymbol{p}$ and unit cost vector $\boldsymbol{c}$, to obtain the following LP problem:

$$\begin{aligned} &[\text{Profit}] \quad && \boldsymbol{p}\boldsymbol{y}^* - \boldsymbol{c}\boldsymbol{x}^* = \max_{\boldsymbol{x},\boldsymbol{y},\boldsymbol{\lambda}} \boldsymbol{p}\boldsymbol{y} - \boldsymbol{c}\boldsymbol{x} \\ &\text{subject to} && \boldsymbol{x} = X\boldsymbol{\lambda} \leq \boldsymbol{x}_o \\ &&& \boldsymbol{y} = Y\boldsymbol{\lambda} \geq \boldsymbol{y}_o \\ &&& L \leq \boldsymbol{e}\boldsymbol{\lambda} \leq U \\ &&& \boldsymbol{\lambda} \geq \boldsymbol{0}. \end{aligned} \tag{8.8}$$

This formulation extends (8.1) with an additional constraint on scale ($L \leq \boldsymbol{e}\boldsymbol{\lambda} \leq U$). So, substitutions in inputs or outputs are not allowed in this case. Here, the purpose is to find a profit-maximization mix in the production possibility set $P = \{(\boldsymbol{x}, \boldsymbol{y}) | \boldsymbol{x} \geq X\boldsymbol{\lambda}, \boldsymbol{y} \leq Y\boldsymbol{\lambda}, L \leq \boldsymbol{e}\boldsymbol{\lambda} \leq U, \boldsymbol{\lambda} \geq \boldsymbol{0}\}$. Based on an optimal solution $(\boldsymbol{x}^*, \boldsymbol{y}^*)$, the *profit efficiency* can be defined in ratio form by

$$E_P \text{(Profit Efficiency)} = \frac{\boldsymbol{p}\boldsymbol{y}_o - \boldsymbol{c}\boldsymbol{x}_o}{\boldsymbol{p}\boldsymbol{y}^* - \boldsymbol{c}\boldsymbol{x}^*}. \tag{8.9}$$

where $\boldsymbol{y}^*$, $\boldsymbol{x}^*$ are optimal for (8.8) and $\boldsymbol{y}_o, \boldsymbol{x}_o$ are the vectors of observed values for DMU_o.

Under the assumption $\boldsymbol{p}\boldsymbol{y}_o > \boldsymbol{c}\boldsymbol{x}_o$, we have $0 < E_P \leq 1$ and DMU $(\boldsymbol{x}_o, \boldsymbol{y}_o)$ is *profit efficient* if and only if $E_P = 1$. The differences between $\boldsymbol{x}^*$ and $\boldsymbol{x}_o$ and between $\boldsymbol{y}^*$ and $\boldsymbol{y}_o$ may suggest directions for managerial improvement and this can be analyzed, constraint by constraint, in (8.8).

8.2.4 An Example

The data in the following table will provide examples to illustrate the use of these models. Here each of three DMUs produces a single output in the amount y, shown in the column for y under "Output," by using two inputs in the amounts x_1 and x_2 shown in the two columns headed by x_1 and x_2. The common unit costs and price are exhibited in the columns for c_1, c_2 and p, respectively.

Table 8.1. Sample Data for Allocative Efficiency

	Input				Output	
DMU	x_1	c_1	x_2	c_2	y	p
A	3	4	2	2	3	6
B	1	4	3	2	5	6
C	4	4	6	2	6	6

We solved this data set under the constant returns-to-scale assumption so the constraint $L \leq \boldsymbol{e}\boldsymbol{\lambda} \leq U$ was omitted. We then obtained the results shown in Table 8.2. DMU B is the only one that is efficient and is the best performer in all efficiency measures.

Table 8.2. Efficiencies

	Efficiency				
DMU	Technical	Cost	Allocative	Revenue	Profit
A	0.9	0.375	0.417	0.9	0.15
B	1	1	1	1	1
C	0.6	0.429	0.715	0.6	0.2

8.3 NEW COST EFFICIENCY UNDER DIFFERENT UNIT PRICES

Firstly we observe an unacceptable property of the traditional Farrell-Debreu cost efficiency models described in the preceding section which can occur when the unit prices of input are not identical among DMUs.

Suppose that DMUs A and B have the same amount of inputs and outputs, i.e., $\boldsymbol{x}_A = \boldsymbol{x}_B$ and $\boldsymbol{y}_A = \boldsymbol{y}_B$. Assume further that the unit cost of DMU A is twice that of DMU B for each input, i.e., $\boldsymbol{c}_A = 2\boldsymbol{c}_B$. Under these assumptions, we have the following theorem:

Theorem 8.1 (Tone(2002)[4]) *Using the Farrell-Debreu cost efficiency model both DMUs A and B have the same cost (overall) and allocative efficiencies even when the latter is more costly than the former.*

Proof: Since DMUs A and B have the same inputs and outputs, they have the same technical efficiency, i.e., $\theta_A^* = \theta_B^*$. The Farrell measure of cost efficiency for DMU A (or DMU B) can be obtained by solving the following LP (see (8.1)):

$$\min \ \boldsymbol{c}_A\boldsymbol{x} (= 2\boldsymbol{c}_B\boldsymbol{x}) \tag{8.10}$$

$$\text{subject to} \quad x_i \geq \sum_{j=1}^{n} x_{ij}\lambda_j \ (i = 1, \ldots, m) \tag{8.11}$$

$$y_{rA} (= y_{rB}) \leq \sum_{j=1}^{n} y_{rj}\lambda_j \ (r = 1, \ldots, s) \tag{8.12}$$

$$\lambda_j \geq 0. \ (\forall j) \tag{8.13}$$

Apparently, DMUs A and B have the same optimal solution (inputs) $\boldsymbol{x}_A^* = \boldsymbol{x}_B^*$, and hence the same cost efficiency, since we have:

$$\gamma_A^* = \boldsymbol{c}_A\boldsymbol{x}_A^*/\boldsymbol{c}_A\boldsymbol{x}_A = 2\boldsymbol{c}_B\boldsymbol{x}_B^*/2\boldsymbol{c}_B\boldsymbol{x}_B = \boldsymbol{c}_B\boldsymbol{x}_B^*/\boldsymbol{c}_B\boldsymbol{x}_B = \gamma_B^*.$$

By definition, they also have the same allocative efficiency. □

This is not acceptable, since DMUs A and B have the same cost and allocative efficiencies but the cost of DMU B is half that of DMU A.

8.3.1 *A New Scheme for Evaluating Cost Efficiency*

The previous example reveals a serious shortcoming in the traditional Farrell-Debreu cost and allocative efficiency measures. These shortcomings are caused by the structure of the supposed production possibility set P as defined by:

$$P = \{(\boldsymbol{x}, \boldsymbol{y}) | \boldsymbol{x} \geq X\boldsymbol{\lambda}, \boldsymbol{y} \leq Y\boldsymbol{\lambda}, \boldsymbol{\lambda} \geq \boldsymbol{0}\}. \tag{8.14}$$

P is defined only by using technical factors $X = (\boldsymbol{x}_1, \ldots, \boldsymbol{x}_n) \in R^{m\times n}$ and $Y = (\boldsymbol{y}_1, \ldots, \boldsymbol{y}_n) \in R^{s\times n}$, but excludes consideration of the unit input costs $C = (\boldsymbol{c}_1, \ldots, \boldsymbol{c}_n)$.

Let us define another cost-based production possibility set P_c as:

$$P_c = \left\{(\overline{\boldsymbol{x}}, \boldsymbol{y}) | \overline{\boldsymbol{x}} \geq \overline{X}\boldsymbol{\lambda}, \boldsymbol{y} \leq Y\boldsymbol{\lambda}, \boldsymbol{\lambda} \geq \mathbf{0}\right\}, \tag{8.15}$$

where $\overline{X} = (\overline{\boldsymbol{x}}_1, \ldots, \overline{\boldsymbol{x}}_n)$ with $\overline{\boldsymbol{x}}_j = (c_{1j}x_{1j}, \ldots, c_{mj}x_{mj})^T$.

Here we assume that the matrices X and C are non-negative. We also assume that the elements of $\overline{x}_{ij} = (c_{ij}x_{ij})$ $(\forall(i,j))$ are denominated in homogeneous units, viz., dollars, so that adding up the elements of $\overline{x}_{ij}$ has a well defined meaning.

Based on this new production possibility set P_c, a new "technical efficiency" measure, $\overline{\theta}^*$, is obtained as the optimal solution of the following LP problem:

$$\begin{aligned} \text{[NTec]} \qquad & \overline{\theta}^* = \min_{\overline{\theta}, \boldsymbol{\lambda}} \ \overline{\theta} && (8.16) \\ \text{subject to} \qquad & \overline{\theta}\overline{\boldsymbol{x}}_o \geq \overline{X}\boldsymbol{\lambda} && (8.17) \\ & \boldsymbol{y}_o \leq Y\boldsymbol{\lambda} && (8.18) \\ & \boldsymbol{\lambda} \geq \mathbf{0}. && (8.19) \end{aligned}$$

The new cost efficiency $\overline{\gamma}^*$ is defined as

$$\overline{\gamma}^* = \boldsymbol{e}\overline{\boldsymbol{x}}_o^* / \boldsymbol{e}\overline{\boldsymbol{x}}_o, \tag{8.20}$$

where $\boldsymbol{e} \in R^m$ is a row vector with all elements being equal to 1, and $\overline{\boldsymbol{x}}_o^*$ is the optimal solution of the LP given below:

$$\begin{aligned} \text{[NCost]} \qquad & \boldsymbol{e}\overline{\boldsymbol{x}}_o^* = \min_{\overline{\boldsymbol{x}}, \boldsymbol{\lambda}} \ \boldsymbol{e}\overline{\boldsymbol{x}} && (8.21) \\ \text{subject to} \qquad & \overline{\boldsymbol{x}} \geq \overline{X}\boldsymbol{\lambda} && (8.22) \\ & \boldsymbol{y}_o \leq Y\boldsymbol{\lambda} && (8.23) \\ & \boldsymbol{\lambda} \geq \mathbf{0}. && (8.24) \end{aligned}$$

Theorem 8.2 *The new "cost efficiency", $\overline{\gamma}^*$ in (8.20), is not greater than the new "technical efficiency" $\overline{\theta}^*$ in (8.16).*

Proof : Let an optimal solution for (8.16)-(8.19) be $(\overline{\theta}^*, \boldsymbol{\lambda}^*)$. Then, $(\overline{\theta}^*\overline{\boldsymbol{x}}_o, \boldsymbol{\lambda}^*)$ is feasible for (8.21)-(8.24). Hence, it follows that $\boldsymbol{e}\overline{\theta}^*\overline{\boldsymbol{x}}_o \geq \boldsymbol{e}\overline{\boldsymbol{x}}_o^*$. This leads to $\overline{\theta}^* \geq \boldsymbol{e}\overline{\boldsymbol{x}}_o^*/\boldsymbol{e}\overline{\boldsymbol{x}}_o = \overline{\gamma}^*$. □

The new *allocative* efficiency $\overline{\alpha}^*$ is then defined as the ratio of $\overline{\gamma}^*$ to $\overline{\theta}^*$, i.e.,

$$\overline{\alpha}^* = \overline{\gamma}^* / \overline{\theta}^*. \tag{8.25}$$

We note that the new efficiency measures $\overline{\theta}^*$, $\overline{\gamma}^*$ and $\overline{\alpha}^*$ are all units invariant so long as $\overline{X}$ has a common unit of cost, e.g., dollars, cents or pounds.

On monotonicity of the new measures with respect to unit cost, we have the following theorem.

Theorem 8.3 (Tone(2002)) *If $\boldsymbol{x}_A = \boldsymbol{x}_B$, $\boldsymbol{y}_A = \boldsymbol{y}_B$ and $\boldsymbol{c}_A \geq \boldsymbol{c}_B$, then we have the following inequalities: $\overline{\theta}_A^* \leq \overline{\theta}_B^*$ and $\overline{\gamma}_A^* \leq \overline{\gamma}_B^*$. Furthermore, strict inequalities hold if $\boldsymbol{c}_A > \boldsymbol{c}_B$.*

Proof. Since $\overline{\boldsymbol{x}}_A \geq \overline{\boldsymbol{x}}_B$ and $\boldsymbol{y}_A = \boldsymbol{y}_B$, the new technical measure $\overline{\theta}_A^*$ is less than or equal to $\overline{\theta}_B^*$ and a strict inequality holds if $\boldsymbol{c}_A > \boldsymbol{c}_B$. Regarding the new cost efficiency, we note that the optimal solution of [NCost] depends only on $\boldsymbol{y}_o$. Hence, DMUs $(\overline{\boldsymbol{x}}_A, \boldsymbol{y}_A)$ and $(\overline{\boldsymbol{x}}_B, \boldsymbol{y}_B)$ with $\boldsymbol{y}_A = \boldsymbol{y}_B$ have a common optimal solution $\overline{\boldsymbol{x}}^*$. Therefore, we have $\overline{\gamma}_A^* = \boldsymbol{e}\overline{\boldsymbol{x}}^* / \boldsymbol{e}\overline{\boldsymbol{x}}_A \leq \boldsymbol{e}\overline{\boldsymbol{x}}^* / \boldsymbol{e}\overline{\boldsymbol{x}}_B = \overline{\gamma}_B^*$ and strict inequality holds if $\boldsymbol{c}_A > \boldsymbol{c}_B$. □

Thus, the new measure eliminates the possible occurrence of the phenomenon observed at the beginning of this section.

8.3.2 Differences Between the Two Models

We now comment on the differences existing between the traditional "Farrell-Debreu" and the new models. In the traditional model, keeping the unit cost of DMU$_o$ fixed at c_o, the optimal input mix $\boldsymbol{x}^*$ that produces the output $\boldsymbol{y}_o$ is found. In the new model, we search for the optimal input mix $\overline{\boldsymbol{x}}^*$ for producing $\boldsymbol{y}_o$ (or more). More concretely, the optimal mix is described as:

$$\overline{x}_i^* = \sum_{j=1}^{n} c_{ij} x_{ij} \lambda_j^*. \ (i = 1, \ldots, m) \tag{8.26}$$

Hence, it is assumed that, for a given output $\boldsymbol{y}_o$, the optimal input mix can be found (and realized) independently of the current unit cost $\boldsymbol{c}_o$ of DMU$_o$.

These are fundamental differences between the two models. Using the traditional "Farrell-Debreu" model we can fail to recognize the existence of other cheaper input mixes, as we have demonstrated earlier. We demonstrate this with a simple example involving three DMUs A, B and C with each using two inputs (x_1, x_2) to produce one output (y) along with input costs (c_1, c_2). The data and the resulting measures are exhibited in Table 8.3.

For DMUs A and B, the traditional model gives the same technical (θ^*), cost (γ^*) and allocative (α^*) efficiency scores — as expected from Theorem 1. DMU C is found to be the only efficient performer in this framework.

The new scheme devised as in Tone (2002) — see footnote 4 — distinguishes DMU A from DMU B by according them different technical and cost efficiency scores. (See New Scheme in Table 8.3). This is due to the difference in their unit costs. Moreover, DMU B is judged as technically, cost and allocatively efficient with improvement in cost efficiency score from $0.35(\gamma_B^*)$ to $1(\overline{\gamma}_B^*)$ as exhibited in these two tables. As shown in Table 8.3, this cost difference produces a drop in DMU A's cost efficiency score from $0.35(\gamma_A^*)$ to $0.1(\overline{\gamma}_A^*)$. This drop in DMU A's performance is explained by its higher cost structure. Lastly, DMU C is no longer efficient in any of its technical, cost or allocative efficiency performances.

Table 8.3. Comparison of Traditional and New Scheme

						Traditional Efficiency		
						Tech.	Cost	Alloc.
	x_1	c_1	x_2	c_2	y	θ^*	γ^*	α^*
A	10	10	10	10	1	0.5	0.35	0.7
B	10	1	10	1	1	0.5	0.35	0.7
C	5	3	2	6	1	1	1	1
						New Scheme Efficiency		
						Tech.	Cost	Alloc.
	$\overline{x}_1$	e_1	$\overline{x}_2$	e_2	y	$\overline{\theta}^*$	$\overline{\gamma}^*$	$\overline{\alpha}^*$
A	100	1	100	1	1	0.1	0.1	1
B	10	1	10	1	1	1	1	1
C	15	1	12	1	1	0.8333	0.7407	0.8889

8.3.3 An Empirical Example

In this section, we apply our new method to a set of hospital data. Table 8.4 records the performances of 12 hospitals in terms of two inputs, number of doctors and nurses, and two outputs identified as number of outpatients and inpatients (each in units of 100 persons/month). Relative unit costs of doctors and nurses for each hospital are also recorded in columns 4 and 6.

Multiplying the number of doctors and nurses by their respective unit costs we obtain the new data set $(\overline{X}, Y)$ exhibited in Table 8.5. The results of efficiency scores: CCR(θ^*), New technical $(\overline{\theta}^*)$, New cost $(\overline{\gamma}^*)$ and New allocative $(\overline{\alpha}^*)$, are also recorded.

From the results, it is seen that the best performer is Hospital B with all its efficiency scores being equal to one. Regarding the cost-based measures, Hospitals E and L received full efficiency marks even though they fell short in their CCR efficiency score. Conversely, although E has the worst CCR score (0.763), its lower unit costs are sufficient to move its cost-based performance to the top rank. This information obtained from $\theta^* = 0.763$ shows that this hospital still has room for input reductions compared with other technically efficient hospitals. Hospital L, on the other hand, may be regarded as positioned in the best performer group. These two DMUs show that the usual assumption does not hold and thus technical efficiency (CCR $\theta^* = 1$) being achieved is not a necessary condition for the new cost and allocative efficiencies. This is caused by the difference between the two production possibility sets, i.e., the technology-based (X, Y) and the cost-based $(\overline{X}, Y)$. On the other hand, Hospital D is rated worst with respect to cost-based measures, although it receives full efficiency marks in terms of its CCR score. This gap is due to its

Table 8.4. Data for 12 Hospitals

		Inputs				Outputs	
		Doctor		Nurse		Outpat.	Inpat.
No.	DMU	Number	Cost	Number	Cost	Number	Number
1	A	20	500	151	100	100	90
2	B	19	350	131	80	150	50
3	C	25	450	160	90	160	55
4	D	27	600	168	120	180	72
5	E	22	300	158	70	94	66
6	F	55	450	255	80	230	90
7	G	33	500	235	100	220	88
8	H	31	450	206	85	152	80
9	I	30	380	244	76	190	100
10	J	50	410	268	75	250	100
11	K	53	440	306	80	260	147
12	L	38	400	284	70	250	120
	Average	33.6	435.8	213.8	85.5	186.3	88.2

Table 8.5. New Data Set and Efficiencies

		Data				Efficiency			
		$\overline{X}$		Y		CCR	Tech.	Cost	Alloc.
No.	DM	Doctor	Nurse	Inp.	Outp.	θ^*	$\overline{\theta}^*$	$\overline{\gamma}^*$	$\overline{\alpha}^*$
1	A	10000	15100	100	90	1	.994	.959	.965
2	B	6650	10480	150	50	1	1	1	1
3	C	11250	14400	160	55	.883	.784	.724	.923
4	D	16200	20160	180	72	1	.663	.624	.941
5	E	6600	11060	94	66	.763	1	1	1
6	F	24750	20400	230	90	.835	.831	.634	.764
7	G	16500	23500	220	88	.902	.695	.693	.997
8	H	13950	17510	152	80	.796	.757	.726	.959
9	I	11400	18544	190	100	.960	.968	.953	.984
10	J	20500	20100	250	100	.871	.924	.776	.841
11	K	23320	24480	260	147	.955	.995	.863	.867
12	L	15200	19880	250	120	.958	1	1	1

high cost structure. Hospital D needs reductions in its unit costs to attain good cost-based scores.

Hospital F has the worst allocative efficiency score, and hence needs a change in input-(cost)mix. This hospital has the current input-(cost)mix, $\overline{x}_F = (24750, 20400)$, while the optimal mix $\overline{x}_F^*$ is $(11697, 16947)$. So, under its current costs, F needs to reduce the number of doctors from 55 to 26 (=11697/450), and nurses from 255 to 212 (=16947/80). Otherwise, if F retains its current input numbers, it needs to reduce the unit cost of doctors from 450 to 213 (=11697/55), and that of nurses from 80 to 66 (=16947/255). Of course, there are many other adjustment plans as well. In any case our proposed new measures provide much more information than the traditional ones.

8.3.4 Extensions

We can also extend this new cost efficiency model to three other situations as follows.

1. **Revenue Efficiency**
   Given the unit price $\boldsymbol{p}_j$ for each output $\boldsymbol{y}_j$ $(j = 1, \ldots, n)$, the conventional revenue efficiency ρ_o^* of DMU$_o$ is evaluated by $\rho_o^* = \boldsymbol{p}_o\boldsymbol{y}_o/\boldsymbol{p}_o\boldsymbol{y}_o^*$. Here, $\boldsymbol{p}_o\boldsymbol{y}_o^*$ is obtained from the optimal objective value of LP problem (8.6).

   However, in the situation of different unit prices, this revenue efficiency ρ_o^* suffers from shortcomings similar to the traditional cost efficiency measure shortcomings described in the previous section. We can eliminate such shortcomings by introducing the price-based output $\overline{Y} = (\overline{\boldsymbol{y}}_1, \ldots, \overline{\boldsymbol{y}}_n)$ with $\overline{\boldsymbol{y}}_j = (p_{1j}y_{1j}, \ldots, p_{sj}y_{sj})$ into the following LP:

$$[\text{NRevenue}] \quad \boldsymbol{e}\overline{\boldsymbol{y}}_o^* = \max_{\overline{\boldsymbol{y}},\boldsymbol{\lambda}} \boldsymbol{e}\overline{\boldsymbol{y}} \tag{8.27}$$

$$\begin{aligned} \text{subject to} \quad & \boldsymbol{x}_o \geq X\boldsymbol{\lambda} \\ & \overline{\boldsymbol{y}} \leq \overline{Y}\boldsymbol{\lambda} \\ & L \leq \boldsymbol{e}\boldsymbol{\lambda} \leq U \\ & \boldsymbol{\lambda} \geq \mathbf{0}. \end{aligned}$$

   The new revenue efficiency measure $\overline{\rho}_o$ is defined by

$$\text{New Revenue Efficiency } (\overline{\rho}_o) = \boldsymbol{e}\overline{\boldsymbol{y}}_o/\boldsymbol{e}\overline{\boldsymbol{y}}_o^*. \tag{8.28}$$

2. **Profit Efficiency**
   Using the new cost and revenue efficiency models, we can also define a new profit efficiency model as follows:

$$[\text{NProfit}] \quad \boldsymbol{e}\overline{\boldsymbol{y}}_o^* - \boldsymbol{e}\overline{\boldsymbol{x}}_o^* = \max_{\overline{\boldsymbol{x}},\overline{\boldsymbol{y}},\boldsymbol{\lambda}} \boldsymbol{e}\overline{\boldsymbol{y}} - \boldsymbol{e}\overline{\boldsymbol{x}} \tag{8.29}$$

$$\begin{aligned} \text{subject to} \quad & \overline{\boldsymbol{x}} = \overline{X}\boldsymbol{\lambda} \leq \overline{\boldsymbol{x}}_o \\ & \overline{\boldsymbol{y}} = \overline{Y}\boldsymbol{\lambda} \geq \overline{\boldsymbol{y}}_o \\ & L \leq \boldsymbol{e}\boldsymbol{\lambda} \leq U \\ & \boldsymbol{\lambda} \geq \mathbf{0}, \end{aligned}$$

where $\overline{X}$ and $\overline{Y}$ are defined in the new cost and revenue models, respectively. The new profit efficiency is defined as:

$$\text{New Profit Efficiency } (\overline{\pi}_o) = (e\overline{y}_o - e\overline{x})/(e\overline{y}_o^* - e\overline{x}_o^*). \tag{8.30}$$

3. **Profit Ratio Model**
We also propose a model for maximizing the revenue vs. cost ratio,

$$\frac{\text{revenue}}{\text{expenses}},$$

instead of maximizing profit (revenue − expenses), since in some situations the latter gives a negative value that is awkward to deal with. This new profit ratio model can be formulated as a problem of maximizing the *revenue/expenses* ratio to obtain the following fractional programming problem,[5]

$$\begin{aligned}
\text{[Profit Ratio]} \quad & \max_{\boldsymbol{x},\boldsymbol{y},\boldsymbol{\lambda}} \frac{\boldsymbol{p}_o\boldsymbol{y}}{\boldsymbol{c}_o\boldsymbol{x}} && (8.31)\\
\text{subject to} \quad & \boldsymbol{x} = X\boldsymbol{\lambda} \leq \boldsymbol{x}_o \\
& \boldsymbol{y} = Y\boldsymbol{\lambda} \geq \boldsymbol{y}_o \\
& L \leq \boldsymbol{e\lambda} \leq U \\
& \boldsymbol{\lambda} \geq \boldsymbol{0}.
\end{aligned}$$

We can transform this program to the linear programming problem below, by introducing a variable $t \in R$ and use the Charnes-Cooper transformation of fractional programming which sets $\widehat{\boldsymbol{x}} = t\boldsymbol{x}$, $\widehat{\boldsymbol{y}} = t\boldsymbol{y}$, $\widehat{\boldsymbol{\lambda}} = t\boldsymbol{\lambda}$. Then multiplying all terms by $t > 0$, we change (8.31) to

$$\begin{aligned}
& \max_{\widehat{\boldsymbol{x}},\widehat{\boldsymbol{y}},\widehat{\boldsymbol{\lambda}},t} \boldsymbol{p}_o\widehat{\boldsymbol{y}} && (8.32)\\
\text{subject to} \quad & \boldsymbol{c}_o\widehat{\boldsymbol{x}} = 1 \\
& t\boldsymbol{x}_o \geq X\widehat{\boldsymbol{\lambda}} = \widehat{\boldsymbol{x}} \\
& t\boldsymbol{y}_o \leq Y\widehat{\boldsymbol{\lambda}} = \widehat{\boldsymbol{y}} \\
& Lt \leq \boldsymbol{e}\widehat{\boldsymbol{\lambda}} \leq Ut \\
& \widehat{\boldsymbol{\lambda}} \geq \boldsymbol{0}.
\end{aligned}$$

Let an optimal solution of this LP problem be $(t^*, \widehat{\boldsymbol{x}}^*, \widehat{\boldsymbol{y}}^*, \widehat{\boldsymbol{\lambda}}^*)$. Since $t^* > 0$ we can reverse this transformation and obtain an optimal solution to the fractional program in (8.31) from

$$\boldsymbol{x}^* = \widehat{\boldsymbol{x}}^*/t^*,\ \boldsymbol{y}^* = \widehat{\boldsymbol{y}}^*/t^*,\ \boldsymbol{\lambda}^* = \widehat{\boldsymbol{\lambda}}^*/t^*. \tag{8.33}$$

The *revenue/cost* efficiency E_{RC} of DMU$_o$ can then be related to actual revenue and costs by the following "ratio of ratios,"

$$E_{RC} = \frac{\boldsymbol{p}_o\boldsymbol{y}_o/\boldsymbol{c}_o\boldsymbol{x}_o}{\boldsymbol{p}_o\boldsymbol{y}^*/\boldsymbol{c}_o\boldsymbol{x}^*}. \tag{8.34}$$

As noted in the following remark, this efficiency index is related to profit efficiency but is applicable even when there are many deficit-DMUs, i.e. DMU$_o$s for which $\boldsymbol{p}_o\boldsymbol{y}_o - \boldsymbol{c}_o\boldsymbol{x}_o < 0$.

[Remark] We can follow Cooper *et al.* (2005)[6] and obtain a profit-to-cost return ratio by noting that an optimal solution to (8.31) is not altered by replacing the objective with

$$\max_{\boldsymbol{x},\boldsymbol{y},\boldsymbol{\lambda}} \frac{\boldsymbol{p}_o\boldsymbol{y}}{\boldsymbol{c}_o\boldsymbol{x}} - 1.$$

This gives

$$\frac{\boldsymbol{p}_o\boldsymbol{y}^*}{\boldsymbol{c}_o\boldsymbol{x}^*} - \frac{\boldsymbol{c}_o\boldsymbol{x}^*}{\boldsymbol{c}_o\boldsymbol{x}^*} = \frac{\boldsymbol{p}_o\boldsymbol{y}^* - \boldsymbol{c}_o\boldsymbol{x}^*}{\boldsymbol{c}_o\boldsymbol{x}^*}$$

which is the commonly used profit-to-cost ratio measure of performance, after it has been adjusted to eliminate inefficiencies. Finally, if the observed profit-to-cost ratio is positive we can use the following

$$0 \leq \frac{\boldsymbol{p}_o\boldsymbol{y}_o - \boldsymbol{c}_o\boldsymbol{x}_o}{\boldsymbol{c}_o\boldsymbol{x}_o} \bigg/ \frac{\boldsymbol{p}_o\boldsymbol{y}^* - \boldsymbol{c}_o\boldsymbol{x}^*}{\boldsymbol{c}_o\boldsymbol{x}^*} \leq 1$$

as a measure of efficiency with unity achieved if and only if $\frac{\boldsymbol{p}_o\boldsymbol{y}_o - \boldsymbol{c}_o\boldsymbol{x}_o}{\boldsymbol{c}_o\boldsymbol{x}_o} = \frac{\boldsymbol{p}_o\boldsymbol{y}^* - \boldsymbol{c}_o\boldsymbol{x}^*}{\boldsymbol{c}_o\boldsymbol{x}^*}$.

8.4 DECOMPOSITION OF COST EFFICIENCY

Technology and cost are the wheels that drive modern enterprises. Some enterprises have advantages in terms of technology and others in cost. Hence, a management may want to know how and to what extent their resources are being effectively and efficiently utilized, compared to similar enterprises in the same or a similar field.

In an effort to address this subject, Tone and Tsutsui (2004)[7] developed a scheme for decomposing actual observed cost into the sum of the minimum cost and the loss due to input inefficiency. Furthermore, the loss due to input inefficiency can be expressed as the sum of the loss due to input technical, price and allocative inefficiencies.

8.4.1 Loss due to Technical Inefficiency

We consider n DMUs, each having m inputs for producing s outputs. We utilize the notations for denoting observed inputs ($\boldsymbol{x}_o$), outputs ($\boldsymbol{y}_o$) and input unit prices ($\boldsymbol{c}_o$). We also assume that unit input prices are not identical among DMUs. The actual (observed) input cost for DMU ($\boldsymbol{x}_o, \boldsymbol{y}_o$) can be calculated as follows:

$$C_o = \sum_{i=1}^{m} c_{io}x_{io}. \ (o = 1, \ldots, n) \tag{8.35}$$

We postulate the production possibility set P defined by

$$P = \{(\boldsymbol{x}, \boldsymbol{y}) | \boldsymbol{x} \geq X\boldsymbol{\lambda},\ \boldsymbol{y} \leq Y\boldsymbol{\lambda},\ \boldsymbol{\lambda} \geq \mathbf{0}\}. \tag{8.36}$$

Let the (technically) efficient input for DMU$_o$ be $\boldsymbol{x}_o^*$, which can be obtained by solving the CCR, SBM or Hybrid models depending on the situation. The technically efficient input cost for DMU$_o$ is calculated as

$$C_o^* = \sum_{i=1}^{m} c_{io} x_{io}^*. \ (o = 1, \ldots, n) \tag{8.37}$$

Then, the loss in input cost due to technical inefficiency is expressed as follows:

$$L_o^* = C_o - C_o^* (\geq 0). \tag{8.38}$$

8.4.2 *Loss due to Input Price Inefficiency*

We now construct a cost-based production possibility set analogous to that in the preceding section as follows:

$$\overline{P}_c = \{(\overline{\boldsymbol{x}}, \boldsymbol{y}) | \overline{\boldsymbol{x}} \geq \overline{X}\boldsymbol{\lambda},\ \boldsymbol{y} \leq Y\boldsymbol{\lambda},\ \boldsymbol{\lambda} \geq \mathbf{0}\}, \tag{8.39}$$

where $\overline{X} = (\overline{\boldsymbol{x}}_1, \ldots, \overline{\boldsymbol{x}}_n) \in R^{m \times n}$, $\overline{\boldsymbol{x}}_j = (\overline{x}_{ij}, \ldots, \overline{x}_{mj})$, and $\overline{x}_{ij} = c_{ij} x_{ij}^*$. It should be noted that $\boldsymbol{x}_j^*$ represents the technically efficient input for producing $\boldsymbol{y}_j$. Hence, we utilize $c_{ij} x_{ij}^*$ instead of $c_{ij} x_{ij}$ in the preceding section in order to eliminate technical inefficiency to the maximum possible extent. Then we solve the CCR model on $\overline{P}_c$ in a manner similar to that of [NTec] in (8.16):

$$[\text{NTec-2}] \qquad \rho^* = \min_{\rho, \boldsymbol{\mu}, \boldsymbol{t}^-, \boldsymbol{t}^+} \rho \tag{8.40}$$

$$\text{subject to} \qquad \rho \overline{\boldsymbol{x}}_o = \overline{X}\boldsymbol{\mu} + \boldsymbol{t}^- \tag{8.41}$$

$$\boldsymbol{y}_o = Y\boldsymbol{\mu} - \boldsymbol{t}^+ \tag{8.42}$$

$$\boldsymbol{\mu} \geq \mathbf{0},\ \boldsymbol{t}^- \geq \mathbf{0},\ \boldsymbol{t}^+ \geq \mathbf{0}. \tag{8.43}$$

Let $(\rho^*, \boldsymbol{\mu}^*, \boldsymbol{t}^{-*}, \boldsymbol{t}^{+*})$ be an optimal solution for [NTec-2]. Then, $\rho^* \overline{\boldsymbol{x}}_o = (\rho^* c_{1o} x_{1o}^*, \ldots, \rho^* c_{mo} x_{mo}^*)$ indicates the radially reduced input vector on the (weakly) efficient frontier of the cost-based production set $\overline{P}_c$ in (8.39). Now we define

$$\boldsymbol{c}_o^* = \rho^* \boldsymbol{c}_o = (\rho^* c_{1o}, \ldots, \rho^* c_{mo}). \tag{8.44}$$

$\boldsymbol{c}_o^*$ is the radially reduced input factor price vector for the technically efficient input $\boldsymbol{x}_o^*$ that can produce $\boldsymbol{y}_o$. The [NTec-2] projection is given by

$$[\text{NTec-2 Projection}]\ \overline{\boldsymbol{x}}_o^* = \rho^* \overline{\boldsymbol{x}}_o - \boldsymbol{t}^{-*},\ \boldsymbol{y}_o^* = \boldsymbol{y}_o + \boldsymbol{t}^{+*}. \tag{8.45}$$

We define the strongly efficient cost C_o^{**}, which is the technical and price efficient cost, and the loss L_o^{**} due to the difference of the input price as follows:

$$C_o^{**} = \sum_{i=1}^{m} \overline{x}_{io}^{*} = \sum_{i=1}^{m} (\rho^* \overline{x}_{io} - t_{io}^{-*}) \leq \rho^* \sum_{i=1}^{m} \overline{x}_{io} = \rho^* C_o^* \leq C_o^* \quad (8.46)$$

$$L_o^{**} = C_o^* - C_o^{**} (\geq 0). \quad (8.47)$$

8.4.3 Loss due to Allocative Inefficiency

Furthermore, we solve the [NCost] model in (8.21) for $\overline{P}_c$ as follows:

$$\text{[NCost-2]} \quad C_o^{***} = \min_{\overline{\boldsymbol{x}}, \boldsymbol{\mu}} \boldsymbol{e}\overline{\boldsymbol{x}} \quad (8.48)$$

$$\text{subject to} \quad \overline{\boldsymbol{x}} \geq \overline{X}\boldsymbol{\mu}$$

$$\boldsymbol{y}_o \leq Y\boldsymbol{\mu}$$

$$\boldsymbol{\mu} \geq \mathbf{0}.$$

Let $(\overline{\boldsymbol{x}}_o^{**}, \boldsymbol{\mu}^*)$ be an optimal solution. Then, the cost-based pair $(\overline{\boldsymbol{x}}_o^{**}, \boldsymbol{y}_o)$ is the minimum production cost in the assumed production possibility set $\overline{P}_c$. This set can differ substantially from P if the unit prices of the inputs vary from DMU to DMU. The (global) allocative efficiency α^* of DMU$_o$ is defined as follows:

$$\alpha^* = \frac{C_o^{***}}{C_o^{**}} (\leq 1). \quad (8.49)$$

We also define the loss L_o^{***} due to the suboptimal cost mix as

$$L_o^{***} = C_o^{**} - C_o^{***} (\geq 0). \quad (8.50)$$

8.4.4 Decomposition of the Actual Cost

From (8.38), (8.47) and (8.50), we can derive at the following theorem.

Theorem 8.4 (Tone and Tsutsui (2004))

$$C_o \geq C_o^* \geq C_o^{**} \geq C_o^{***}. \quad (8.51)$$

Furthermore, we can obtain the relationship among the optimal cost and losses, and the actual cost (C_o) can be decomposed into three losses and the minimum cost (C_o^{***}):

$$L_o^* = C_o - C_o^* (\geq 0) \text{ Loss due to Technical Inefficiency} \quad (8.52)$$

$$L_o^{**} = C_o^* - C_o^{**} (\geq 0) \text{ Loss due to Price Inefficiency} \quad (8.53)$$

$$L_o^{***} = C_o^{**} - C_o^{***} (\geq 0) \text{ Loss due to Allocative Inefficiency} \quad (8.54)$$

$$C_o = L_o^* + L_o^{**} + L_o^{***} + C_o^{***}. \quad (8.55)$$

For further developments of this scheme, see Problems 8.1-8.4 at the end of this chapter.

8.4.5 An Example of Decomposition of Actual Cost

We applied the above procedure to the data set exhibited in Table 8.4 and obtained the results listed in Table 8.6. We then utilized the input-oriented CCR model and the projection formulas in Chapter 3 for finding the technical efficient inputs ($\boldsymbol{x}_o^*$).

Table 8.6. Decomposition of Actual Cost

DMU	Actual C	Cost Minimum C^{***}	C^{***}/C	Loss Tech. L^*	Loss Price L^{**}	Loss Alloc. L^{***}
A	25100	18386	0.73	0	5959	754
B	17130	17130	1	0	0	0
C	25650	18404	0.72	3557	2658	1032
D	36360	21507	0.59	0	13470	1383
E	17660	13483	0.76	4177	0	0
F	45150	27323	0.61	12911	1767	3149
G	40000	26287	0.66	4256	8931	526
H	31460	19684	0.63	6407	4230	1139
I	29944	24605	0.82	3254	1617	468
J	40600	29871	0.74	7725	0	3004
K	47800	34476	0.72	5367	5030	2927
L	35080	31457	0.90	2348	0	1275
Total	391934	282615	0.72	50002	43661	15656

As can be seen from the results, Hospital B is again the most efficient DMU in the sense that it has no loss due to technical, price or allocative inefficiencies, while Hospital D has the worst ratio $C^{***}/C(=0.59)$ caused by losses due to price and allocative inefficiencies. Hospital A has inefficiency in cost-based aspects, i.e., price and allocation but not in technical-physical aspects, whereas E has inefficiency due to technical-physical aspects but not in cost-based aspects. Figure 8.2 exhibits the decomposition graphically.

8.5 SUMMARY OF CHAPTER 8

This chapter has covered approaches which have been studied in DEA for evaluations of efficiencies such as "allocative" and "overall" efficiencies. Problems in the standard approaches were identified with cases in which different prices or different unit costs may be associated with the performance of different firms producing the same outputs and utilizing the same inputs. Types of price-cost efficiencies were therefore identified and related to extensions of the customary production possibility sets that reflected the unit price and cost differences.

Standard approaches were also extended in models that permit substitutions so that worsening of some inputs or outputs may be made in order to improve

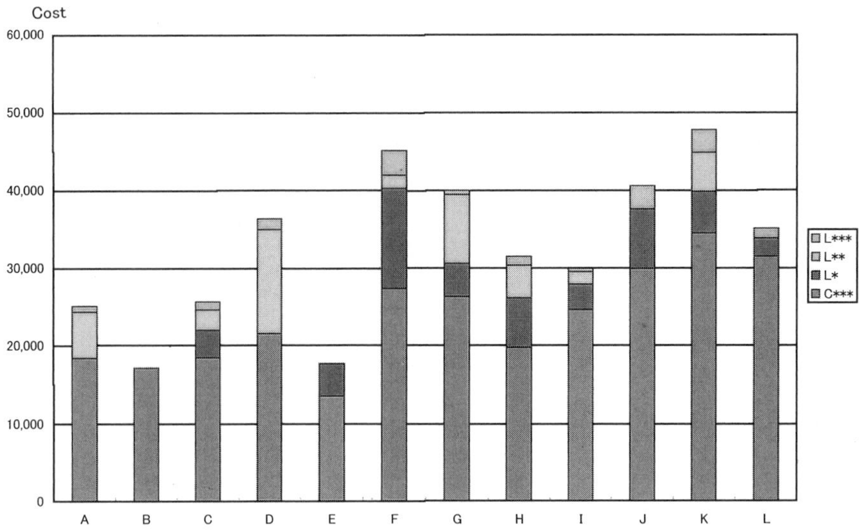

Figure 8.2. Decomposition of Actual Cost

other inputs or outputs. We also extended the traditional Farrell-Debreu cost efficiency measures and introduced new ones that can deal with non-identical cost and price situations. Furthermore, we provided a decomposition of the actual cost into the minimum cost and into losses due to other inefficiencies.

Problems in the use of these concepts may be encountered because many companies are unwilling to disclose their unit costs. As noted by Farrell, unit prices may also be subject to large fluctuations. One may, of course, use averages or other summaries of such prices and also deduce or infer which unit costs are applicable. However, this may not be satisfactory because in many cases accurate costs and prices may not really reflect criteria that are being used. Cases in point include attempts to evaluate public programs such as education, health, welfare, or military and police activities.

Earlier in this book a variety of approaches were suggested that can be applied to these types of problems. This includes the case of assurance regions and like concepts, as treated in Chapter 6, which can replace exact knowledge of prices and costs with corresponding bounds on their values. When this is done, however, the precise relations between allocative, overall and technical efficiencies may become blurred.

8.6 NOTES AND SELECTED BIBLIOGRAPHY

The concepts of cost efficiency related subjects were introduced by M.J. Farrell (1957) and G. Debreu (1951) and developed into implementable form by Färe, Grosskopf and Lovell (1985) using linear programming technologies. Cooper, Park and Pastor (1999)[8] extended these treatments to the Additive models with a new "translation invariant" measure named "RAM."

Inadequacies in attempts to move from "technical" to price based or cost based efficiencies were identified by Tone (2002). In response a new approach to cost efficiency was developed by Tone (2002) and further extended to decompositions of cost efficiency by Tone and Tsutsui (2004) in a form they applied to Japan-US electric utilities comparisons. Tone and Sahoo (2005)[9] applied the new cost efficiency model to examine the performance of Life Insurance Corporation (LIC) of India and found a significant heterogeneity in the cost efficiency scores over the course of 19 years. See also Tone and Sahoo (2005)[10] in which the issues of cost elasticity are extensively discussed based on the new cost efficiency model. Fukuyama and Weber (2004)[11] developed a variant of the new cost efficiency model using "directional distance functions" introduced in Chambers, Chung and Färe (1996)[12] to measure inefficiency. See Färe, Grosskopf and Whittaker (2004)[13] for an updated survey. See, however, Ray (2004), [14] p.95, who identified a deficiency in the failure of "directional distance functions" to account for nonzero slacks in the measures of efficiency.

8.7 RELATED DEA-SOLVER MODELS FOR CHAPTER 8

(New-)Cost-C(V) Cost-C(V) code evaluates the *cost efficiency* of each DMU as follows. First we solve the LP problem below:

$$\begin{aligned}
\min \quad & \sum_{i=1}^{m} c_i x_i \\
\text{subject to} \quad & x_i \geq \sum_{j=1}^{n} x_{ij}\lambda_j \quad (i = 1, \ldots, m) \\
& y_{ro} \leq \sum_{i=1}^{n} y_{rj}\lambda_j \quad (r = 1, \ldots, s) \\
& L \leq \sum_{j=1}^{n} \lambda_j \leq U \\
& \lambda_j \geq 0 \ \forall j,
\end{aligned} \tag{8.56}$$

where c_i is the unit cost of the input i. This model allows substitutions in inputs. Based on an optimal solution $(\boldsymbol{x}^*, \boldsymbol{\lambda}^*)$ of the above LP, the cost efficiency of DMU$_o$ is defined as

$$E_C = \frac{\boldsymbol{c}\boldsymbol{x}^*}{\boldsymbol{c}\boldsymbol{x}_o}. \tag{8.57}$$

The code "Cost-C" solves the case $L = 0$. $U = \infty$ (the case of constant returns to scale) and "Cost-V" for the case $L = U = 1$ (the variable returns to scale case).

The data set (X, Y, C) should be prepared in an Excel Workbook under an appropriate Worksheet name, e.g., ".Data", prior to execution of this code. See the sample format displayed in Figure B.6 in Section B.5 of Appendix B and refer to explanations above the figure.

The results will be obtained in the Worksheets of the selected Workbook: "Score", "Projection" (projection onto the efficient frontier), "Graph1", "Graph2" and "Summary."

New-Cost-C(V) solves the model described in [NCost] (8.21). Data format is the same with the Cost-C(V) model.

(New-)Revenue-C(V) Revenue-C(V) code solves the following revenue maximization program for each DMU:

$$\begin{aligned} \max \quad & \boldsymbol{p}\boldsymbol{y} \\ \text{subject to} \quad & \boldsymbol{x}_o \geq X\boldsymbol{\lambda} \\ & \boldsymbol{y} \leq Y\boldsymbol{\lambda} \\ & L \leq \boldsymbol{e}\boldsymbol{\lambda} \leq U \\ & \boldsymbol{\lambda} \geq \boldsymbol{0}, \end{aligned}$$

where the vector $\boldsymbol{p} = (p_1, \ldots, p_s)$ expresses the unit prices of the output. This model allows substitutions in outputs.

The code "Revenue-C" solves the case $L = 0$. $U = \infty$ (the case of constant returns to scale) and "Revenue-V", the case $L = U = 1$ (the variable returns to scale case).

Based on an optimal solution $\boldsymbol{y}^*$ of this program, the *revenue efficiency* is defined as

$$E_R = \frac{\boldsymbol{p}\boldsymbol{y}_o}{\boldsymbol{p}\boldsymbol{y}^*}. \tag{8.58}$$

E_R satisfies $0 < E_R \leq 1$, provided $\boldsymbol{p}\boldsymbol{y}_o > 0$. See the sample data format displayed in Figure B.7 in Section B.5 of Appendix B and refer to the explanations above the figure.

The results will be obtained in the Worksheets of the selected Workbook: "Score", "Projection" (projection onto the efficient frontier), "Graph1," "Graph2," and "Summary."

New-Revenue-C(V) solves the model described in [NRevenue] (8.27). Data format is the same with Revenue-C(V).

(New-)Profit-C(V) Profit-C(V) code solves the LP problem defined in (8.8) for each DMU. Based on an optimal solution ($\boldsymbol{x}^*$, $\boldsymbol{y}^*$), the *profit efficiency* is defined as

$$E_P = \frac{\boldsymbol{p}\boldsymbol{y}_o - \boldsymbol{c}\boldsymbol{x}_o}{\boldsymbol{p}\boldsymbol{y}^* - \boldsymbol{c}\boldsymbol{x}^*}. \tag{8.59}$$

Under the assumption $\boldsymbol{p}\boldsymbol{y}_o > \boldsymbol{c}\boldsymbol{x}_o$, we have $0 < E_P \leq 1$ and DMU$_o$ is *profit efficient* if $E_P = 1$.

The data format is a combination of *Cost* and *Revenue* models. The cost columns are headed by (C) for input names and the price column are headed by (P) for output names.

The results will be obtained in the Worksheets of the selected Workbook: "Score", "Projection" (projection onto the efficient frontier), "Graph1," "Graph2," and "Summary."

New-Profit-C(V) solves the model described in [NProfit] (8.29). Data format is the same with Profit-C(V).

Ratio-C(V) This code solves the LP problem defined in (8.31). Based on the optimal solution $(\boldsymbol{x}^*, \boldsymbol{y}^*)$, the *ratio* (revenue/cost) *efficiency* is defined as

$$E_{RC} = \frac{\boldsymbol{p}_o\boldsymbol{y}_o/\boldsymbol{c}_o\boldsymbol{x}_o}{\boldsymbol{p}_o\boldsymbol{y}^*/\boldsymbol{c}_o\boldsymbol{x}^*},$$

which satisfies $0 < E_{RC} \leq 1$ and DMU$_o$ is *ratio efficient* if $E_{RC} = 1$.

The data format is a combination of *Cost* and *Revenue* models. The cost columns are headed by (C) for input names and the price column are headed by (P) for output names.

The results will be obtained in the Worksheets of the selected Workbook: "Score", "Projection" (projection onto the efficient frontier), "Graph1," "Graph2," and "Summary."

8.8 PROBLEM SUPPLEMENT FOR CHAPTER 8

Problem 8.1

In Section 8.4, the actual cost is decomposed into the *sum* of the minimum cost and losses due to technical, price and allocative inefficiencies.

Can you decompose it in the *productive* form (not in the *sum* form)?

Suggested Response : Define the following efficiency measures:

- C^{***}/C = cost efficiency (CE)
- C^{*}/C = technical efficiency (TE)
- C^{**}/C^{*} = price efficiency (PE)
- C^{***}/C^{**} = allocative efficiency (AE)

Then, we have:

$$\text{CE} = \text{TE} \times \text{PE} \times \text{AE}.$$

Problem 8.2

Write out the decomposition of the actual profit in the same vein as described in Section 8.4.

Suggested Response : The actual profit of DMU $(\boldsymbol{x}_o, \boldsymbol{y}_o)$ is calculated as:

$$E_o = \boldsymbol{p}_o\boldsymbol{y}_o - \boldsymbol{c}_o\boldsymbol{x}_o. \tag{8.60}$$

Using radial or non-radial technical efficiency models, e.g., the CCR and SBM, we project the DMU onto the efficient frontier and obtain the technically efficient $(\boldsymbol{x}_o^*, \boldsymbol{y}_o^*)$ with profit given by

$$E_o^* = \boldsymbol{p}_o \boldsymbol{y}_o^* - \boldsymbol{c}_o \boldsymbol{x}_o^* \ (\geq \boldsymbol{p}_o \boldsymbol{y}_o - \boldsymbol{c}_o \boldsymbol{x}_o = E_o). \tag{8.61}$$

Thus, the *loss due to technical inefficiency* is evaluated as

$$L_o^* = E_o^* - E_o. \tag{8.62}$$

We formulate the new cost-price based production possibility set as

$$P_{cp} = \left\{ (\overline{\boldsymbol{x}}, \overline{\boldsymbol{y}}) | \ \overline{\boldsymbol{x}} \geq \overline{X}\boldsymbol{\lambda}, \overline{\boldsymbol{y}} \leq \overline{Y}\boldsymbol{\lambda}, \boldsymbol{\lambda} \geq \boldsymbol{0} \right\}, \tag{8.63}$$

where $\overline{X} = (\overline{\boldsymbol{x}}_1, \ldots, \overline{\boldsymbol{x}}_n)$ with $\overline{\boldsymbol{x}}_j = (c_{1j}x_{1j}^*, \ldots, c_{mj}x_{mj}^*)$ and $\overline{Y} = (\overline{\boldsymbol{y}}_1, \ldots, \overline{\boldsymbol{y}}_n)$ with $\overline{\boldsymbol{y}}_j = (p_{1j}y_{1j}^*, \ldots, p_{sj}y_{sj}^*)$. On this PPS we form a technical efficiency model similar to [NTec-2] (8.40) as follows:

$$\begin{aligned} \text{[NTec-3]} \qquad & \rho^* = \min_{\overline{\boldsymbol{x}}, \overline{\boldsymbol{y}}, \rho, \boldsymbol{\mu}} \rho \\ \text{subject to} \qquad & \rho \overline{\boldsymbol{x}}_o \geq \overline{X}\boldsymbol{\mu} = \overline{\boldsymbol{x}} \\ & \boldsymbol{y}_o \leq \overline{Y}\boldsymbol{\mu} = \overline{\boldsymbol{y}} \\ & \boldsymbol{\mu} \geq \boldsymbol{0}. \end{aligned} \tag{8.64}$$

Let the optimal solution of [NTec-3] be $(\overline{\boldsymbol{x}}_o^*, \overline{\boldsymbol{y}}_o^*, \boldsymbol{\mu}^*, \rho^*)$. (Note that, instead of [NTec-3], we can apply the non-radial and non-oriented SBM for obtaining $(\overline{\boldsymbol{x}}_o^*, \overline{\boldsymbol{y}}_o^*)$.)

We then have the technical and cost-price efficient profit given by

$$E_o^{**} = \boldsymbol{e}\overline{\boldsymbol{y}}_o^* - \boldsymbol{e}\overline{\boldsymbol{x}}_o^* \ (\geq E_o^*). \tag{8.65}$$

The loss due to cost-price inefficiency is estimated by

$$L_o^{**} = E_o^{**} - E_o^* \ (\geq 0). \tag{8.66}$$

Lastly, we solve the following max profit model on P_{cp}.

$$\begin{aligned} \text{[NProfit-2]} \qquad & \boldsymbol{e}\overline{\boldsymbol{y}}_o^{**} - \boldsymbol{e}\overline{\boldsymbol{x}}_o^{**} = \max_{\overline{\boldsymbol{x}}, \overline{\boldsymbol{y}}, \boldsymbol{\lambda}} \boldsymbol{e}\overline{\boldsymbol{y}} - \boldsymbol{e}\overline{\boldsymbol{x}} \\ \text{subject to} \qquad & \overline{\boldsymbol{x}}_o \geq \overline{X}\boldsymbol{\lambda} = \overline{\boldsymbol{x}} \\ & \boldsymbol{y}_o \leq \overline{Y}\boldsymbol{\lambda} = \overline{\boldsymbol{y}} \\ & \boldsymbol{\lambda} \geq \boldsymbol{0}. \end{aligned} \tag{8.67}$$

Let the optimal solution be $(\overline{\boldsymbol{x}}_o^{**}, \overline{\boldsymbol{y}}_o^{**})$. The allocative efficient profit is given by

$$E_o^{***} = \boldsymbol{e}\overline{\boldsymbol{y}}_o^{**} - \boldsymbol{e}\overline{\boldsymbol{x}}_o^{**}. \tag{8.68}$$

The loss due to allocative inefficiency is evaluated by

$$L_o^{***} = E_o^{***} - E_o^{**} \ (\geq 0). \tag{8.69}$$

Summing up, we have the decomposition of the actual profit into the maximum profit and the losses due to technical, cost-price and allocative inefficiencies as follows:

$$E_o = E_o^{***} - L_o^* - L_o^{**} - L_o^{***}. \tag{8.70}$$

Problem 8.3

Most of the models in this Chapter are developed under the constant returns-to-scale (CRS) assumption. Can you develop them under the variable returns-to-scale assumption?

Suggested Response : This can be done by adding the convex constraint on the intensity variable $\boldsymbol{\lambda}$ as follows:

$$\sum_{j=1}^{n} \lambda_j = 1.$$

Problem 8.4

In connection with the preceding problem, can you incorporate the scale inefficiency effect, i.e., loss due to scale inefficiency, in the model?

Suggested Response : In the cost efficiency case, we first solve the input-oriented technical efficiency model under the VRS assumption. Let the optimal solution be $(\boldsymbol{x}_o^{VRS}, \boldsymbol{y}_o^{VRS})$. The cost and loss due to this projection are, respectively:

$$C_o^{VRS} = \boldsymbol{c}_o \boldsymbol{x}_o^{VRS}, \; L_o^{VRS} = C_o - C_o^{VRS}, \tag{8.71}$$

where C_o is the actual observed cost of DMU $(\boldsymbol{x}_o, \boldsymbol{y}_o)$. Then we construct the data set (X^{VRS}, Y^{VRS}) consisting of $(\boldsymbol{x}_j^{VRS}, \boldsymbol{y}_j^{VRS})$ $j = 1, \ldots, n$. We next evaluate the technical efficiency of $(\boldsymbol{x}_o^{VRS}, \boldsymbol{y}_o^{VRS})$ with respect to (X^{VRS}, Y^{VRS}) under the constant returns-to-scale (CRS) assumption. Let the optimal solution be $(\boldsymbol{x}_o^*, \boldsymbol{y}_o^*)$, with its cost $C_o^* = \boldsymbol{c}_o \boldsymbol{x}_o^*$ $(\leq C_o^{VRS})$. Thus, we obtain the loss due to scale inefficiency as follows:

$$L_o^{Scale} = C_o^{VRS} - C_o^*. \tag{8.72}$$

Referring to (8.52)-(8.55), we can decompose the actual cost into four losses with minimum cost as follows:

$$L_o^{VRS} = C_o - C_o^{VRS} (\geq 0) \text{ Loss due to Pure Tech. Inefficiency} \tag{8.73}$$
$$L_o^{Scale} = C_o^{VRS} - C_o^* (\geq 0) \text{ Loss due to Scale Inefficiency} \tag{8.74}$$
$$L_o^{**} = C_o^* - C_o^{**} (\geq 0) \text{ Loss due to Price Inefficiency} \tag{8.75}$$
$$L_o^{***} = C_o^{**} - C_o^{***} (\geq 0) \text{ Loss due to Allocative Inefficiency} \tag{8.76}$$
$$C_o = L_o^{VRS} + L_o^{Scale} + L_o^{**} + L_o^{***} + C_o^{***}. \tag{8.77}$$

Problem 8.5

The concluding part of Section 5.5, Chapter 5, quoted Dr. Harold Wein, a steel industry consultant, who believed that the concept of returns to scale, as formulated in economics, is useless because increases in plant size are generally accompanied by mix changes in outputs or inputs — or both.

Assignment : Formulate the responses an economist might make to this criticism.

Suggested Response : Under the assumption of profit maximization, as employed in economics, both scale and mix are determined simultaneously. This is consistent with Dr. Wein's observation from his steel industry experience.

One response to Dr. Wein's criticism of the scale concept as employed in economics (which holds mix constant) is to note that economists are interested in being able to distinguish between scale and mix changes when treating empirical data *after* the decisions have been made. Economists like Farrell[15] and Debreu[16] contributed concepts and methods for doing this, as cited in this chapter (and elsewhere in this text), and these have been further extended by economists like Färe, Grosskopf and Lovell whose works have also been cited at numerous points in this text.[17]

The above response is directed to *ex post* analyses. Another response is directed to whether economics can contribute to these decisions in an *ex ante* fashion. The answer is that there is a long-standing literature which provides guidance to scale decisions by relating marginal (incremental) costs to marginal (incremental) receipts. Generally speaking, returns to scale will be increasing as long as marginal costs are below average (unit) costs. The reverse situation applies when returns to scale is decreasing. Equating marginal costs to marginal receipts will determine the best (most profitable) scale size.

This will generally move matters into the region of decreasing returns to scale. Reasons for this involve issues of stability of solutions which we cannot treat here. Marginal cost lying below average unit cost in regions of increasing returns to scale means that average unit cost can be decreased by incrementing outputs. Hence if marginal receipts equal or exceed average unit cost it is possible to increase total profit by incrementing production.

The above case refers to single output situations. Modifications are needed to allow for complementary and substitution interactions when multiple outputs are involved. This has been accomplished in ways that have long been available which show that the above rule continues to supply general guidance. Indeed, recent years have seen this extended to ways for determining "economics of scope" in order to decide whether to add or delete product lines while simultaneously treating mix and scale decisions. See W.S. Baumol, J.C. Panzar and R.D. Willig (1982) *Contestable Markets and the Theory of Industry Structure* (New York: Harcourt Brace Jovanovich).

Much remains to be done in giving the above concepts implementable form — especially when technical inefficiencies are involved and errors and uncertainties are also involved. A start has been made in the form of what are referred to as "stochastic frontier" approaches to statistical estimation. However, these approaches are, by and large, confined to the use of single output regressions.

Comment : As already noted, profit maximization requires a simultaneous determination of the best (i.e., most profitable) combination of scale, scope, technical and mix efficiencies. The models and methods described in earlier chapters can then be used to determine what was done and whether and where any inefficiencies occurred.

Problem 8.6

Prove that the *profit ratio* model in (8.31) does not suffer from the inadequacies pointed out in Section 8.3.

Suggested Response : Since the profit ratio efficiency is defined as ratio of ratios (between revenue and cost), it is invariant when we double both the unit cost and price.

Notice that the traditional profit efficiency model [Profit] (8.8) gives the same efficiency value when we double both the unit cost and price. This is unacceptable.

Notes

1. M.J. Farrell (1957), "The Measurement of Productive Efficiency," *Journal of the Royal Statistical Society* Series A, 120, III, pp.253-281. G. Debreu (1951), "The Coefficient of Resource Utilization," *Econometrica* 19, pp.273-292.

2. R. Färe, S. Grosskopf and C.A.K. Lovell, *Measurement of Efficiency of Production* (Boston: Kluwer-Nijhoff Publishing Co., Inc., 1985).

3. This means that P lies below the production possibility surface and hence is inefficient.

4. K. Tone (2002), "A Strange Case of the Cost and Allocative Efficiencies in DEA," *Journal of the Operational Research Society* 53, pp.1225-1231.

5. This ratio model was introduced in K. Tone (1993), *Data Envelopment Analysis* (in Japanese)(Tokyo: JUSE Press, Ltd.).

6. W.W. Cooper, Z. Huang, S. Li and J.T. Pastor (2005) "Aggregation with Enhanced Russell Measure in DEA," *European Journal of Operational Research* (forthcoming).

7. K. Tone and M. Tsutsui (2004), "Decomposition of Cost Efficiency and its Application to Japan-US Electric Utility Comparisons," Research Report Series I-2004-0004, GRIPS (National Graduate Institute for Policy Studies), also forthcoming in *Socio-Economic Planning Sciences.*

8. W.W. Cooper, K.S. Park and J.T. Pastor (1999), "RAM: A Range Adjusted Measure of Inefficiency for Use with Additive Models and Relations to Other Models and Measures in DEA," *Journal of Productivity Analysis* 11, pp.5-42.

9. K. Tone and B.K. Sahoo (2005), "Evaluating Cost Efficiency and Returns to Scale in the Life Insurance Corporation of India Using Data Envelopment Analysis," *Socio-Economic Planning Sciences* 39, pp.261-285.

10. K. Tone and B.K. Sahoo (2005), "Cost-Elasticity: A Re-Examination in DEA," *Annals of Operations Research* (forthcoming).

11. H. Fukuyama and W.L. Weber (2004), "Economic Inefficiency Measurement of Input Spending When Decision-making Units Face Different Input Prices," *Journal of the Operational Research Society* 55, pp.1102-1110.

12. R.G. Chambers, Y. Chung and R. Färe (1996) "Benefit and Distance Functions," *Journal of Economic Theory* 70, pp.407-418.

13. R. Färe, S. Grosskopf and G. Whittaker (2004) "Distance Functions," Chapter 5 in W.W. Cooper, L.M. Seiford and J. Zhu, eds., *Handbook on Data Envelopment Analysis* (Norwell Mass., Kluwer Academic Publishers).

14. S. Ray (2004) *Data Envelopment Analysis: Theory and Techniques for Economics and Operations Research* (Cambridge University Press).

15. See Note 1. See also M.J. Farrell and M. Fieldhouse (1962), "Estimating Efficient Production Functions Under Increasing Returns to Scale," *Journal of the Royal Statistical Society* Series A, 125, Part 2, pp.252-267.

16. See the Note 1 reference.

17. See the Note 2 above. See also R. Färe, S. Grosskopf and C.A.K. Lovell (1994), *Production Frontiers* (Cambridge: Cambridge University Press).

9 DATA VARIATIONS

9.1 INTRODUCTION

We have now covered a wide variety of topics ranging from straightforward DEA models and their uses and extending to modifications such as are incorporated in assurance regions or in the treatment of variables with values that have been exogenously fixed to varying degrees. This, however, does not end the possibilities. New uses of DEA with accompanying new developments and extensions continue to appear.

An attempt to cover all of these topics would be beyond the scope of this text — and perhaps be impossible of achievement as still more uses and new developments continue to appear. Hence the course we follow is to provide relatively brief introductions to some of these topics. This will include sensitivity analysis, stochastic-statistical characterizations and probabilistic formulations with their associated approaches. Dynamic extensions of DEA will also be indicated in the form of window analysis.

9.2 SENSITIVITY ANALYSIS

9.2.1 Degrees of Freedom

The topic of sensitivity (= stability or robustness) analysis has taken a variety of forms in the DEA literature. One part of this literature studies the responses when DMUs are deleted or added to the set being considered or when outputs or inputs are added or withdrawn from consideration. See Wilson (1995).[1] See also the discussion of "window analysis" later in this chapter. Another part of this literature deals with the increases or decreases in the number of inputs and outputs to be treated. Analytically oriented treatments of these topics are not lacking[2] but most of this literature has taken the form of simulation studies, as in Banker *et al.* (1996).[3]

Comment: As in statistics or other empirically oriented methodologies, there is a problem involving degrees of freedom, which is compounded in DEA because of its orientation to *relative* efficiency. In the envelopment model, the number of degrees of freedom will increase with the number of DMUs and decrease with the number of inputs and outputs. A rough rule of thumb which can provide guidance is as follows.

$$n \geq \max\{m \times s, 3(m + s)\}$$

where n= number of DMUs, m= number of inputs and s = number of outputs.

9.2.2 *Algorithmic Approaches*

Attention to this topic of sensitivity analysis in DEA was initiated in Charnes *et al.* (1985)[4] which built on the earlier work in Charnes and Cooper (1968)[5] after noting that variations in the data for the DMU_o under evaluation could alter the inverse matrix used to generate solutions in the usual simplex algorithm computer codes. (See expressions (3.52)-(3.53) in Chapter 3.) Proceeding along the path opened by the latter publication (by Charnes and Cooper) this work is directed to the use of algorithms that avoid the need for additional matrix inversions. Originally confined to treating a single input or output this line of work was extended and improved in a series of papers published by Charnes and Neralic.[6]

We do not pursue these algorithmic approaches here. We turn instead to other approaches where attention is confined to transitions from efficient to inefficient status.

9.2.3 *Metric Approaches*

Another avenue for sensitivity analysis opened by Charnes *et al.* (1992)[7] bypasses the need for these kinds of algorithmic forays by turning to metric concepts. The basic idea is to use concepts such as "distance" or "length" (= norm of a vector) in order to determine "radii of stability" within which the occurrence of data variations will not alter a DMU's classification from efficient to inefficient status (or *vice versa*).

The resulting classifications can range from "unstable" to "stable" with the latter identified by a radius of some finite value within which no reclassification will occur. Points like E or F in Figure 9.1 provide examples identified as stable. A point like A, however, is unstable because an infinitesimal perturbation to the left of its present position would alter its status from inefficient to efficient.

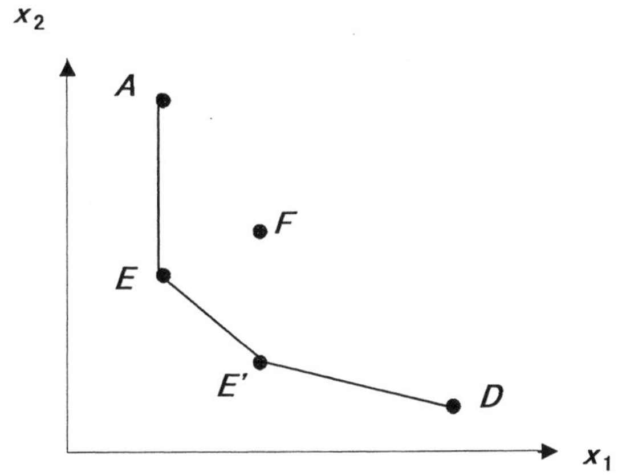

Figure 9.1. Stable and Unstable DMUs

A variety of metrics and models are examined but here attention will be confined to the Chebychev ($= l_\infty$) norm, as in the following model taken from Charnes, Haag, *et al.* (1992, p.795),[7]

$$\max \ \delta \tag{9.1}$$

$$\text{subject to} \quad y_{ro} = \sum_{j=1}^{n} y_{rj}\lambda_j - s_r^+ - \delta d_r^+, \quad r = 1, \ldots, s$$

$$x_{io} = \sum_{j=1}^{n} x_{ij}\lambda_j + s_i^- + \delta d_i^-, \quad i = 1, \ldots, m$$

$$1 = \sum_{j=1}^{n} \lambda_j$$

with all variables (including δ) constrained to be nonnegative while the d_r^+ and d_i^- are fixed constants (to serve as weights) which we now set to unity.

With all $d_i^- = d_r^+ = 1$ the solution to (9.1) may be written

$$\sum_{j=1}^{n} y_{rj}\lambda_j^* - s_r^{+*} = y_{ro} + \delta^*, \quad r = 1, \ldots, s \tag{9.2}$$

$$\sum_{j=1}^{n} x_{ij}\lambda_j^* + s_i^{-*} = x_{io} - \delta^*, \quad i = 1, \ldots, m$$

This shows that we are improving all outputs and inputs to the maximum that this metric allows consistent with the solution on the left.

The formulation in (9.2) is for an inefficient DMU which continues to be inefficient for all data alteration from y_{ro} to $y_{ro} + \delta^*$ and from x_{io} to $x_{io} - \delta^*$. This is interpreted to mean that no reclassification to efficient status will occur within the open set defined by the value of $0 \leq \delta^*$ — which is referred to as a "radius of stability." See, for example, the point F in Figure 9.2 which is centered in the square (or box) defined by this C (=Chebyshev) norm which is referred to as a "unit ball."[8]

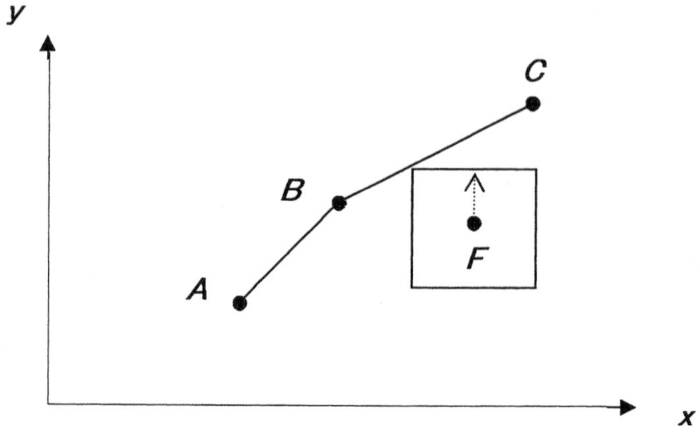

Figure 9.2. A Radius of Stability

The above model dealt with improvements in both inputs and outputs that could occur for an inefficient point before its status would change to efficient — as in the upper left hand corner of the box surrounding F in Figure 9.2. The treatment of efficient points proceeds in the direction of "worsening" outputs and inputs as in the following model.

$$\min \ \delta \tag{9.3}$$

$$\text{subject to} \quad y_{ro} = \sum_{j=1, j\neq o}^{n} y_{rj}\lambda_j - s_r^+ - \delta, \quad r = 1, \ldots, s$$

$$x_{io} = \sum_{j=1, j\neq o}^{n} x_{ij}\lambda_j + s_i^- + \delta, \quad i = 1, \ldots, m$$

$$1 = \sum_{j=1, j\neq o}^{n} \lambda_j$$

where, again, all variables are constrained to be nonnegative.

In this case $j \neq o$ refers to the efficient DMU_o that is being analyzed. Otherwise, as in the following definition, the result will always be unstable.[9]

Definition 9.1 *The coordinates of the point associated with an efficient DMU will always have both efficient and inefficient points within a radius of $\varepsilon > 0$ however small the value of ε.*

Definition 9.2 *Any point with the above property is unstable.*

To see that this property is not confined to points associated with efficient DMUs, note that A in Figure 9.1 has this property since a slight variation to the left will change its status from inefficient to efficient. In any case, a solution, δ^*, provides a radius in the Chebychev norm that is to be attained before an efficient DMU is changed to inefficient status.

To see what is happening in the case of an efficient point refer to B in Figure 9.2. The radius of stability for this point would be determined by the "worsenings" allowed in (9.3) until the line connecting A and C is reached. This follows from the fact that worsenings which only move B to a new point which is on the left of this line will not affect its efficiency status.

Comment : The above formulations are recommended only as the *start* for a sensitivity analysis by Charnes *et al.*[10] because, *inter alia*, this norm does not reflect any nonzero slacks which may be present.[11] It might be supposed that omitting the DMU to be evaluated from the right-hand side of (9.3) could lead to non-solution possibilities. This is not the case. Solutions to (9.3) always exist, as is proved in W.W. Cooper, S. Li, L.M. Seiford, K. Tone, R.M. Thrall and J. Zhu (2001) "Sensitivity and Stability Analysis in DEA: Some Recent Developments," *Journal of Productivity Analysis* 15, pp.217-246. See also L.M. Seiford and J. Zhu (1998) "Sensitivity Analysis of DEA Models for Simultaneous Changes in All of the Data," *Journal of the Operational Research Society* 49, pp.1060-1071 as well as Seiford and Zhu (1999) "Infeasibility of Super-Efficiency Data Envelopment Analysis Models," *INFOR* 37, pp.174-187.

9.2.4 Multiplier Model Approaches

The above approaches treat one DMU at a time. However, this needs to be extended for treating cases where the DMUs are numerous and it is not clear which ones require attention. Ideally it should be possible to vary all data simultaneously until the status of at least one DMU is changed from inefficient to efficient or *vice versa*. A third approach initiated by R.G. Thompson and R.M. Thrall[12] and their associates moves in this direction in a manner that we now describe.

For this purpose we record the following dual pair from Thompson *et al.* (1996).[13]

Envelopment Model	Multiplier Model
$\text{minimize}_{\theta,\lambda}\ \theta$	$\text{maximize}_{\boldsymbol{u},\boldsymbol{v}}\ z = \boldsymbol{u}\boldsymbol{y}_o$
subject to	subject to
$Y\boldsymbol{\lambda} \geq \boldsymbol{y}_o$	$\boldsymbol{u} \geq \boldsymbol{0}$
$\theta\boldsymbol{x}_o - X\boldsymbol{\lambda} \geq \boldsymbol{0}$	$\boldsymbol{v} \geq \boldsymbol{0}$
$\boldsymbol{\lambda} \geq \boldsymbol{0}$	$\boldsymbol{u}Y - \boldsymbol{v}X \leq \boldsymbol{0}$
θ unrestricted	$\boldsymbol{v}\boldsymbol{x}_o = 1,$

(9.4)

where Y, X and $\boldsymbol{y}_o, \boldsymbol{x}_o$ are data matrices and vectors of outputs and inputs, respectively, and $\boldsymbol{\lambda}, \boldsymbol{u}, \boldsymbol{v}$ are vectors of variables ($\boldsymbol{\lambda}$: a column vector, $\boldsymbol{u}$ and $\boldsymbol{v}$: row vectors). θ, a scalar, which can be positive, negative or zero in the envelopment model is the source of the condition $\boldsymbol{v}\boldsymbol{x}_o = 1$ in the multiplier model.

No allowance for nonzero slacks is made in the objective of the above envelopment model. Hence the variables in the multiplier model are constrained only to be nonnegative. That is, the positivity requirement associated with the non-Archimedean element, ε, is absent from both members of this dual pair. Thompson *et al.* refer to Charnes, Cooper and Thrall (1991) [14] to justify this omission of non-Archimedean elements. For present purposes, however, we note only that these sensitivity analyses are centered around the set, E, of efficient extreme points and these points always have a unique optimum with nonzero slacks.

We also note that the analysis is carried forward via the multiplier models[15] by Thompson, *et al.* This makes it possible to exploit the fact that the values $\boldsymbol{u}^*, \boldsymbol{v}^*$ which are optimal for the DMU being evaluated will remain valid over some (generally positive) range of variation in the data.[16]

Following Thompson, *et al.* we try to exploit this property by defining a new vector $\boldsymbol{w} = (\boldsymbol{u}, \boldsymbol{v})$ and a function $h_j(\boldsymbol{w})$ as follows

$$h_j(\boldsymbol{w}) = \frac{f_j(\boldsymbol{w})}{g_j(\boldsymbol{w})} = \frac{\sum_{r=1}^{s} u_r y_{rj}}{\sum_{i=1}^{m} v_i x_{ij}}. \tag{9.5}$$

Next, let

$$h_o(\boldsymbol{w}) = \max_{j=1,\ldots,n} h_j(\boldsymbol{w}) \tag{9.6}$$

so that

$$h_o(\boldsymbol{w}) \geq h_j(\boldsymbol{w}) \ \ \forall j. \tag{9.7}$$

It is now to be noted that (9.5) returns matters to the CCR ratio form which was introduced as early as Chapter 2, Section 2.2. Hence we need not be concerned with continued satisfaction of the norm condition, $\boldsymbol{v}\boldsymbol{x}_o = 1$ in (9.4), as we study variations in the data.

When an *optimal* $\boldsymbol{w}^*$ does not satisfy (9.7), the DMU_o being evaluated is said to be "radial inefficient." The term is appropriate because this means that $\theta^* < 1$ will occur in the envelopment model. The full panoply of relations between the CCR ratio, multiplier and envelopment models is thus brought into play without any need for extensive computations or analyses.

Among the frontier points for which $\theta^* = 1$, attention is directed by Thompson *et al.* to "extreme efficient points" — i.e., points in the set E which, for some multiplier w^*,

$$h_o(w^*) > h_j(w^*) \;\; \forall j \neq o. \tag{9.8}$$

This (strict) inequality will generally remain valid over some range of variation in the data. Hence, in more detail we will have

$$h_o(w^*) = \frac{\sum_{r=1}^{s} u_r^* y_{ro}}{\sum_{i=1}^{m} v_i^* x_{io}} > \frac{\sum_{r=1}^{s} u_r^* y_{rj}}{\sum_{i=1}^{m} v_i^* x_{ij}} = h_j(w^*) \;\; \forall j \neq o, \tag{9.9}$$

which means that DMU_o is more efficient than any other DMU_j and hence will be rated as fully efficient by DEA.

Thompson, *et al.* employ a ranking principle which they formulated as: "If DMU_o is more efficient than any other DMU_j *relative* to the vector w^*, then DMU_o is said to be top ranked." Holding w^* fixed, the data are then varied and DMU_o is said to be "top ranked" if (9.8) continues to hold.

Thompson, *et al.* allow the data to vary in several different ways which include allowing the data variations to occur at random. Among these possibilities we examine only the following one. For DMU_o, which is extreme efficient, the outputs are all decreased and the inputs are all increased by a stipulated amount (or percentage). This same treatment is accorded to the other DMUs which are efficient. For the other DMU_j, the reverse adjustment is made: All outputs are increased and all inputs are decreased in these same amounts (or percentages). In this way the ratio in (9.8) for DMU_o will be decreased along with the other extreme efficient DMUs while the ratios for the other DMU_j will be increased. Continuing in this manner a reversal can be expected to occur at some point — in which case DMU_o will no longer be "top ranked" and it will then lose the status of being fully (DEA) efficient.

Table 9.1 taken from Thompson *et al.* (1994) will be used to illustrate the procedure in a simple manner by varying only the data for the inputs x_1, x_2 in this table. To start this sensitivity analysis, Table 9.2 records the initial

Table 9.1. Data for a Sensitivity Analysis

	E-Efficient*			Not Efficient		
DMU	1	2	3	4	5	6
Output: y	1	1	1	1	1	1
Input: x_1	4	2	1	2	3	4
Input: x_2	1	2	4	3	2	4

* E-Efficient = Extreme Point Efficient

solutions by applying the multiplier model for (9.8) to these data for each of DMU_1, DMU_2 and DMU_3 which are all extreme point efficient.[17] As can be

Table 9.2. Initial Solutions

DMU	DMU_1 $h_j(w^1)$	DMU_2 $h_j(w^2)$	DMU_3 $h_j(w^3)$
1	1.000	0.800	0.400
2	0.714	1.000	0.714
3	0.400	0.800	1.000
4	0.500	0.800	0.667
5	0.667	0.800	0.550
6	0.357	0.500	0.357

seen these solutions show DMU_1, DMU_2 and DMU_3 to be top ranked in their respective columns.

The gaps between the top and other ranks from these results show that some range of data variation can be undertaken without changing this top-ranked status. To start we therefore introduce 5% increases in each of x_1 and x_2 for DMU_1, DMU_2 and DMU_3 and thereby worsen their performances. Simultaneously we decrease these inputs by 5% for the other DMUs to obtain Table 9.3.

Table 9.3. Results of 5% Data Variations

DMU	DMU_1 $h_j(w^1)$	DMU_2 $h_j(w^2)$	DMU_3 $h_j(w^3)$
1	0.952	0.762	0.381
2	0.680	0.952	0.680
3	0.381	0.762	0.952
4	0.526	0.842	0.702
5	0.702	0.842	0.526
6	0.376	0.526	0.376

The values of the $h_j(w)$ resulting from these data variations are portrayed in Table 9.3. As can be seen, each of DMU_1, DMU_2 and DMU_3 maintain their "top ranked status" and hence continue to be DEA fully efficient (relatively). Nor is this the end of the line. Continuing in this 5% increment-decrement fashion, as Thompson, *et al.* (1994) report, a 15% increment-decrement is needed for a first displacement in which DMU_2 is replaced by DMU_4 and DMU_5. Continuing further, a 20% increment-decrement is needed to replace DMU_1 with DMU_5 and, finally, still further incrementing and decrementing is needed to replace DMU_3 with DMU_4 as top ranked.

Comment : Note that the $h_j(\boldsymbol{w})$ values for all of the efficient DMUs decrease in every column when going from Table 9.2 to Table 9.5 and, simultaneously, the $h_j(\boldsymbol{w})$ values increase for the inefficient DMUs. The same behavior occurs for the other data variations, including the random choices of data changes, used by Thompson, Thrall and their associates in other studies. As noted on page 401 of Thompson *et al.* (1994) this robust behavior is obtained for extreme efficient DMUs which are identified by their satisfaction of the Strong Complementary Slackness Condition (described in Section A.8 of our Appendix A) for which a gap will appear like ones between the top and second rank shown in every column of Table 9.2. In fact, the choice of $\boldsymbol{w}^*$ can affect the degree of robustness as reported in Thompson *et al.* (1996) where use of an interior point algorithm produces a $\boldsymbol{w}^*$ closer to the analytic center and this considerably increases the degree of robustness for the above example. For a more detailed treatment see Cooper, Li, Seiford and Zhu. [18]

9.3 STATISTICAL APPROACHES

Treatment of data variations by statistical methods has taken a variety of forms in DEA and related literatures. More precisely, Banker (1993)[19] and Banker and Natarasan (2004)[20] show that DEA provides a consistent estimator of arbitrary monotone and concave production functions when the (one-sided) deviations from such a production function are regarded as stochastic variations in technical inefficiency.[21] Convergence is slow, however, since, as is shown by Korostolev *et al.* (1995),[22] the DEA likelihood estimator in the single output - m input case converges at the rate $n^{-2/(1+m)}$ and no other estimator can converge at a faster rate.[23]

The above approaches treat only the single output - multiple input case. Simar and Wilson (1998)[24] turn to "bootstrap methods" which enable them to deal with the case of multiple outputs and inputs. In this manner, the sensitivity of θ^*, the efficiency score obtained from the BCC model, can be tested by repeatedly sampling from the original samples. A sampling distribution of θ^* values is then obtained from which confidence intervals may be derived and statistical tests of significance developed.

All of this work represents significant new developments. More remains to be done, however, since neither Banker nor Simar and Wilson make any mention of how to treat nonzero slacks. Thus, it is not even clear that they are estimating efficiency frontiers.

Another line of research proceeds through what are referred to as "stochastic frontier regressions." This line of work has a longer history which can (in a sense) by traced all the way back to Farrell (1957).[25] Subsequently extended by Aigner and Chu (1968)[26] this approach was given its first statistical formulation in Aigner, Lovell and Schmidt (1977) in a form that is now called the "composed error" approach.[27]

To see what is involved in this "composed error" approach we start with the usual formulation of a statistical regression model as in

$$y = f(\boldsymbol{x}) + \varepsilon. \tag{9.10}$$

Here $f(\boldsymbol{x})$ is a prescribed (known) function with parameters to be estimated and ε represents random errors which occur in the dependent (regressand) variable, a scalar, and not in the independent (regressor) variables represented by the vector $\boldsymbol{x}$. The components of $\boldsymbol{x}$, we emphasize, are assumed to be known without error.

The concept of a "composed error" is represented by replacing ε with a 2-component term which we can represent as

$$\varepsilon = \nu - \tau. \tag{9.11}$$

Here ν represents the random error component which may be positive, negative or zero while τ is restricted to nonnegative ranges that represent values of y that fail to achieve the efficient frontier. The term $\tau \geq 0$ is usually assumed to have statistical distributions, such as the exponential or half normal, which are confined to nonnegative ranges that represent inefficiencies.

Following Jondrow, Lovell, Materov and Schmidt (1982),[28] the estimates of technical efficiency are obtained from

$$\widehat{\tau} = -\left[\mu_\tau - \sigma \frac{f^*(\mu_\tau/\sigma)}{F^*(-\mu_\tau/\sigma)}\right] \tag{9.12}$$

where

$$\mu_\tau = \frac{\sigma_\tau^2}{\sigma_\nu^2 + \sigma_\tau^2} \quad \text{and} \quad \sigma^2 = \frac{\sigma_\nu^2 \sigma_\tau^2}{\sigma_\nu^2 + \sigma_\tau^2}$$

and where $f^*(\cdot)$ and $F^*(\cdot)$ represent the standard normal density and cumulative normal distribution functions, respectively, with mean μ and variance σ^2. The efficiency corresponding to specified values for the components of $\boldsymbol{x}$ are then estimated from

$$0 \leq e^{-\widehat{\tau}} \leq 1 \tag{9.13}$$

which is equal to unity when $\widehat{\tau} = 0$ and becomes 0 as $\widehat{\tau} \to \infty$.

To see how this measure of efficiency is to be used we employ (9.10) and (9.11) in the following simple (two-input) version of a log-linear (=Cobb-Douglas) production function

$$y = \beta_0 x_1^{\beta_1} x_2^{\beta_2} e^{\varepsilon} = \beta_0 x_1^{\beta_1} x_2^{\beta_2} e^{\nu - \tau} \tag{9.14}$$

so that

$$y e^{\tau} = \beta_0 x_1^{\beta_1} x_2^{\beta_2} e^{\nu}. \tag{9.15}$$

Hence $ye^\tau = \widehat{y}$ with $\widehat{y} \geq y$ is estimated stochastically with inefficiency embodied in an output shortfall and not in overuses of either input.

It is possible to view these stochastic frontier regressions as competing with DEA and to study them from this standpoint as is done in Gong and Sickles

(1990),[29] for example, who bring a simulation approach to this task. Carried to an extreme the two approaches, DEA vs. Stochastic Frontier Regressions, can be regarded as mutually exclusive — as in Schmidt (1985).[30] An alternative view is also possible in which the two approaches can be used in complementary fashion. Ferrier and Lovell (1990),[31] for example, use the two approaches to cross-check each other. In this approach, the objective is to avoid what Charnes, Cooper and Sueyoshi (1988)[32] refer to as "methodological bias" when large issues of policy are being addressed. Indeed, it is possible to go a step further and join the two approaches in a common effort as in the example we now discuss.[33]

Arnold *et al.* (1994)[34] describe an experience in which the use of Cobb-Douglas regressions yielded unsatisfactory results in an attempt to use this kind of regression approach in a study conducted under legislative mandate to develop methods for evaluating the performances of public schools in Texas. Using this same functional form, however, and applying it to the same body of data, Arnold *et al.* reported satisfactory results from a two-stage DEA-regression approach which proceeded in the following manner: In stage one all of the 640 schools in this study were submitted to treatment by DEA. The original Cobb-Douglas form was then extended to incorporate these results in the form of "dummy variables." In this approach a school which had been found to be DEA efficient was associated with a value of unity for the dummy variables. A school which had been found to be DEA inefficient was assigned a value of zero. The regression was then recalculated and found to yield very satisfactory results.

The above study was followed by a simulation experiment which we now review for added insight.[35] For this purpose we replaced (9.10) with

$$y = 0.75x_1^{0.65}x_2^{0.55}e^{\varepsilon}. \tag{9.16}$$

In short, the above expression is used to generate all observations with the Cobb-Douglas form having known parameter values

$$\begin{aligned} \beta_0 &= 0.75 \\ \beta_1 &= 0.65 \\ \beta_2 &= 0.55 \end{aligned} \tag{9.17}$$

and e^{ε} is used to generate random variables which are then used to adjust the thus generated y values to new values which contain these random terms. This procedure conforms to the assumptions of both SF (=Stochastic Frontier) and OLS (=Ordinary Least Squares) regression uses.

The input values for x_1 and x_2 in (9.16) are generated randomly, as a bias avoiding mechanism, and these values are inserted in (9.16) to provide the truly efficient values of y after which the values of $\widehat{y}$ defined in (9.15) are then generated in the previously described manner.

The inputs are then adjusted to new values

$$\widehat{x}_1 = x_1 e^{\tau_1} \text{ and } \widehat{x}_2 = x_2 e^{\tau_2} \text{ with } \tau_1, \tau_2 \geq 0 \tag{9.18}$$

where τ_1 and τ_2 represent input-specific technical inefficiencies drawn at random to yield the corresponding input inefficiencies embedded in $\widehat{x}_1$ and $\widehat{x}_2$.

This procedure, we note, violates the SF assumption that all inefficiencies are impounded only in the regressand, y. See the discussion immediately following (9.15). It also violates OLS since these $\widehat{x}_1$, $\widehat{x}_2$ are not the input amounts used to generate the y values. Nevertheless it reproduces a situation in which the observed y (or $\widehat{y}$) will tend to be too low for these inputs. Finally, to complete the experimental design, a subset of the observations, chosen at random, used the original x_1, x_2 values rather than the $\widehat{x}_1$, $\widehat{x}_2$ generated as in (9.18). This was intended to conform to the assumption that some DMUs are wholly efficient and it also made it possible (a) to determine whether the first-stage DEA identified the efficient DMUs in an approximately correct manner as well as (b) to examine the effects of such an efficient subset on the derived statistical estimates.

Further details on the experimental design may be found in Bardhan *et al.* (1998).

We therefore now turn to the estimating relations which took a logarithmic form as follows,

$$\ln y = \ln \beta_0 + \beta_1 \ln \widehat{x}_1 + \beta_2 \ln \widehat{x}_2 \tag{9.19}$$

and

$$\ln y = \ln \beta_0 + \beta_1 \ln \widehat{x}_1 + \beta_2 \ln \widehat{x}_2 + \delta D + \delta_1 D \ln \widehat{x}_1 + \delta_2 D \ln \widehat{x}_2 + \varepsilon \tag{9.20}$$

Table 9.4. OLS Regression Estimates without Dummy Variables

Case 1: *EXPONENTIAL* distribution of input inefficiencies

Parameter Estimates	*Case A* $\sigma_\varepsilon^2 = 0.04$ (1)	*Case B* $\sigma_\varepsilon^2 = 0.0225$ (2)	*Case C* $\sigma_\varepsilon^2 = 0.01$ (3)	*Case D* $\sigma_\varepsilon^2 = 0.005$ (4)
β_0	1.30*	1.58*	1.40*	1.43*
	(0.19)	(0.15)	(0.13)	(0.10)
β_1	0.46*	0.43*	0.45*	0.46*
	(0.024)	(0.02)	(0.016)	(0.013)
β_2	0.48*	0.47*	0.47*	0.46*
	(0.02)	(0.013)	(0.01)	(0.01)

The asterisk "*" denotes statistical significance at 0.05 significance level or better. Standard errors are shown in parentheses.
The values for σ_ε^2 shown in the top row of the table represent the true variances for the statistical error distributions.

Table 9.5. Stochastic Frontier Regression Estimates without Dummy Variables

Case 1: *EXPONENTIAL* distribution of input inefficiencies

Parameter Estimates	*Case A* $\sigma_\varepsilon^2 = 0.04$ (1)	*Case B* $\sigma_\varepsilon^2 = 0.0225$ (2)	*Case C* $\sigma_\varepsilon^2 = 0.01$ (3)	*Case D* $\sigma_\varepsilon^2 = 0.005$ (4)
β_0	1.42*	1.62*	1.25*	1.28*
	(0.19)	(0.14)	(0.14)	(0.11)
β_1	0.46*	0.43*	0.48*	0.46*
	(0.024)	(0.02)	(0.017)	(0.01)
β_2	0.48*	0.47*	0.48*	0.47*
	(0.017)	(0.013)	(0.01)	(0.01)
σ_τ	0.15*	0.11*	0.15*	0.15*
	(0.035)	(0.01)	(0.01)	(0.01)
σ_ν	0.15*	0.13*	0.08*	0.04
	(0.01)	(0.02)	(0.01)	(0.025)

The asterisk "*" denotes statistical significance at 0.05 significance level or better. Standard errors are shown in parentheses. The values for σ_ε^2 shown in the top row of the table represent the true variances for the statistical error distributions.

where D represents a dummy variable which is assigned the following values

$$D = \begin{cases} 1 & : \text{if a DMU is identified as 100\% efficient in stage 1} \\ 0 & : \text{if a DMU is not identified as 100\% efficient in stage 1.} \end{cases} \tag{9.21}$$

Tables 9.4 and 9.5 exhibit the results secured when (9.19) was used as the estimating relation to obtain parameter values for both OLS and SF regressions. As might be expected, all parameter estimates are wide of the true values represented in (9.17)and significantly so in all cases as recorded in Tables 9.4 and 9.5. Somewhat surprising, however, is the fact that the OLS and SF estimates are very close and, in many cases they are identical.

When (9.20) is used — which is the regression with dummy variable values described in (9.21) — the situation is reversed for the efficient, but not for the inefficient DMUs. When the estimates are formed in the manner noted at the bottoms of Tables 9.6 and 9.7, none of the estimate of β_1 and β_2 differ significantly from their true values as given in (9.17). These are the estimates to be employed for $D = 1$. For $D = 0$, the case of inefficient DMUs, the previous result is repeated. All of the estimates differ significantly from their true values in both the empirical and simulation studies we described as can be seen in both of Tables 9.6 and 9.7.

Table 9.6. OLS Regression Estimates without Dummy Variables on DEA-efficient DMUs

Case 1: *EXPONENTIAL* distribution of input inefficiencies

Parameter Estimates	*Case A* $\sigma_\varepsilon^2 = 0.04$ (1)	*Case B* $\sigma_\varepsilon^2 = 0.0225$ (2)	*Case C* $\sigma_\varepsilon^2 = 0.01$ (3)	*Case D* $\sigma_\varepsilon^2 = 0.005$ (4)
β_0	1.07* (0.21)	1.47* (0.17)	1.28* (0.14)	1.34* (0.11)
β_1	0.49* (0.03)	0.43* (0.02)	0.46* (0.02)	0.47* (0.01)
β_2	0.48* (0.02)	0.48* (0.015)	0.48* (0.01)	0.46* (0.01)
δ	-1.57* (0.64)	-2.30* (0.43)	-1.50* (0.35)	-1.50* (0.21)
δ_1	0.155* (0.075)	0.26* (0.05)	0.16* (0.04)	0.16* (0.03)
δ_2	0.12* (0.05)	0.12* (0.04)	0.10* (0.03)	0.09* (0.02)
Combining Parameters with *Dummy Variables*				
$H_0 : \beta_1 + \delta_1 = 0.65$ $H_a : \beta_1 + \delta_1 \neq 0.65$	$t_1 = 0.07$	$t_1 = 0.87$	$t_1 = -0.72$	$t_1 = 0.82$
$H_0 : \beta_2 + \delta_2 = 0.55$ $H_a : \beta_2 + \delta_2 \neq 0.55$	$t_2 = 1.09$	$t_2 = 1.76$	$t_2 = 1.02$	$t_2 \simeq 0$

The asterisk "*" denotes statistical significance at 0.05 significance level or better. Standard errors are shown in parentheses.
The values for σ_ε^2 shown in the top row of the table represent the true variances for the statistical error distributions.

The above tables report results for an exponential distribution of the inefficiencies associated with $\widehat{x}_1, \widehat{x}_2$ as defined in (9.18). However, uses of other statistical distributions and other forms of production functions did not alter these kinds of results for either the efficient or the inefficient DMUs. Thus this two-stage approach provided a new way of evaluating efficient and inefficient behaviors in both the empirical and simulation studies where it was used. It also provides an OLS regression as an alternative to the SF regression and this alternative is easier to use (or at least is more familiar) for many uses. See Brockett *et al.* (2004) [36] for an application to advertising strategy and a comparison with other types of statistical regression.

Comment : There are shortcomings and research challenges that remain to be met. One such challenge is to expand these uses to include multiple outputs as well as multiple inputs. Another challenge is to develop ways for identifying and estimating input specific as well as output specific inefficiencies. In order to meet such challenges it will be necessary to develop an analytically based theory in order to extend what can be accomplished by empirical applications and simulation studies.

Table 9.7. Stochastic Frontier Regression Estimates without Dummy Variables on DEA-efficient DMUs

Case 1: *EXPONENTIAL* distribution of input inefficiencies

Parameter Estimates	*Case A* $\sigma_\varepsilon^2 = 0.04$ (1)	*Case B* $\sigma_\varepsilon^2 = 0.0225$ (2)	*Case C* $\sigma_\varepsilon^2 = 0.01$ (3)	*Case D* $\sigma_\varepsilon^2 = 0.005$ (4)
β_0	1.18*	1.50*	0.80*	1.40*
	(0.23)	(0.16)	(0.16)	(0.13)
β_1	0.50*	0.44*	0.53*	0.49*
	(0.03)	(0.02)	(0.02)	(0.01)
β_2	0.48*	0.49*	0.50*	0.47*
	(0.02)	(0.02)	(0.01)	(0.02)
δ	-1.60*	-2.4*	-1.25*	-1.55*
	(0.57)	(0.56)	(0.38)	(0.23)
δ_1	0.16*	0.26*	0.13*	0.15*
	(0.07)	(0.06)	(0.04)	(0.03)
δ_2	0.11*	0.13*	0.086*	0.09*
	(0.05)	(0.04)	(0.04)	(0.03)
σ_ν	0.13(0.01)*	0.09(0.01)*	0.05(0.01)*	0.04(0.01)*
Combining Parameters with *Dummy Variables*				
$H_0 : \beta_1 + \delta_1 = 0.65$ $H_a : \beta_1 + \delta_1 \neq 0.65$	$t_1 = 0.20$	$t_1 = 0.93$	$t_1 = 0.28$	$t_1 = -0.4$
$H_0 : \beta_2 + \delta_2 = 0.55$ $H_a : \beta_2 + \delta_2 \neq 0.55$	$t_2 = 1.03$	$t_2 = 1.90$	$t_2 = 1.16$	$t_2 = 0.45$

The asterisk "*" denotes statistical significance at 0.05 significance level or better. Standard errors are shown in parentheses.
The values for σ_ε^2 shown in the top row of the table represent the true variances for the statistical error distributions.

Fortunately, excellent texts dealing with stochastic frontier and other approaches to efficiency evaluation have become available in the following two books,

1. T. Coelli, D.S.P. Rao and G.E. Battese (1998) *An Introduction to Efficiency and Productivity Analysis* (Boston: Kluwer Academic Publishers).

2. S.C. Kumbhakar and C.A.K. Lovell (2000) *Stochastic Frontier Analysis* (Cambridge: Cambridge University Press).

9.4 CHANCE-CONSTRAINED PROGRAMMING AND SATISFICING IN DEA

9.4.1 Introduction

S. Thore's (1987)[37] paper initiated a series of joint efforts with R. Land, and C.A.K. Lovell[38] directed to joining chance-constrained programming (CCP) with DEA as a third method for treating data uncertainties in DEA. Here we turn to Cooper, Huang and Li (1996)[39] to show how this approach can also be used to make contact with the concept of "satisficing" as developed in the psychology literature by H.A. Simon[40] as an alternative to the assumption of "optimizing" behavior which is extensively used in economics.

9.4.2 Satisficing in DEA

We start with the following CCP formulation that extends the CCR (ratio) model of DEA which was introduced in Section 2.3 of Chapter 2,

$$\max \; P\left(\frac{\sum_{r=1}^{s} u_r \tilde{y}_{ro}}{\sum_{i=1}^{m} v_i \tilde{x}_{io}} \geq \beta_o\right) \tag{9.22}$$

$$\text{subject to} \quad P\left(\frac{\sum_{r=1}^{s} u_r \tilde{y}_{rj}}{\sum_{i=1}^{m} v_i \tilde{x}_{ij}} \leq \beta_j\right) \geq 1 - \alpha_j, \quad j = 1, \ldots, n$$

$$u_r,\; v_i \geq 0 \;\; \forall r,\; i.$$

Here "P" means "probability" and "$\sim$" identifies these outputs and inputs as random variables with a known probability distribution while $0 \leq \alpha_j \leq 1$ is a scalar, specified in advance, which represents an allowable chance (=risk) of failing to satisfy the constraints with which it is associated.

For "satisficing," the values of β_o is interpreted as an "aspiration level" specified as an efficiency rating which is to be attained. The β_j are also prescribed constants imposed by the individual, or by outside conditions including superior levels of management.

To exhibit another aspect of satisficing behavior we might consider the case of inconsistent constraints. The problem will then have no solution. In such cases, according to Simon, an individual must either quit or else he must revise his aspiration level — or the risk of not achieving this level (or both). Thus, probabilistic (chance-constrained programming) formulations allow for types of behavior which are not present in the customary deterministic models of satisficing.

Now, however, we want to make contact with the deterministic DEA models which we discussed earlier. For this purpose we select the CCR model which was introduced as early as Chapter 2. This model always has a solution and the same is true for (9.22). This can be exhibited by choosing $u_r = 0\ \forall r$ and $v_i > 0$ for some i. Although not minimal for the objective, this choice satisfies all constraints with a probability of unity.

9.4.3 *Deterministic Equivalents*

To align the development more closely with our earlier versions of the CCR ratio model we note that

$$P\left(\frac{\sum_{r=1}^{s} u_r \tilde{y}_{ro}}{\sum_{i=1}^{m} v_i \tilde{x}_{io}} \le \beta_o\right) + P\left(\frac{\sum_{r=1}^{s} u_r \tilde{y}_{ro}}{\sum_{i=1}^{m} v_i \tilde{x}_{io}} \ge \beta_o\right) = 1 \tag{9.23}$$

where, for simplicity, we restrict attention to the class of continous distributions. We therefore replace (9.22) with

$$\begin{aligned} & \max\ P\left(\frac{\sum_{r=1}^{s} u_r \tilde{y}_{ro}}{\sum_{i=1}^{m} v_i x_{io}} \ge \beta_o\right) \\ \text{subject to} \quad & P\left(\frac{\sum_{r=1}^{s} u_r \tilde{y}_{rj}}{\sum_{i=1}^{m} v_i x_{ij}} \le \beta_j\right) \ge 1 - \alpha_j, \quad j = 1, \ldots, n \\ & u_r,\ v_i \ge 0\ \ \forall r,\ i. \end{aligned} \tag{9.24}$$

Here we follow Land, Lovell and Thore (1993) and omit the symbol "~" from the x_{ij} (and x_{io}) in order to represent the inputs as deterministic. This model corresponds to a situation in which DMU managers choose the inputs without being able to completely control the outputs. Moreover if we also remove the symbol "~" from the y_{rj} (and y_{ro}), set $\beta_j = 1,\ j = 1, \ldots, n$ and remove β_o from the objective we will reproduce the CCR model that was first encountered in expression (2.3)-(2.6) in Chapter 2.

This identification having been made, we restore the symbol "~" to the y_{rj} (and y_{ro}) — thereby characterizing them as random variables — but we continue to treat the x_{ij} (and x_{io}) as deterministic. Then using vector-matrix notation we restate the constraints in (9.24) via the following development,

$$P\left(\frac{\boldsymbol{u}^T \tilde{\boldsymbol{y}}_j}{\boldsymbol{v}^T \boldsymbol{x}_j} \le \beta_j\right) = P\left(\boldsymbol{u}^T \tilde{\boldsymbol{y}}_j \le \beta_j \boldsymbol{v}^T \boldsymbol{x}_j\right). \tag{9.25}$$

Now let $\bar{\boldsymbol{y}}_j$ be the vector of output means and let Σ_j represent the variance-covariance matrix. We assume that this matrix is positive definite so we can represent the variance by $\boldsymbol{u}^T \Sigma_j \boldsymbol{u}$, a scalar, which is also positive for all choices of $\boldsymbol{u} \ne \boldsymbol{0}$. We then subtract $\boldsymbol{u}^T \bar{\boldsymbol{y}}_j$ from both sides of the right-hand inequality in (9.25) and divide through by $\sqrt{\boldsymbol{u}^T \Sigma_j \boldsymbol{u}}$ to obtain

$$P\left(\frac{\boldsymbol{u}^T \tilde{\boldsymbol{y}}_j - \boldsymbol{u}^T \bar{\boldsymbol{y}}_j}{\sqrt{\boldsymbol{u}^T \Sigma_j \boldsymbol{u}}} \le \frac{\beta_j \boldsymbol{v}^T \boldsymbol{x}_j - \boldsymbol{u}^T \bar{\boldsymbol{y}}_j}{\sqrt{\boldsymbol{u}^T \Sigma_j \boldsymbol{u}}}\right) \ge 1 - \alpha_j \tag{9.26}$$

for each $j = 1, \ldots, n$. Now we note that the expression on the right in the parenthesis does not contain any random elements.

To simplify our notation we introduce a new random variable defined by

$$\tilde{z}_j = \frac{\boldsymbol{u}^T \tilde{\boldsymbol{y}}_j - \boldsymbol{u}^T \bar{\boldsymbol{y}}_j}{\sqrt{\boldsymbol{u}^T \Sigma_j \boldsymbol{u}}}. \tag{9.27}$$

We then replace (9.26) with

$$P\left(\tilde{z}_j \le k_j(\boldsymbol{u}^T, \boldsymbol{v}^T)\right) \ge 1 - \alpha_j, \; j = 1, \ldots, n \tag{9.28}$$

where

$$k_j(\boldsymbol{u}^T, \boldsymbol{v}^T) = \frac{\beta_j \boldsymbol{v}^T \boldsymbol{x}_j - \boldsymbol{u}^T \bar{\boldsymbol{y}}_j}{\sqrt{\boldsymbol{u}^T \Sigma_j \boldsymbol{u}}}$$

so we can write

$$\int_{-\infty}^{k_j(\boldsymbol{u}^T, \boldsymbol{v}^T)} f(z_j) dz_j = \Phi\left(\frac{\beta_j \boldsymbol{v}^T \boldsymbol{x}_j - \boldsymbol{u}^T \bar{\boldsymbol{y}}_j}{\sqrt{\boldsymbol{u}^T \Sigma_j \boldsymbol{u}}}\right) \ge 1 - \alpha_j \tag{9.29}$$

in place of (9.26).

We now assume that Φ is the normal distribution which has been standardized via

$$\tilde{z}_j = \frac{\boldsymbol{u}^T (\tilde{\boldsymbol{y}}_j - \bar{\boldsymbol{y}}_j)}{\sqrt{\boldsymbol{u}^T \Sigma_j \boldsymbol{u}}}. \tag{9.30}$$

Assuming $\alpha_j \le 0.5$ we can utilize the property of invertibility associated with this distribution and apply it to (9.29) to obtain

$$\frac{\beta_j \boldsymbol{v}^T \boldsymbol{x}_j - \boldsymbol{u}^T \bar{\boldsymbol{y}}_j}{\sqrt{\boldsymbol{u}^T \Sigma_j \boldsymbol{u}}} \ge \Phi^{-1}(1 - \alpha_j) \tag{9.31}$$

where Φ^{-1} is the fractile function associated with the standard normal distribution. Hence also

$$\beta_j \boldsymbol{v}^T \boldsymbol{x}_j - \boldsymbol{u}^T \bar{\boldsymbol{y}}_j \ge \Phi^{-1}(1 - \alpha_j)\sqrt{\boldsymbol{u}^T \Sigma_j \boldsymbol{u}}. \tag{9.32}$$

We now employ what Charnes and Cooper (1963)[41] refer to as "splitting variables" which we symbolize by η_j in

$$\beta_j \boldsymbol{v}^T \boldsymbol{x}_j - \boldsymbol{u}^T \bar{\boldsymbol{y}}_j \ge \eta_j \ge \Phi^{-1}(1 - \alpha_j)\sqrt{\boldsymbol{u}^T \Sigma_j \boldsymbol{u}}. \tag{9.33}$$

For every $j = 1, \ldots, n$ this variable is nonnegative by virtue of the expression on the right. Provided this nonnegativity is preserved we can therefore use this variable to split the expression in (9.33) into the following pair

$$\begin{aligned} &\beta_j \boldsymbol{v}^T \boldsymbol{x}_j - \boldsymbol{u}^T \bar{\boldsymbol{y}}_j \geq \eta_j \geq 0 \\ &K^2_{(1-\alpha_j)} \boldsymbol{u}^T \Sigma_j \boldsymbol{u} \leq \eta_j^2 \end{aligned} \tag{9.34}$$

where

$$\begin{aligned} &K_{(1-\alpha_j)} = \Phi^{-1}(1-\alpha_j) \\ &j = 1, \ldots, n. \end{aligned}$$

We have thus separated the conditions in (9.33) into a pair for which the first relation refers to a valuation effected by multipliers assigned to the inputs and outputs while the second relation treats the "risks" as in a portfolio analysis of the Markowitz-Sharpe type used in finance.[42]

In place (9.24) we now have

$$\begin{aligned} &\max\ P\left(\frac{\boldsymbol{u}^T \widetilde{\boldsymbol{y}}_o}{\boldsymbol{v}^T \boldsymbol{x}_o} \geq \beta_o\right) \\ \text{subject to}\quad &\beta_j \boldsymbol{v}^T \boldsymbol{x}_j - \boldsymbol{u}^T \bar{\boldsymbol{y}}_j - \eta_j \geq 0 \\ &K^2_{(1-\alpha_j)} \boldsymbol{u}^T \Sigma_j \boldsymbol{u} - \eta_j^2 \leq 0 \\ &\boldsymbol{v},\ \boldsymbol{u} \geq \boldsymbol{0},\ \eta_j \geq 0,\ j = 1, \ldots, n. \end{aligned} \tag{9.35}$$

The constraints, but not the objective, are now deterministic. To bring our preceding development to bear we therefore replace (9.35) with

$$\begin{aligned} &\max\ \gamma_o \\ \text{subject to}\quad &P\left(\frac{\boldsymbol{u}^T \widetilde{\boldsymbol{y}}_o}{\boldsymbol{v}^T \boldsymbol{x}_o} \geq \beta_o\right) \geq \gamma_o \\ &\beta_j \boldsymbol{v}^T \boldsymbol{x}_j - \boldsymbol{u}^T \bar{\boldsymbol{y}}_j - \eta_j \geq 0 \\ &K^2_{(1-\alpha_j)} \boldsymbol{u}^T \Sigma_j \boldsymbol{u} - \eta_j^2 \leq 0 \\ &\boldsymbol{v},\ \boldsymbol{u} \geq \boldsymbol{0},\ \eta_j \geq 0,\ j = 1, \ldots, n. \end{aligned} \tag{9.36}$$

Proceeding as before we then have

$$\begin{aligned} &\max\ \gamma_o \\ \text{subject to}\quad &\boldsymbol{u}^T \bar{\boldsymbol{y}}_o - \beta_o \boldsymbol{v}^T \boldsymbol{x}_o \geq \Phi^{-1}(\gamma_o)\sqrt{\boldsymbol{u}^T \Sigma_j \boldsymbol{u}} \\ &\eta_j + \boldsymbol{u}^T \boldsymbol{y}_j - \beta_j \boldsymbol{v}^T \boldsymbol{x}_j \leq 0 \\ &\eta_j^2 - K^2_{(1-\alpha_j)} \boldsymbol{u}^T \Sigma_j \boldsymbol{u} \geq 0 \\ &\boldsymbol{u},\ \boldsymbol{v} \geq \boldsymbol{0},\ \eta_j \geq 0,\ j = 1, \ldots, n \\ &0 \leq \gamma_o \leq 1. \end{aligned} \tag{9.37}$$

This is a "deterministic equivalent" for (9.24) in that the optimal values of $\boldsymbol{u}^*$, $\boldsymbol{v}^*$ in (9.37) will also be optimal for (9.24).

9.4.4 Stochastic Efficiency

Although entirely deterministic, this problem is difficult to solve because, by virtue of the first constraint, it is nonconvex as well as nonlinear. As shown in Cooper, Huang and Li (1996), it is possible to replace (9.37) with a convex programming problem but we do not undertake the further development needed to show this. Instead, we assume that we have a solution with

$$\gamma_o^* = P\left(\frac{\boldsymbol{u}^{*T}\widetilde{\boldsymbol{y}}_o}{\boldsymbol{v}^{*T}\boldsymbol{x}_o} \geq \beta_o\right), \tag{9.38}$$

where γ_o^* is obtained from (9.36).

To develop the concepts of stochastic efficiency we assume that $\beta_o = \beta_{j_o}$ so the level prescribed for DMU_o in its constraints is the same as the β_o level prescribed in the objective. Then we note that $\gamma_o^* > \alpha_{j_o}$ is not possible because this would fail to satisfy this constraint. To see that this is so we note, as in (9.23), that

$$P\left(\frac{\boldsymbol{u}^{*T}\widetilde{\boldsymbol{y}}_o}{\boldsymbol{v}^{*T}\boldsymbol{x}_o} \geq \beta_o\right) + P\left(\frac{\boldsymbol{u}^{*T}\widetilde{\boldsymbol{y}}_o}{\boldsymbol{v}^{*T}\boldsymbol{x}_o} \leq \beta_o\right) = 1 \tag{9.39}$$

because $P\left(\frac{\boldsymbol{u}^{*T}\widetilde{\boldsymbol{y}}_o}{\boldsymbol{v}^{*T}\boldsymbol{x}_o} = \beta_o\right) = 0$ for a continuous distribution. Hence

$$\begin{aligned} P\left(\frac{\boldsymbol{u}^{*T}\widetilde{\boldsymbol{y}}_o}{\boldsymbol{v}^{*T}\boldsymbol{x}_o} \leq \beta_o\right) &= 1 - P\left(\frac{\boldsymbol{u}^{*T}\widetilde{\boldsymbol{y}}_o}{\boldsymbol{v}^{*T}\boldsymbol{x}_o} \geq \beta_o\right) \\ &= 1 - \gamma_o^* \ < \ 1 - \alpha_{j_o} \end{aligned} \tag{9.40}$$

which fails to satisfy the constraint

$$P\left(\frac{\boldsymbol{u}^{*T}\widetilde{\boldsymbol{y}}_o}{\boldsymbol{v}^{*T}\boldsymbol{x}_o} \leq \beta_o\right) \geq 1 - \alpha_{j_o}.$$

Now if $\beta_{j_o} = \beta_o = 1$, which will usually be the case of interest, the above development leads to the following,

Theorem 9.1 (Cooper, Huang and Li (1996)) *If $\beta_{j_o} = \beta_o = 1$ then DMU_o will have performed in a stochastically efficient manner if and only if $\gamma_o^* = \alpha_{j_o}$.*

This leaves the case of $\gamma_o^* < \alpha_{j_o}$ to be attended to. In this case the risk of failing to satisfy the constraints for DMU_{j_o} falls below the level which was specified as satisfactory. To restate this in a more positive manner, we return to (9.40) and reorient it to

$$\begin{aligned} P\left(\frac{\boldsymbol{u}^{*T}\widetilde{\boldsymbol{y}}_o}{\boldsymbol{v}^{*T}\boldsymbol{x}_o} \leq \beta_o\right) &= 1 - P\left(\frac{\boldsymbol{u}^{*T}\widetilde{\boldsymbol{y}}_o}{\boldsymbol{v}^{*T}\boldsymbol{x}_o} \geq \beta_o\right) \\ &= 1 - \gamma_o^* \ > \ 1 - \alpha_{j_o}. \end{aligned} \tag{9.41}$$

This leads to the following corollary to the above theorem,

Corollary 9.1 *If $\beta_{j_o} = \beta_o = 1$ then DMU$_o$'s performance is stochastically inefficient with probability $1 - \gamma_o^*$ if and only if $\gamma_o^* < \alpha_{j_o}$.*

We justify this by noting that (9.41) means that the probability of falling below β_o exceeds the probability that was specified as being satisfactory.

To return to the satisficing model and concepts discussed in Section 9.4.2, we assume that $0 \le \beta_o = \beta_{j_o} < 1$. The above theorem and corollary then translate into: "satisficing was attained or not according to whether $\gamma_o^* = \alpha_{j_o}$ or $\gamma_o^* < \alpha_{j_o}$." To see what this means we note that (9.39) consists of opposed probability statements except for the case $P\left(\frac{\boldsymbol{u}^{*T}\tilde{\boldsymbol{y}}_o}{\boldsymbol{v}^{*T}\boldsymbol{x}_o} = \beta_o\right) = 0$. Hence failure to attain satisficing occurs when the goal specified in the objective can be attained only at a level below the risk that is specified for being wrong in making this inference.

Returning to our earlier choice of $u_r = 0\ \forall r$ and some $v_i > 0$ for (9.22) we note that this assigns a positive value to some inputs and a zero value for all outputs. The objective and the associated constraints in (9.22) will then be achieved with probability one because of refusal to play and hence there will be a zero probability of achieving the objective in (9.24). See (9.39). This is an important special case of Simon's "refusal to play" behavior that was noted in the discussion following (9.22).

Comment : This is as far as we carry our analyses of these chance constrained programming approaches to DEA. We need to note, however, that this analysis has been restricted to what is referred to as the "P-model" in the chance-constrained programming literature. Most of the other DEA literature on this topic has utilized the "E-model," so named because its objective is stated in terms of optimizing "expected values." None has yet essayed a choice of "V-models" for which the objective is to minimize "variance" or "mean-square error." See A. Charnes and W.W. Cooper (1963) "Deterministic Equivalents for Optimizing and Satisficing under Chance Constraints," *Operations Research* 11, pp.18-39.

The formulations here (and elsewhere in the DEA literature) are confined to a use of "conditional chance constraints." A new chapter was opened for this research, however, by Olesen and Petersen (1995)[43] who used "joint chance constraints" in their CCP models. In a subsequent paper, Cooper, Huang, Lelas, Li and Olesen (1998)[44] utilize such joint constraints to extend the concept of "stochastic efficiency" to a measure called "α-stochastic efficiency" for use in problems where "efficiency dominance" is of interest.

Like the rest of the DEA literature dealing with CCP, we have restricted attention to the class of "zero order decision rules." This corresponds to a "here and now" approach to decision making in contrast to the "wait and see" approach that is more appropriate to dynamic settings in which it may be better to delay some parts of a decision until more information is available. To go further in this direction leads to the difficult problem of choosing a best decision rule from an admissible class of decision rules. To date, this problem has only been addressed in any detail for the class of linear (first order) decision rules even in the general literature on CCP and even this treatment was conducted under restrictive assumptions on the nature of the random variables and their statistical distributions.[45] See also Charnes, Cooper and Symods (1958), [46] an article which originated the (as-yet-to-be named) "chance constraint programming." This topic is important and invites further treatments as an example of the "information processing" that plays an important role in studying use of the term "satisficing behavior." See Gigerenzer (2004). [47] Finally we come to the treatments which, by and large, have assumed that these probability distributions are known. Hence there is a real need and interest in relaxing this assumption. Only bare beginnings on this topic have been made as in R. Jagannathan (1985) "Use of Sample Information in Stochastic Recourse and Chance Constrained Programming Models," *Management Science* 31, pp.96-108.

9.5 WINDOW ANALYSIS

9.5.1 An Example

Although now used in many other contexts we here revert to the applications in army recruiting efforts that gave rise to "window analysis." This example is built around 3 outputs consisting of recruits such as male and female high school graduates and degree candidates (such as high school seniors not yet graduated). The 10 inputs consist of number of recruiters (such as recruiting sergeants) and their aids, amount spent on local advertising, as well as qualities such as recruiter experience which are all discretionary variables as well as non-discretionary variables such as local unemployment, number of high school seniors in the local market and their propensity to enlist.[48]

These variables are used (among others) by the U.S. Army Recruiting Command (USAREC) to evaluate performance in its various organization units which Klopp (1985, p.115)[48] describes as follows. USAREC recruits for the entire United States. To facilitate control of the recruiting process USAREC divides the U.S. into 5 regions managed by entities referred to as "Recruiting Brigades." Each Recruiting Brigade is responsible for some of the 56 "Recruiting Battalions" that operate in the U.S. and the latter are responsible, in turn, for the "Recruiting Stations" where recruiters are assigned specific missions. For the example we employ, it was decided to use the Recruiting Battalions each of which was designated as a DMU.

The DEA analyses took a variety of forms. One consisted of standard (static) DEA reports in the manner of Table 3.7 expanded to include detailed information on the inefficiencies present in each DMU. Both discretionary and non-discretionary variables were included along with the members of the peer group used to effect the evaluation.

Something more was wanted in the form of trend analyses of the quarterly reports that USAREC received. A use of statistical regressions and time series analyses of the efficiency scores proved unsatisfactory. Experimentation was therefore undertaken which led to the window analysis we now describe.

9.5.2 *Application*

Table 9.8, as adapted from Klopp (1985) will be used for guidance in this discussion. The basic idea is to regard each DMU as if it were a different DMU in each of the reporting dates represented by Q1, Q2, etc., at the top of the table. For instance, the results in row 1 for Battalion 1A represent four values obtained by using its results in Q1, Q2, Q3 and Q4 by bringing this DMU into the objective for each of these quarters. Because these 56 DMUs are regarded as different DMUs in each quarter, these evaluations are conducted by reference to the entire set of 4$\times$ 56 = 224 DMUs that are used to form the data matrix. Thus the values of $\theta^* = 0.83$, 1.00, 0.95 and 1.00 in row 1 for Battalion 1A represent its quarterly performance ratings as obtained from this matrix with 13 $\times$ 224 = 2,912 entries.

Similar first-row results for other DMUs are obtained for each of the 10 DMUs in Table 9.8 which we have extracted from a larger tabulation of 56 DMUs to obtain compactness. After first row values have been similarly obtained, a new 4-period window is obtained by dropping the data for Q1 and adding the data for Q5. Implementing the same procedure as before produces the efficiency ratings in row 2 for each Battalion. The process is then continued until no further quarters are added — as occurs (here) in "row" 5 with its 4 entries.

To help interpret the results we note that the "column views" enable us to examine the stability of results across the different data sets that occur with these removal and replacement procedures. "Row views" make it possible to determine trends and/or observed behavior with the same data set. Thus, the column view for Battalion 1A shows stability and the row view shows steady behavior after the improvement over Q1. At the bottom of Table 9.8, however, Battalion 1K exhibits deteriorating behavior and this same deterioration continues to be manifested with different data sets. Moreover, to reenforce this finding the values in each column for Battalion 1K do not change very much and this stability reenforces this finding.

Finally we augment this information by the statistical values noted on the right side of Table 9.8. Here the average is used as a representative measure obtained from the θ^* values for each DMU and matched against its variance. Medians might also be used and matched with ranges and so on. Other summary statistics and further decompositions may also be used, of course, but a good deal of information is supplied in any case.

Weaknesses are also apparent, of course, such as the absence of attention to nonzero slacks. However, the principle and formulas to be supplied in the next section of this chapter may be applied to slack portrayals, too, if desired and other similar measures may be used such as the SBM (Slacks Based Measure) given in Section 4.4 of Chapter 4.

Another deficiency is apparent in that the beginning and ending period DMUs are not tested as frequently as the others. Thus the DMUs in Q1 are examined in only one window and the same is true for Q8. In an effort to address this problem Sueyoshi (1992)[49] introduced a "round robin" proce-

Table 9.8. Window Analysis: 56 DMUs in U.S. Army Recruitment Battalions 3 Outputs - 10 Inputs

Battalion	Efficiency Scores Q1	Q2	Q3	Q4	Q5	Q6	Q7	Q8	Summary Measures Mean	Var	Column Range	Tot Ra
1A	0.83	1.00	0.95	1.00								
		1.00	0.95	1.00	1.00				0.99	0.03	.05	.17
			1.00	1.00	1.00	1.00						
				1.00	1.00	1.00	1.00					
					1.00	1.00	1.00	1.00				
1B	0.71	1.00	0.91	0.89								
		1.00	0.91	0.89	0.77				0.92	0.21	.11	.29
			1.00	1.00	0.76	1.00						
				1.00	0.75	1.00	1.00					
					0.80	1.00	1.00	1.00				
1C	0.72	1.00	0.86	1.00								
		1.00	0.87	1.00	0.74				0.93	0.21	.26	.28
			0.87	1.00	0.74	1.00						
				1.00	1.00	1.00	1.00					
					0.80	1.00	1.00	1.00				
1D	0.73	1.00	1.00	1.00								
		1.00	1.00	1.00	1.00				0.99	0.07	.00	.27
			1.00	1.00	1.00	1.00						
				1.00	1.00	1.00	1.00					
					1.00	1.00	1.00	1.00				
1E	0.69	1.00	0.83	1.00								
		0.83	0.83	1.00	0.77				0.88	0.17	.17	.31
			0.83	1.00	0.79	0.93						
				1.00	0.77	0.95	0.83					
					0.82	0.88	0.88	1.00				
1F	1.00	1.00	1.00	1.00								
		1.00	1.00	1.00	1.00				0.99	0.04	.20	.20
			1.00	1.00	1.00	1.00						
				1.00	1.00	1.00	1.00					
					0.80	1.00	1.00	1.00				
1G	0.65	0.73	0.68	0.77								
		0.73	0.68	0.77	0.68				0.79	0.32	.32	.35
			0.68	0.77	0.68	0.72						
				0.76	0.74	0.75	1.00					
					1.00	1.00	1.00	1.00				
1H	0.82	1.00	1.00	1.00								
		1.00	1.00	1.00	0.71				0.90	0.32	.29	.3
			1.00	1.00	0.71	0.70						
				1.00	1.00	0.72	0.72					
					0.79	0.81	1.00	1.00				
1I	1.00	1.00	1.00	1.00								
		1.00	1.00	1.00	1.00				0.99	0.02	.14	.1
			0.86	1.00	1.00	1.00						
				1.00	1.00	1.00	1.00					
					1.00	1.00	1.00	1.00				
1K	0.81	1.00	1.00	1.00								
		1.00	1.00	1.00	0.73				0.86	0.37	.11	.3
			1.00	1.00	0.73	0.67						
				1.00	0.85	0.67	0.67					
					0.67	0.67	0.67	0.67				

dure which proceeds as follows: First each period is examined independently. This is followed by a 2-period analysis after which a three-period analysis is used. And so on. However, this analysis becomes unwieldy since the number of combinations grows to $2^p - 1$ so that some care is needed with this approach.

9.5.3 *Analysis*

The following formulas adapted from D.B. Sun (1988)[50] can be used to study the properties of these window analyses. For this purpose we introduce the following symbols

$$\begin{aligned} n &= \text{number of DMUs} \\ k &= \text{number of periods} \\ p &= \text{length of window } (p \leq k) \\ w &= \text{number of windows.} \end{aligned} \tag{9.42}$$

We then reduce the number of DMUs to the 10 in Table 9.8 so we can use it for numerical illustrations.

	Formula	Application (to Table 9.8)
no. of windows:	$w = k - p + 1$	$8 - 4 + 1 = 5$
no. of DMUs in each window:	$np/2$	$10 \times 4/2 = 20$
no. of "different" DMUs:	npw	$10 \times 4 \times 5 = 200$
Δ no. of DMUs:	$n(p-1)(k-p)$	$10 \times 3 \times 4 = 120$

Here "Δ" represents an increase compared to the $8 \times 10 = 80$ DMUs that would have been available if the evaluation had been separately effected for each of the 10 DMUs in each of the 8 quarters.

An alternate formula for deriving the total number of DMUs is given in Charnes and Cooper (1990)[51] as follows.

$$\text{Total no. of "different" DMUs}: \; n(k-p+1)p = 10\times(8-4+1)\times 4 = 200 \tag{9.43}$$

Differentiating this last function and equating to zero gives

$$p = \frac{k+1}{2} \tag{9.44}$$

as the condition for a maximum number of DMUs. This result need not be an integer, however, so we utilize the symmetry of (9.43) and (9.44) and modify the latter to the following,

$$p = \begin{cases} \frac{k+1}{2} & \text{when } k \text{ is odd} \\ \frac{k+1}{2} \pm \frac{1}{2} & \text{when } k \text{ is even.} \end{cases} \tag{9.45}$$

To see how to apply this formula when k is even we first note that

$$n(k-p+1)p = n\left[(k+1)p - p^2\right].$$

Hence by direct substitution we obtain

$$\begin{aligned} & n\left[(k+1)\left(\frac{k+1}{2}-\frac{1}{2}\right)-\left(\frac{k+1}{2}-\frac{1}{2}\right)^2\right] \\ = \; & n\left[(k+1)\left(\frac{k+1}{2}+\frac{1}{2}\right)-\left(\frac{k+1}{2}+\frac{1}{2}\right)^2\right] = \frac{n}{4}\left[(k+1)^2-1\right] \end{aligned}$$

Then, for $k = 8$, as in our example, we find from (9.44) that

$$p = \frac{8+1}{2} = 4.5$$

which is not an integer. Hence using $[p]$ to mean "the integer closest to p" we apply the bottom expression in (9.45) to obtain

$$[p] = \begin{cases} 4 = 4.5 - 0.5 \\ 5 = 4.5 + 0.5 \end{cases}$$

and note that substitution in (9.43) produces 200 "different" DMUs as the maximum number in either case.

9.6 SUMMARY OF CHAPTER 9

In this chapter we have treated the topic of data variability in the following manner. Starting with the topic of sensitivity and stability analysis we moved to statistical regression approaches which we aligned with DEA in various ways. We then went on to probabilistic formulations using the P-model of chance-constrained programming which we could relate to our CCR and BCC models which we had previously treated in deterministic manners. Finally we turned to window analysis which allowed us to study trends as well as stability of results when DMUs are systematically dropped and added to the collection to be examined.

The topics treated in this chapter are all under continuing development. Hence we do not provide problems and suggested answers like those we presented in preceding chapters. Instead we have provided comments and references that could help to point up issues for further research. We hope that readers will respond positively and regard this as an opportunity to join in the very active research that is going on in these (and other) areas.

We here note that research on the topics treated in this chapter were prompted by problems encountered in attempts to bring a successful conclusion to one or many attempts to use DEA in actual applications. Indeed a good deal of the very considerable progress in DEA has emanated from actual attempts to apply it to different problems. This, too, has been an important source of progress in that new applications as well as new developments in DEA are being simultaneously reported.

To help persons who want to pursue additional topics and uses we have supplied an extensive bibliography in the disk that accompanies this book as well as in this chapter. We hope this will be helpful and we hope readers of our text will experience some of the fun and exhilaration that we have experienced as we watch the rapid pace of developments in DEA.

9.7 RELATED DEA-SOLVER MODELS FOR CHAPTER 9

Window-I(O)-C(V) These codes execute Window Analysis in Input (Output) orientation under constant (CRS) or variable (VRS) returns-to-scale assumptions. See the sample data format in Section B.5.10 and explanation on results in Section B.7 of Appendix B.

Notes

1. P.W. Wilson (1995), "Detecting Influential Observations in Data Envelopment Analysis," *Journal of Productivity Analysis* 6, pp.27-46.

2. See, for instance, R.M. Thrall (1989), "Classification of Transitions under Expansion of Inputs and Outputs," *Managerial and Decision Economics* 10, pp.159-162.

3. R.D. Banker, H. Chang and W.W. Cooper (1996), "Simulation Studies of Efficiency, Returns to Scale and Misspecification with Nonlinear Functions in DEA," *Annals of Operations Research* 66, pp.233-253.

4. A. Charnes, W.W. Cooper, A.Y. Lewin, R.C. Morey and J.J. Rousseau (1985), "Sensitivity and Stability Analysis in DEA," *Annals of Operations Research* 2, pp.139-156.

5. A. Charnes and W.W. Cooper (1968), "Structural Sensitivity Analysis in Linear Programming and an Exact Product Form Left Inverse," *Naval Research Logistics Quarterly* 15, pp.517-522.

6. For a summary discussion see A. Charnes and L. Neralic (1992), "Sensitivity Analysis in Data Envelopment Analysis 3," *Glasnik Matematicki* 27, pp.191-201. A subsequent extension is L. Neralic (1997), "Sensitivity in Data Envelopment Analysis for Arbitrary Perturbations of Data," *Glasnik Matematicki* 32, pp.315-335. See also L. Neralic (2004), "Preservation of Efficiency and Inefficiency Classification in Data Envelopment Analysis," *Mathematical Communications* 9, pp.51-62.

7. A. Charnes, S. Haag, P. Jaska and J. Semple (1992), "Sensitivity of Efficiency Calculations in the Additive Model of Data Envelopment Analysis," *International Journal of System Sciences* 23, pp.789-798. Extensions to other classes of models may be found in A. Charnes, J.J. Rousseau and J.H.Semple (1996) "Sensitivity and Stability of Efficiency Classifications in DEA," *Journal of Productivity Analysis* 7, pp.5-18.

8. The shape of this "ball" will depend on the norm that is used. For a discussion of these and other metric concepts and their associated geometric portrayals see Appendix A in A. Charnes and W.W. Cooper (1961), *Management Models and Industrial Applications of Linear Programming* (New York: John Wiley & Sons).

9. This omission of DMU_o is also used in developing a measure of "super efficiency" as it is called in P. Andersen and N.C. Petersen (1993), "A Procedure for Ranking Efficient Units in DEA," *Management Science* 39, pp.1261-1264. Their use is more closely associated with stability when a DMU is omitted, however, so we do not cover it here. See the next chapter, Chapter 10, in this text. See also R.M. Thrall (1996) "Duality Classification and Slacks in DEA," *Annals of Operations Research* 66, pp.104-138.

10. A. Charnes, J.J. Rousseau and J.H. Semple (1996), "Sensitivity and Stability of Efficiency Classification in Data Envelopment Analysis," *Journal of Productivity Analysis* 7, pp.5-18.

11. In fact, Seiford and Zhu propose an iterative approach to assemble an exact stability region in L. Seiford and J. Zhu, "Stability Regions for Maintaining Efficiency in Data Envelopment Analysis," *European Journal of Operational Research* 108, 1998, pp.127-139.

12. R.G. Thompson, P.S. Dharmapala and R.M. Thrall (1994), "Sensitivity Analysis of Efficiency Measures with Applications to Kansas Farming and Illinois Coal Mining," in A. Charnes, W.W. Cooper, A.Y. Lewin and L.M. Seiford, eds., *Data Envelopment Analysis: Theory, Methodology and Applications* (Norwell, Mass., Kluwer Academic Publishers) pp.393-422.

13. R.G. Thompson, P.S. Dharmapala, J. Diaz, M.D. Gonzales-Lina and R.M. Thrall (1996), "DEA Multiplier Analytic Center Sensitivity Analysis with an Illustrative Application to Independent Oil Cos.," *Annals of Operations Research* 66, pp.163-180.

14. A. Charnes, W.W. Cooper and R.M.Thrall (1991), "A Structure for Classifying and Characterizing Efficiency in Data Envelopment Analysis," *Journal of Productivity Analysis* 2, pp.197-237.

15. An alternate approach to simultaneous variations in all data effected by the envelopment model is available in L.M. Seiford and J. Zhu (1998), "Sensitivity Analysis of DEA Models for Simultaneous Changes in All Data," *Journal of the Operational Research Society* 49, pp.1060-1071. See also the treatments of simultaneous changes in all data for additive models in L. Neralic (2004), "Preservation of Efficiency and Inefficiency Classification in Data Envelopment Analysis," *Mathematical Communications* 9, pp.51-62.

16. The ranges within which these dual variable values do not change generally form part of the printouts in standard linear programming computer codes.

17. This status is easily recognized because $\theta_1^* = \theta_2^* = \theta_3^* = 1$ are all associated with uniquely obtained solutions with zero slacks. See A. Charnes, W.W. Cooper and R.M. Thrall (1991) "A Structure for Classifying and Characterizing Efficiency in Data Envelopment Analysis," *Journal of Productivity Analysis* 2, pp.197-237.

18. W.W. Cooper, S. Li, L.M. Seiford and J. Zhu (2004), Chapter 3 in W.W. Cooper, L.M. Seiford and J. Zhu, eds., *Handbook on Data Envelopment Analysis* (Norwell, Mass., Kluwer Academic Publishers).

19. R.D. Banker (1993), "Maximum Likelihood, Consistency and Data Envelopment Analysis: A Statistical Foundation," *Management Science* 39, pp.1265-1273.

20. R. Banker and R. Natarasan (2004), "Statistical Tests Based on DEA Efficiency Scores," Chapter 11 in in W.W. Cooper, L.M. Seiford and J. Zhu, eds., *Handbook on Data Envelopment Analysis* (Norwell, Mass., Kluwer Academic Publishers).

21. Quoted from p.139 in R.D Banker (1996), "Hypothesis Tests Using Data Envelopment Analysis," *Journal of Productivity Analysis* pp.139-159.

22. A.P. Korostolev, L. Simar and A.B. Tsybakov (1995), "On Estimation of Monotone and Convex Boundaries," *Public Institute of Statistics of the University of Paris,* pp.3-15. See also Korostolev, Simar and Tsybakov (1995), "Efficient Estimation of Monotone Boundaries," *Annals of Statistics* 23, pp.476-489.

23. See the discussion in L. Simar (1996), "Aspects of Statistical Analysis in DEA-Type Frontier Models," *Journal of Productivity Analysis* 7, pp.177-186.

24. L. Simar and P.W. Wilson (1998), "Sensitivity Analysis of Efficiency Scores: How to Bootstrap in Nonparametric Frontier Models," *Management Science* 44, pp.49-61. See also Simar and Wilson (2004) "Performance of the Bootstrap for DEA Estimators and Iterating the Principle," Chapter 10 in W.W. Cooper, L.M. Seiford and J. Zhu, eds. *Handbook on Data Envelopment Analysis* (Norwell, Mass: Kluwer Academic Publishers).

25. M.J. Farrell (1951), "The Measurement of Productive Efficiency," *Journal of the Royal Statistical Society* Series A, 120, pp.253-290.

26. D.J. Aigner and S.F. Chu (1968), "On Estimating the Industry Production Frontiers," *American Economic Review* 56, pp.826-839.

27. D.J. Aigner, C.A.K. Lovell and P. Schmidt (1977), "Formulation and Estimation of Stochastic Frontier Production Models," *Journal of Econometrics* 6, pp.21-37. See also W. Meeusen and J. van den Broeck (1977) "Efficiency Estimation from Cobb-Douglas Functions with Composed Error," *International Economic Review* 18, pp.435-444.

28. J. Jondrow, C.A.K. Lovell, I.S. Materov and P. Schmidt (1982), "On the Estimation of Technical Inefficiency in the Stochastic Frontier Production Model," *Journal of Econometrics* 51, pp.259-284.

29. B.H. Gong and R.C. Sickles (1990), "Finite Sample Evidence on the Performance of Stochastic Frontiers and Data Envelopment Analysis Using Panel Data," *Journal of Econometrics* 51, pp.259-284.

30. P. Schmidt (1985-1986), "Frontier Production Functions," *Econometric Reviews* 4, pp.289-328. See also P.W. Bauer (1990), "Recent Development in Econometric Estimation of Frontiers," *Journal of Econometrics* 46, pp.39-56.

31. G.D. Ferrier and C.A.K. Lovell (1990), "Measuring Cost Efficiency in Banking — Econometric and Linear Programming Evidence," *Journal of Econometrics* 6, pp.229-245.

32. A. Charnes, W.W. Cooper and T. Sueyoshi (1988), "A Goal Programming/Constrained Regression Review of the Bell System Breakup," *Management Science* 34, pp.1-26.

33. See R.S. Barr, L.M. Seiford and T.F. Siems (1994), "Forcasting Bank Failure: A Non-Parametric Frontier Estimation Approach," *Recherches Economiques de Louvain* 60, pp. 417-429. for an example of a different two-stage DEA regression approach in which the DEA scores from the first stage served as an independent variable in the second stage regression model.

34. V. Arnold, I.R. Bardhan, W.W. Cooper and S.C. Kumbhakar (1994), "New Uses of DEA and Statistical Regressions for Efficiency Evaluation and Estimation — With an Illustrative Application to Public Secondary Schools in Texas," *Annals of Operations Research* 66, pp.255-278.

35. I.R. Bardhan, W.W. Cooper and S.C. Kumbhakar (1998), "A Simulation Study of Joint Uses of Data Envelopment Analysis and Stochastic Regressions for Production Function Estimation and Efficiency Evaluation," *Journal of Productivity Analysis* 9, pp.249-278.

36. P.L. Brockett, W.W. Cooper, S.C. Kumbhakar, M.J. Kwinn Jr. and D. McCarthy (2004), "Alternative Statistical Regression Studies of the Effects of Joint and Service-Specific Advertising on Military Recruitment," *Journal of the Operational Research Society* 55, pp.1039-1048.

37. S. Thore (1987), "Chance-Constrained Activity Analysis," *European Journal of Operational Research* 30, pp.267-269.

38. See the following three papers by K.C. Land, C.A.K. Lovell and S. Thore: (1) "Productive Efficiency under Capitalism and State Socialism: the Chance Constrained Programming Approach" in Pierre Pestieau, ed. in *Public Finance in a World of Transition* (1992) supplement to *Public Finance* 47, pp.109-121; (2) "Chance-Constrained Data Envelopment Analysis," *Managerial and Decision Economics* 14, 1993, pp.541-554; (3) "Productive Efficiency under Capitalism and State Socialism: An Empirical Inquiry Using Chance-Constrained Data Envelopment Analysis," *Technological Forecasting and Social Change* 46, 1994, pp.139-152. In "Four Papers on Capitalism and State Socialism" (Austin Texas: The University of Texas, IC^2 Institute) S. Thore notes that publication of (2) was delayed because it was to be presented at a 1991 conference in Leningrad which was cancelled because of the Soviet Crisis.

39. W.W. Cooper, Z. Huang and S. Li (1996), "Satisficing DEA Models under Chance Constraints," *Annals of Operations Research* 66, pp.279-295. For a survey of chance constraint programming uses in DEA, see W.W. Cooper, Z. Huang and S. Li (2004), "Chance Constraint DEA," in W.W. Cooper, R.M. Seiford and J. Zhu, eds., *Handbook on Data Envelopment Analysis* (Norwell, Mass., Kluwer Academic Publishers).

40. See Chapter 15 in H.A. Simon (1957), *Models of Man* (New York: John Wiley & Sons, Inc.)

41. A. Charnes and W.W. Cooper (1963), "Deterministic Equivalents for Optimizing and Satisficing under Chance Constraints," *Operations Research* 11, pp.18-39.

42. See W.F. Sharpe (1970), *Portfolio Theory and Capital Markets* (New York: McGraw Hill, Inc.)

43. O.B. Olesen and N.C. Petersen (1995), "Chance Constrained Efficiency Evaluation," *Management Science* 41, pp.442-457.

44. W.W. Cooper, Z. Huang, S.X. Li and O.B. Olesen (1998), "Chance Constrained Programming Formulations for Stochastic Characterizations of Efficiency and Dominance in DEA," *Journal of Productivity Analysis* 9, pp.53-79.

45. See A. Charnes and W.W. Cooper (1963) in footnote 41, above.

46. A. Charnes, W.W. Cooper and G.H. Symods (1958), "Cost Horizons and Certainty Equivalents," *Management Science* 4, pp.235-263.

47. G. Gigerenzer (2004), "Striking a Blow for Sanity in Theories of Rationality," in M. Augier and J.G. March, eds., *Models of a Man: Essays in Memory of H.A. Simon* (Cambridge: MIT Press).

48. As determined from YATS (Youth Attitude Tracking Survey) which obtains this information from periodic surveys conducted for the Army. See also G.A. Klopp (1985), "The Analysis of the Efficiency of Productive Systems with Multiple Inputs and Outputs," Ph.D. Dissertation (Chicago: University of Illinois at Chicago). Also available from University Microfilms, Inc., in Ann Arbor, Michigan.

49. T. Sueyoshi (1992), "Comparisons and Analyses of Managerial Efficiency and Returns to Scale of Telecommunication Enterprises by using DEA/WINDOW," (in Japanese) *Communications of the Operations Research Society of Japan* 37, pp.210-219.

50. D.B. Sun (1988), "Evaluation of Managerial Performance in Large Commercial Banks by Data Envelopment Analysis," Ph.D. Thesis (Austin, Texas: The University of Texas, Graduate School of Business). Also available from University Microfilms, Inc.

51. A. Charnes and W.W. Cooper (1991), "DEA Usages and Interpretations" reproduced in *Proceedings of International Federation of Operational Research Societies 12th Triennial Conference* in Athens, Greece, 1990.

10 SUPER-EFFICIENCY MODELS

10.1 INTRODUCTION

In this chapter we introduce a model proposed by Andersen and Petersen (1993),[1] that leads to a concept called "super-efficiency." The efficiency scores from these models are obtained by eliminating the data on the DMUo to be evaluated from the solution set. For the input model this can result in values which are regarded as according DMU_o the status of being "super-efficient." These values are then used to rank the DMUs and thereby eliminate some (but not all) of the ties that occur for efficient DMUs.

Other uses of this approach have also been proposed. Wilson (1993)[2], for example, suggests two uses of these measures in which each DMU_o is ranked according to its "influence" in either (or both) of the following two senses: (1) the number of observations that experience a change in their measure of technical efficiency as a result of these eliminations from the solution set and (2) the magnitude of these changes. Still other interpretations and uses are possible, and we shall add further to such possibilities. See also Ray (2000).[3]

There are troubles with these "super-efficiency" measures, as we shall see. These troubles can range from a lack of units invariance for these measures and extend to non-solution possibilities when convexity constraints are to be dealt with — as in the BCC model. However, the underlying concept is important, so we shall first review this approach to ranking in the form suggested by Andersen and Petersen (1993) and then show how these deficiencies may be eliminated by using other non-radial models that we will suggest.

10.2 RADIAL SUPER-EFFICIENCY MODELS

We start with the model used by Andersen and Petersen which takes the form of a CCR model and thereby avoids the possibility of non-solution that is associated with the convexity constraint in the BCC model. In vector form this model is

$$\begin{aligned} \text{[Super Radial-I-C]} \quad \theta^* &= \min_{\theta,\boldsymbol{\lambda},\boldsymbol{s}^-,\boldsymbol{s}^+} \theta - \varepsilon \boldsymbol{e}\boldsymbol{s}^+ \qquad (10.1) \\ \text{subject to} \quad \theta \boldsymbol{x}_o &= \sum_{j=1,\neq o}^{n} \lambda_j \boldsymbol{x}_j + \boldsymbol{s}^- \\ \boldsymbol{y}_o &= \sum_{j=1,\neq o}^{n} \lambda_j \boldsymbol{y}_j - \boldsymbol{s}^+ \end{aligned}$$

where all components of the $\boldsymbol{\lambda}$, $\boldsymbol{s}^-$ and $\boldsymbol{s}^+$ are constrained to be non-negative, $\varepsilon > 0$ is the usual non-Archimedean element and $\boldsymbol{e}$ is a row vector with unity for all elements.

We refer to (10.1) as a "Radial Super-Efficiency" model and note that the vectors, $\boldsymbol{x}_o$, $\boldsymbol{y}_o$ are omitted from the expression on the right in the constraints. The data associated with the DMU$_o$ being evaluated on the left is therefore omitted from the production possibility set. However, solutions will always exist so long as all elements are positive in the matrices X, $Y > 0$. (For weaker conditions see Charnes, Cooper and Thrall (1991)[4]).

The above model is a member of the class of Input Oriented-CCR (CCR-I) models. The output oriented version (Radial Super-O-C) has an optimal $\phi^* = 1/\theta^*$ and $\boldsymbol{\lambda}^*, \boldsymbol{s}^{-*}, \boldsymbol{s}^{+*}$ adjusted by division with θ^*, so we confine our discussion to the input oriented version, after which we will turn to other models, such as the BCC class of models, where this reciprocal relation for optimal solutions does not hold.

We illustrate the use of (10.1) with the data in Table 10.1 that is represented geometrically in Figure 10.1. Taken from Andersen and Petersen, these data are modified by adding DMU F, (which lies halfway between B and C), to portray all possibilities.

Table 10.1. Test Data

DMU	A	B	C	D	E	F
Input 1	2.0	2.0	5.0	10.0	10.0	3.5
Input 2	12.0	8.0	5.0	4.0	6.0	6.5
Output 1	1.0	1.0	1.0	1.0	1.0	1.0

Source: Andersen and Petersen (1993)

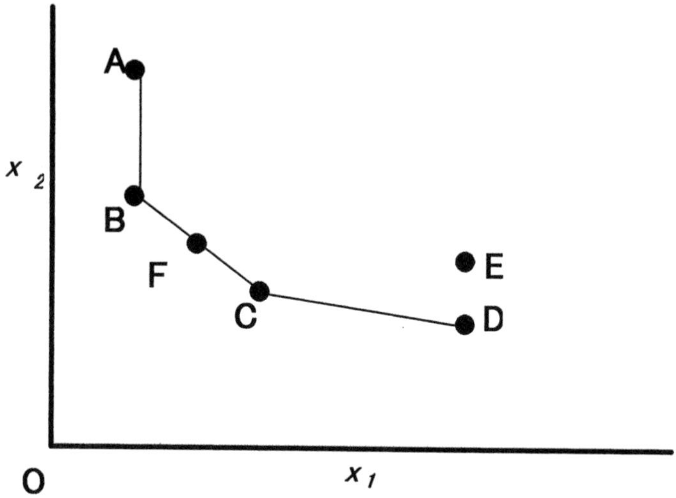

Figure 10.1. The Unit Isoquant Spanned by the Test Data in Table 10.1

Consider, first, the evaluation of DMU A as determined from the following adaptation of (10.1) where, as can be seen, the data for A are represented on the left but not on the right.

$$\begin{aligned}
&z^* = \min \theta - \varepsilon(s_1^- + s_2^-) - \varepsilon s^+ \\
&\text{subject to} \\
&2\theta = 2\lambda_B + 5\lambda_C + 10\lambda_D + 10\lambda_E + 3.5\lambda_F + s_1^- \\
&12\theta = 8\lambda_B + 5\lambda_C + 4\lambda_D + 6\lambda_E + 6.5\lambda_F + s_2^- \\
&1 = \lambda_B + \lambda_C + \lambda_D + \lambda_E + \lambda_F + s^+ \\
&0 \leq \lambda_B, \ldots, \lambda_F, s_1^-, s_2^-, s^+.
\end{aligned}$$

The solution to this problem is $\theta^* = 1$, $\lambda_B^* = 1$ $s_2^{-*} = 4$. All other variables are zero so $z^* = 1 - 4\varepsilon$. This is the same solution that would be obtained if the data for DMU A were included on the right. The evaluation of an inefficient point, like A, is not affected by this omission from the production possibility set because the efficient points that enter into the evaluation are unaffected by such a removal.

The latter condition is also present in the case of F. That is, the efficient points used in the evaluation of F are not removed from the production possibility set. Thus, the solution $\theta^* = 1$ and $\lambda_B^* = \lambda_C^* = 1/2$ confirms the efficiency of F both before and after such a removal. The important consideration is that the same efficient points that enter into (or can enter into) the evaluation of a DMU_o are present in both cases. This will also be the case for any inefficient point. For example the evaluation of E is unaffected with $\theta^* = 3/4$ and[5] $\lambda_C^* = \lambda_D^* = 1/2$ both before and after removal of the point E from the production possibility set. A point like A, which is on a part of the frontier that is not efficient will have the value $\theta^* = 1$, as is also true for the efficient point F which is on a part of the efficient frontier but is not an extreme point. See Charnes, Cooper and Thrall (1991) for a discussion of all possible classes of points and their properties. Now consider the rankings. The slacks are represented in the objective of (10.1) so we have the ranking: DMU F > DMU A since $1 > 1 - 4\varepsilon$ while $\theta^* = 3/4$ lies below both values by the definition of $\varepsilon > 0$ so this ranking will be F > A > E.

We next delete C from the production possibility set and obtain the solution $\theta^* = 1.2$ and $\lambda_B^* = \lambda_D^* = 1/2$ so that DMU C ranks as a super-efficient convex combination of B and D. This solution gives the coordinates of C with $x_1 = x_2 = 6$ so the elimination of C results in a replacement that yields the same unit of output with a 1 unit (=20%) increase in each input.

The Andersen-Petersen rankings are exhibited in Table 10.2 with B ranked first followed by D. A possible interpretation of what kind of managerial decision might be involved is to assume that the rankings are directed to studying the consequences of eliminating some DMUs. For this purpose, D might be preferred over B because of the large (7.5 units) reduction in input1 — a reduction that is given the negligible weight of $\varepsilon > 0$ in the rankings associated with the Andersen-Petersen measure.

Table 10.2. Andersen-Petersen Ranking*

Rank Order	B >	D >	C >	F >	A >	E
z^*	1.32	$1.25 - 7.5\varepsilon$	1.2	1	$1 - 4\varepsilon$	0.75
Reference Set	$\lambda_A = 0.79$ $\lambda_C = 0.21$	$\lambda_C = 1$	$\lambda_A = 0.5$ $\lambda_D = 0.5$	$\lambda_B = 0.5$ $\lambda_C = 0.5$	$\lambda_B = 1$	$\lambda_C = 0.5$ $\lambda_D = 0.5$
s	0	$s_1^- = 7.5$	0	0	$s_2^- = 4$	0

* Source: Andersen and Petersen (1993)

For purposes like the possible removal of DMUs the Andersen-Petersen measure can be regarded as deficient in its treatment of the nonzero slacks. It is also deficient because its treatment of the slacks does not yield a measure that is "units invariant." We therefore turn to SBM (the slacks based measure) to eliminate these deficiencies. We will do this in a way that eliminates the non-solution possibilities that are present when the convexity condition $\sum_{j=1}^{n} \lambda_j = 1$ is adjoined to the models we will suggest.

10.3 NON-RADIAL SUPER-EFFICIENCY MODELS

Here, we discuss the super-efficiency issues under the assumption that the DMU $(\boldsymbol{x}_o, \boldsymbol{y}_o)$ is SBM-efficient, i.e., it is strongly efficient. (See Definition 4.6 in Chapter 4.)

Let us define a production possibility set $P \setminus (\boldsymbol{x}_o, \boldsymbol{y}_o)$ spanned by (X, Y) excluding $(\boldsymbol{x}_o, \boldsymbol{y}_o)$, i.e.

$$P \setminus (\boldsymbol{x}_o, \boldsymbol{y}_o) = \left\{ (\bar{\boldsymbol{x}}, \bar{\boldsymbol{y}}) |\ \bar{\boldsymbol{x}} \geq \sum_{j=1, \neq o}^{n} \lambda_j \boldsymbol{x}_j,\ \ \bar{\boldsymbol{y}} \leq \sum_{j=1, \neq o}^{n} \lambda_j \boldsymbol{y}_j,\ \ \bar{\boldsymbol{y}} \geq \boldsymbol{0},\ \ \boldsymbol{\lambda} \geq \boldsymbol{0} \right\}. \tag{10.2}$$

Further, we define a subset $\bar{P} \setminus (\boldsymbol{x}_o, \boldsymbol{y}_o)$ of $P \setminus (\boldsymbol{x}_o, \boldsymbol{y}_o)$ as

$$\bar{P} \setminus (\boldsymbol{x}_o, \boldsymbol{y}_o) = P \setminus (\boldsymbol{x}_o, \boldsymbol{y}_o) \bigcap \{ \bar{\boldsymbol{x}} \geq \boldsymbol{x}_o \text{ and } \bar{\boldsymbol{y}} \leq \boldsymbol{y}_o \}, \tag{10.3}$$

where $P \setminus (\boldsymbol{x}_o, \boldsymbol{y}_o)$ means that point $(\boldsymbol{x}_o, \boldsymbol{y}_o)$ is excluded. By assumption $X > 0$ and $Y > 0$, $\bar{P} \setminus (\boldsymbol{x}_o, \boldsymbol{y}_o)$ is not empty.

As a weighted l_1 distance from $(\boldsymbol{x}_o, \boldsymbol{y}_o)$ and $(\bar{\boldsymbol{x}}, \bar{\boldsymbol{y}}) \in \bar{P} \setminus (\boldsymbol{x}_o, \boldsymbol{y}_o)$, we employ the index δ as defined by

$$\delta = \frac{\frac{1}{m} \sum_{i=1}^{m} \bar{x}_i / x_{io}}{\frac{1}{s} \sum_{r=1}^{s} \bar{y}_r / y_{ro}}. \tag{10.4}$$

From (10.3), this distance is not less than 1 and attains 1 if and only if $(\boldsymbol{x}_o, \boldsymbol{y}_o) \in \bar{P} \setminus (\boldsymbol{x}_o, \boldsymbol{y}_o)$, i.e. exclusion of the DMU $(\boldsymbol{x}_o, \boldsymbol{y}_o)$ has no effect on the original production possibility set P.

We can interpret this index as follows. The numerator is a weighted l_1 distance from $\boldsymbol{x}_o$ to $\bar{\boldsymbol{x}} (\geq \boldsymbol{x}_o)$, and hence it expresses an average expansion rate of $\boldsymbol{x}_o$ to $\bar{\boldsymbol{x}}$ of the point $(\bar{\boldsymbol{x}}, \bar{\boldsymbol{y}}) \in \bar{P} \setminus (\boldsymbol{x}_o, \boldsymbol{y}_o)$. The denominator is a weighted l_1 distance from $\boldsymbol{y}_o$ to $\bar{\boldsymbol{y}} (\leq \boldsymbol{y}_o)$, and hence it is an average reduction rate of $\boldsymbol{y}_o$ to $\bar{\boldsymbol{y}}$ of $(\bar{\boldsymbol{x}}, \bar{\boldsymbol{y}}) \in \bar{P} \setminus (\boldsymbol{x}_o, \boldsymbol{y}_o)$. The smaller the denominator is, the farther $\boldsymbol{y}_o$ is positioned relative to $\bar{\boldsymbol{y}}$. Its inverse can be interpreted as an index of the distance from $\boldsymbol{y}_o$ to $\bar{\boldsymbol{y}}$. Therefore, δ is a product of two indices: one, the distance in the input space, and the other in the output space. Both indices are dimensionless.

10.3.1 Definition of Non-radial Super-efficiency Measure

Based on the above observations, we define the super-efficiency of $(\boldsymbol{x}_o, \boldsymbol{y}_o)$ as the optimal objective function value δ^* from the following program:

$$[\text{SuperSBM-C}] \qquad \delta^* = \min_{\bar{\boldsymbol{x}},\bar{\boldsymbol{y}},\boldsymbol{\lambda}} \frac{\frac{1}{m}\sum_{i=1}^{m} \bar{x}_i/x_{io}}{\frac{1}{s}\sum_{r=1}^{s} \bar{y}_r/y_{ro}} \tag{10.5}$$

$$\text{subject to} \qquad \bar{\boldsymbol{x}} \geq \sum_{j=1,\neq o}^{n} \lambda_j \boldsymbol{x}_j$$

$$\bar{\boldsymbol{y}} \leq \sum_{j=1,\neq o}^{n} \lambda_j \boldsymbol{y}_j$$

$$\bar{\boldsymbol{x}} \geq \boldsymbol{x}_o \text{ and } \bar{\boldsymbol{y}} \leq \boldsymbol{y}_o$$

$$\bar{\boldsymbol{y}} \geq \mathbf{0}, \quad \boldsymbol{\lambda} \geq \mathbf{0}.$$

Let us introduce $\boldsymbol{\phi} \in R^m$ and $\boldsymbol{\psi} \in R^s$ such that

$$\bar{x}_i = x_{io}(1+\phi_i) \ (i = 1,\ldots,m) \text{ and } \bar{y}_r = y_{ro}(1+\psi_r) \ (r = 1,\ldots,s). \tag{10.6}$$

Then, this program can be equivalently stated in terms of $\boldsymbol{\phi}$, $\boldsymbol{\psi}$ and $\boldsymbol{\lambda}$ as follows:

$$[\text{SuperSBM-C}'] \qquad \delta^* = \min_{\boldsymbol{\phi},\boldsymbol{\psi},\boldsymbol{\lambda}} \frac{1+\frac{1}{m}\sum_{i=1}^{m} \phi_i}{1-\frac{1}{s}\sum_{r=1}^{s} \psi_r} \tag{10.7}$$

$$\text{subject to} \qquad \sum_{j=1,\neq o}^{n} x_{ij}\lambda_j - x_{io}\phi_i \leq x_{io} \ (i = 1,\ldots,m)$$

$$\sum_{j=1,\neq o}^{n} y_{rj}\lambda_j + y_{ro}\psi_r \geq y_{ro} \ (r = 1,\ldots,s)$$

$$\phi_i \geq 0 \ (\forall i), \ \psi_r \geq 0 \ (\forall r), \ \lambda_j \geq 0 \ (\forall j)$$

We have the following two propositions (Tone (2002)[6].

Proposition 10.1 *The super-efficiency score δ^* is units invariant, i.e. it is independent of the units in which the inputs and outputs are measured provided these units are the same for every DMU.*

Proof : This proposition holds, since both the objective function and constraints are units invariant. □

Proposition 10.2 *Let $(\alpha\boldsymbol{x}_o, \beta\boldsymbol{y}_o)$ with $\alpha \leq 1$ and $\beta \geq 1$ be a DMU with reduced inputs and enlarged outputs than $(\boldsymbol{x}_o, \boldsymbol{y}_o)$. Then, the super-efficiency score of $(\alpha\boldsymbol{x}_o, \beta\boldsymbol{y}_o)$ is not less than that of $(\boldsymbol{x}_o, \boldsymbol{y}_o)$.*

Proof: The super-efficiency score $(\hat{\delta}^*)$ of $(\alpha\boldsymbol{x}_o, \beta\boldsymbol{y}_o)$ is evaluated by solving the following program:

$$[\text{SuperSBM-C-2}] \qquad \hat{\delta}^* = \min \frac{\frac{1}{m}\sum_{i=1}^{m} \hat{x}_i/(\alpha x_{io})}{\frac{1}{s}\sum_{r=1}^{s} \hat{y}_r/(\beta y_{ro})}$$

$$= \min \frac{\beta}{\alpha} \frac{\frac{1}{m}\sum_{i=1}^{m} \hat{x}_i/x_{io}}{\frac{1}{s}\sum_{r=1}^{s} \hat{y}_r/y_{ro}} \quad (10.8)$$

$$\text{subject to} \quad \hat{\boldsymbol{x}} \geq \sum_{j=1,\neq o}^{n} \lambda_j \boldsymbol{x}_j$$

$$\hat{\boldsymbol{y}} \leq \sum_{j=1,\neq o}^{n} \lambda_j \boldsymbol{y}_j$$

$$\hat{\boldsymbol{x}} \geq \alpha \boldsymbol{x}_o \text{ and } \boldsymbol{0} \leq \hat{\boldsymbol{y}} \leq \beta \boldsymbol{y}_o$$

$$\boldsymbol{\lambda} \geq \boldsymbol{0}.$$

It can be observed that, for any feasible solution $(\hat{\boldsymbol{x}}, \hat{\boldsymbol{y}})$ for [SuperSBM-C-2], $(\hat{\boldsymbol{x}}/\alpha, \hat{\boldsymbol{y}}/\beta)$ is feasible for [SuperSBM-C]. Hence it holds

$$\delta^* \leq \frac{\frac{1}{m}\sum_{i=1}^{m}(\hat{x}_i/\alpha)/x_{io}}{\frac{1}{s}\sum_{r=1}^{s}(\hat{y}_r/\beta)/y_{ro}} = \frac{\beta}{\alpha} \frac{\frac{1}{m}\sum_{i=1}^{m} \hat{x}_i/x_{io}}{\frac{1}{s}\sum_{r=1}^{s} \hat{y}_r/y_{ro}}. \quad (10.9)$$

Comparing (10.8) with (10.9) we see that:

$$\delta^* \leq \hat{\delta}^*.$$

Thus, the super-efficiency score of $(\alpha \boldsymbol{x}_o, \beta \boldsymbol{y}_o)$ ($\alpha \leq 1$ and $\beta \geq 1$) is not less than that of $(\boldsymbol{x}_o, \boldsymbol{y}_o)$. □

10.3.2 Solving Super-efficiency

The fractional program [SuperSBM-C′] can be transformed into a linear programming problem using the Charnes-Cooper transformation as:

$$[\text{LP}] \quad \tau^* = \min t + \frac{1}{m}\sum_{i=1}^{m} \Phi_i \quad (10.10)$$

$$\text{subject to} \quad t - \frac{1}{s}\sum_{r=1}^{s} \Psi = 1$$

$$\sum_{j=1,\neq o}^{n} x_{ij}\Lambda_j - x_{io}\Phi_i - x_{io}t \leq 0 \ (i = 1, \ldots, m)$$

$$\sum_{j=1,\neq o}^{n} y_{rj}\Lambda_j + y_{ro}\Psi_r - y_{ro}t \geq 0 \ (r = 1, \ldots, s)$$

$$\Phi_i \geq 0 \ (\forall i), \ \Psi_r \geq 0 \ (\forall r), \ \Lambda_j \geq 0 \ (\forall j)$$

Let an optimal solution of [LP] be $(\tau^*, \boldsymbol{\Phi}^*, \boldsymbol{\Psi}^*, \boldsymbol{\Lambda}^*, t^*)$. Then we have an optimal solution of [SuperSBM-C′] expressed by

$$\delta^* = \tau^*, \ \boldsymbol{\lambda}^* = \boldsymbol{\Lambda}^*/t^*, \ \boldsymbol{\phi}^* = \boldsymbol{\Phi}^*/t^*, \ \boldsymbol{\psi}^* = \boldsymbol{\Psi}^*/t^*. \quad (10.11)$$

Furthermore, the optimal solution of [Super-SBM-C] is given by:

$$\bar{x}_{io}^* = x_{io}(1 + \phi_i^*) \ (i = 1, \ldots, m) \text{ and } \bar{y}_{ro}^* = y_{ro}(1 - \psi_r^*) \ (r = 1, \ldots, s). \quad (10.12)$$

10.3.3 Input/Output-Oriented Super-efficiency

In order to adapt the non-radial super-efficiency model to input (output) orientation, we can modify the preceding program as follows. For input orientation, we deal with the weighted l_1-distance only in the input space. Thus, the program turns out to be:

$$[\text{SuperSBM-I-C}] \qquad \delta_I^* = \min_{\boldsymbol{\phi},\boldsymbol{\lambda}} 1 + \frac{1}{m}\sum_{i=1}^{m} \phi_i \tag{10.13}$$

$$\text{subject to} \quad \sum_{j=1,\neq o}^{n} x_{ij}\lambda_j - x_{io}\phi_i \leq x_{io} \ (i = 1, \ldots, m)$$

$$\sum_{j=1,\neq o}^{n} y_{rj}\lambda_j \geq y_{ro} \ (r = 1, \ldots, s)$$

$$\phi_i \geq 0 \ (\forall i), \ \lambda_j \geq 0 \ (\forall j)$$

In a similar way we can develop the output-oriented super-efficiency model as follows:

$$[\text{SuperSBM-O-C}] \qquad \delta_O^* = \min_{\boldsymbol{\psi},\boldsymbol{\lambda}} \frac{1}{1 - \frac{1}{s}\sum_{r=1}^{s} \psi_r} \tag{10.14}$$

$$\text{subject to} \quad \sum_{j=1,\neq o}^{n} x_{ij}\lambda_j \leq x_{io} \ (i = 1, \ldots, m)$$

$$\sum_{j=1,\neq o}^{n} y_{rj}\lambda_j + y_{ro}\psi_r \geq y_{ro} \ (r = 1, \ldots, s)$$

$$\psi_r \geq 0 \ (\forall r), \ \lambda_j \geq 0 \ (\forall j)$$

Since the above two models have the same restricted feasible region as [SuperSBM-C], we have:

Proposition 10.3 *$\delta_I^* \geq \delta^*$ and $\delta_O^* \geq \delta^*$, where δ^* is defined in (10.4).*

10.3.4 An Example of Non-radial Super-efficiency

Using [SuperSBM-I-C] model in (10.13), we solved the same data set described in Table 10.1 and obtained the results exhibited in Table 10.3. Since DMUs A and E are enveloped respectively by B and C, their SBM scores are less than unity and SBM-inefficient. Compared with the radial model, DMU A dropped its efficiency from $1 - 4\varepsilon$ to 0.833. This is caused by the slack $s_2^- = 4$. We need to solve [SuperSBM-I-C] for DMUs B, C, D and F which have SBM score unity. DMU B has the optimal solution $\lambda_A^* = 1$, $s_2^* = x_{2A}\phi_2^* = 4$, $(\phi_2^* = 4/8 = 0.5)$ with all other variables zero, and hence its super-efficiency score is $1 + 0.5/2 = 1.25$. Similarly, DMUs C, D and F have super-efficiency scores 1.092, 1.125 and 1, respectively. In this example, the ranking is the same as the Andersen-Petersen model, although they are not always same (see Tone (2002)).

Table 10.3. Non-radial Super-efficiency

DMU	A	B	C	D	E	F
Super-eff.	0.833	1.25	1.092	1.125	0.667	1
Rank	5	1	3	2	6	4
Reference set	$\lambda_B = 1$	$\lambda_A = 1$	$\lambda_D = 0.23$ $\lambda_F = 0.77$	$\lambda_C = 1$	$\lambda_C = 1$	$\lambda_B = 0.5$ $\lambda_C = 0.5$
Slacks						
$s_1^- = x_1\phi_1$	0	0	0	0	5	0
$s_2^- = x_2\phi_2$	4	4	0.923	1	1	0

10.4 EXTENSIONS TO VARIABLE RETURNS-TO-SCALE

In this section, we extend our super-efficiency models to the variable returns-to-scale models and discuss the infeasible LP issues that then arise.

We extend our analysis to the variable returns-to-scale case by adjoining the following convexity constraints to the models:

$$\sum_{j=1}^{n} \lambda_j = 1, \ \lambda_j \geq 0, \ \forall j. \tag{10.15}$$

We observe two approaches as follows:

10.4.1 Radial Super-efficiency Case

We have two models [SuperRadial-I-V] (Input-oriented Variable RTS) and [SuperRadial-O-V] (Output-oriented Variable RTS) with models represented respectively as follows.

[SuperRadial-I-V] $\quad \theta^* = \min \theta \quad (10.16)$

subject to

$$\theta x_{io} \geq \sum_{j=1,\neq o}^{n} \lambda_j x_{ij} \ (i = 1, \ldots, m) \tag{10.17}$$

$$y_{ro} \leq \sum_{j=1,\neq o}^{n} \lambda_j y_{rj} \ (r = 1, \ldots, s) \tag{10.18}$$

$$\sum_{j=1,\neq o}^{n} \lambda_j = 1 \tag{10.19}$$

$$\lambda_j \geq 0 \ (\forall j).$$

$$\text{[SuperRadial-O-V]} \quad 1/\eta^* = \min 1/\eta \tag{10.20}$$

$$\text{subject to} \quad x_{io} \geq \sum_{j=1,\neq o}^{n} \lambda_j x_{ij} \ (i = 1, \ldots, m) \tag{10.21}$$

$$\eta y_{ro} \leq \sum_{j=1,\neq o}^{n} \lambda_j y_{rj} \ (r = 1, \ldots, s) \tag{10.22}$$

$$\sum_{j=1,\neq o}^{n} \lambda_j = 1 \tag{10.23}$$

$$\lambda_j \geq 0 \ (\forall j).$$

The above two programs may suffer from infeasibility under the following conditions. Suppose, for example, y_{1o} is larger than the other y_{1j} $(j \neq o)$, i.e.,

$$y_{1o} > \max_{j=1,\neq o}^{n} \{y_{1j}\}.$$

Then, the constraint (10.18) in [SuperRadial-I-V] is infeasible for $r = 1$ by dint of the constraint (10.19).

Likewise, suppose, for example, x_{1o} is smaller than the other x_{1j} $(j \neq o)$, i.e.,

$$x_{1o} < \min_{j=1,\neq o}^{n} \{x_{1j}\}.$$

Then, the constraint (10.21) in [SuperRadial-O-V] is infeasible for $i = 1$ by dint of the constraint (10.23).

Thus, we have:

Proposition 10.4 [SuperRadial-I-V] *has no feasible solution if there exists* r *such that* $y_{ro} > \max_{j\neq o}\{y_{rj}\}$, *and* [SuperRadial-O-V] *has no feasible solution if there exists* i *such that* $x_{io} < \min_{j\neq o}\{x_{ij}\}$.

We notice that Proposition 10.4 above is not a necessary condition for infeasibility, i.e., infeasibility may occur in other cases.

10.4.2 Non-radial Super-efficiency Case

The non-radial super-efficiency under variable returns-to-scale is evaluated by solving the following program:

$$\text{[SuperSBM-V]} \quad \delta^* = \min \delta = \frac{\frac{1}{m}\sum_{i=1}^{m} \bar{x}_i / x_{io}}{\frac{1}{s}\sum_{r=1}^{s} \bar{y}_r / y_{ro}} \tag{10.24}$$

$$\text{subject to} \quad \bar{\boldsymbol{x}} \geq \sum_{j=1,\neq o}^{n} \lambda_j \boldsymbol{x}_j$$

$$\bar{\boldsymbol{y}} \leq \sum_{j=1,\neq o}^{n} \lambda_j \boldsymbol{y}_j$$

$$\bar{\boldsymbol{x}} \geq \boldsymbol{x}_o \text{ and } \bar{\boldsymbol{y}} \leq \boldsymbol{y}_o$$
$$\sum_{j=1,\neq o}^{n} \lambda_j = 1$$
$$\bar{\boldsymbol{y}} \geq \mathbf{0}, \quad \boldsymbol{\lambda} \geq \mathbf{0}.$$

Under the assumptions $X > O$ and $Y > O$, we demonstrate that the [SuperSBM-V] is always feasible and has a finite optimum in contrast to the radial super-efficiency. This can be shown as follows.

We choose a DMU $j(\neq o)$ with $(\boldsymbol{x}_j, \boldsymbol{y}_j)$ and set $\bar{\lambda}_j = 1$ and $\bar{\lambda}_k = 0$ $(k \neq j)$. Using this DMU $(\boldsymbol{x}_j, \boldsymbol{y}_j)$ we define:

$$\widetilde{x}_i = \max\{x_{io}, x_{ij}\} \ (i = 1, \ldots, m) \tag{10.25}$$
$$\widetilde{y}_r = \min\{y_{ro}, y_{rj}\} \ (r = 1, \ldots, s). \tag{10.26}$$

Thus, the set $(\bar{\boldsymbol{x}} = \widetilde{\boldsymbol{x}}, \bar{\boldsymbol{y}} = \widetilde{\boldsymbol{y}}, \boldsymbol{\lambda} = \bar{\boldsymbol{\lambda}})$ is feasible for the [SuperSBM-V]. Hence, [SuperSBM-V] is always feasible with a finite optimum. Thus, we have:

Theorem 10.1 (Tone (2002)) *The non-radial super-efficiency model under the variable returns-to-scale environment,*[SuperSBM-V]*, is always feasible and has a finite optimum.*

We can define [SuperSBM-I-V] (Input-oriented Variable RTS) and [SuperSBM-O-V] (Output-oriented Variable RTS) models similar to [SuperSBM-I-C] and [SuperSBM-O-C]. We notice that [SuperSBM-I-V] and [SuperSBM-O-V] models confront the same infeasible LP issues as the [SuperRadial-I-V] and [Super Radial-O-V]. See Problem 10.1 for comparisons of super-efficiency models.

10.5 SUMMARY OF CHAPTER 10

This chapter introduced the concept of super-efficiency and presented two types of approach for measuring super-efficiency: radial and non-radial. Super-efficiency measures are widely utilized in DEA applications for many purposes, e.g., ranking efficient DMUs, evaluating the Malmquist productivity index and comparing performances of two groups (the bilateral comparisons model in Chapter 7).

10.6 NOTES AND SELECTED BIBLIOGRAPHY

Andersen and Petersen (1993) introduced the first super-efficiency model for radial models. Tone (2002) introduced non-radial super-efficiency models using the SBM.

10.7 RELATED DEA-SOLVER MODELS FOR CHAPTER 10

Super-efficiency codes have the same data format with the CCR model.

Super-Radial-I(O)-C(V) (Input(Output)-oriented Radial Super-efficiency model).

This code solves the oriented radial super-efficiency model under the constant (variable) returns-to-scale assumption. If the corresponding LP for a DMU is infeasible, we return a score 1 to the DMU.

Super-SBM-I(O)-C(V or GRS) (Input(Output)-oriented Non-radial Super-efficiency model)

This code solves the oriented non-radial super-efficiency model under the constant (variable or general) returns-to-scale assumption. If the corresponding LP for a DMU is infeasible, we return a score 1 to the DMU.

Super-SBM-C(V or GRS) (Non-radial and Non-oriented Super-efficiency model)

This code solves the non-oriented non-radial super-efficiency model under the constant (variable or general) returns-to-scale assumption.

10.8 PROBLEM SUPPLEMENT FOR CHAPTER 10

Problem 10.1

Table 10.4 displays data for 6 DMUs (A, B, C, D, E, F) with two inputs (x_1, x_2) and two outputs (y_1, y_2). Obtain and compare the super-efficiency scores using the attached DEA-Solver.

Table 10.4. Data for Super-efficiency

DMU	A	B	C	D	E	F
Input 1 (x_1)	2	2	5	10	10	3.5
Input 2 (x_2)	12	8	5	4	6	6.5
Output 1 (y_1)	4	3	2	2	1	1
Output 2 (y_2)	1	1	1	1	1	1

Suggested Answer : Tables 10.5 and 10.6 exhibit super-efficiency scores respectively under variable and constant returns-to-scale conditions. In the tables I (O) indicates the Input (Output) orientation and V (C) denotes the Variable (Constant) RTS. "NA" (not available) represents an occurrence of infeasible LP solution.

It is observed that, by dint of Proposition 10.4, Super-BCC-I and Super-SBM-I-V scores of DMU A are not available since it has $y_{1A} = 4$ which is strictly larger than other y_{1j} ($j \neq A$), and Super-BCC-O and Super-SBM-O-V scores of DMU D are not available since it has $x_{2D} = 4$ which is strictly smaller than other x_{2j} ($j \neq D$). The infeasibility of DMUs B and C for Super-BCC-O and Super-SBM-O-V is caused simply by non-existence of feasible solution for the corresponding linear programs. However, Super-SBM-V always has finite solutions as claimed by Theorem 10.1.

Under the constant RTS assumption, all DMUs have a finite score which resulted from exclusion of the convexity constraint on λ_j. The differences in scores between radial and non-radial models are due to non-zero slacks.

Table 10.5. Super-efficiency Scores under Variable RTS

DMU	A	B	C	D	E	F
Super-BCC-I	NA	1.26	1.17	1.25	0.75	1
Super-BCC-O	1.33	NA	NA	NA	1	1
Super-SBM-I-V	NA	1.25	1.12	1.13	0.66	1
Super-SBM-O-V	1.14	NA	NA	NA	0.57	0.57
Super-SBM-V	1.14	1.25	1.12	1.13	0.42	0.57

Table 10.6. Super-efficiency Scores under Constant RTS

DMU	A	B	C	D	E	F
Super-CCR-I	1.33	1.26	1.17	1.25	0.75	1
Super-CCC-O	1.33	1.26	1.17	1.25	0.75	1
Super-SBM-I-C	1.17	1.25	1.12	1.13	0.67	1
Super-SBM-O-C	1.14	1.2	1.09	1.25	0.5	0.57
Super-SBM-C	1.14	1.16	1.09	1.13	0.42	0.57

Notes

1. P. Andersen and N.C. Petersen (1993), "A Procedure for Ranking Efficient Units in Data Envelopment Analysis," *Management Science* 39, pp.1261-1264.

2. P. Wilson (1993), "Detecting Influential Observations in Data Envelopment Analysis," *Journal of Productivity Analysis* 6, pp. 27-46.

3. S. Ray (2004), *Data Envelopment Analysis* (Cambridge: Cambridge University Press).

4. A. Charnes, W.W. Cooper and R.M. Thrall (1991), "A Structure for Classifying and Characterizing Inefficiency in DEA," *Journal of Productivity Analysis* 2, pp. 197-237.

5. Andersen and Petersen report this as $\theta^* = 0.8333$, but their reported value is seen to be incorrect since 0.8333>0.75.

6. K. Tone (2002), "A Slacks-based Measure of Super-efficiency in Data Envelopment Analysis," *European Journal of Operational Research* 143, pp.32-41

Appendix A
Linear Programming and Duality

This appendix outlines linear programming and its duality relations. Readers are referred to text books such as Gass (1985)[1], Charnes and Cooper (1961)[2], Mangasarian (1969)[3] and Tone (1978)[4] for details. More advanced treatments may be found in Dantzig (1963)[5], Spivey and Thrall (1970)[6] and Nering and Tucker (1993).[7] Most of the discussions in this appendix are based on Tone (1978).

A.1 LINEAR PROGRAMMING AND OPTIMAL SOLUTIONS

The following problem, which minimizes a linear functional subject to a system of linear equations in nonnegative variables, is called a *linear programming problem*:

$$
\begin{aligned}
(P) \quad \min \; z &= \boldsymbol{c}\boldsymbol{x} && \text{(A.1)}\\
\text{subject to} \quad A\boldsymbol{x} &= \boldsymbol{b} && \text{(A.2)}\\
\boldsymbol{x} &\geq \boldsymbol{0}, && \text{(A.3)}
\end{aligned}
$$

where $A \in R^{m \times n}$, $\boldsymbol{b} \in R^m$ and $\boldsymbol{c} \in R^n$ are given, and $\boldsymbol{x} \in R^n$ is the vector of variables to be determined optimally to minimize the scalar z in the objective. $\boldsymbol{c}$ is a row vector and $\boldsymbol{b}$ a column vector. (A.3) is called a nonnegativity constraint. Also, we assume that $m < n$ and $\text{rank}(A) = m$.

A nonnegative vector of variables $\boldsymbol{x}$ that satisfies the constraints of (P) is called *a feasible solution* to the linear programming problem. A feasible solution that minimizes the objective function is called an *optimal solution.*

A.2 BASIS AND BASIC SOLUTIONS

We call a nonsingular submatrix $B \in R^{m \times m}$ of A a *basis* of A when it has the following properties: (1) it is of full rank and (2) it spans the space of solutions. We partition A into B and R and write symbolically:

$$A = [B \mid R], \tag{A.4}$$

where R is an $(m \times (n-m))$ matrix. The variable vector $\boldsymbol{x}$ is similarly divided into $\boldsymbol{x}^B$ and $\boldsymbol{x}^R$. $\boldsymbol{x}^B$ is called *basic* and $\boldsymbol{x}^R$ *nonbasic*. (A.2) can be expressed in terms of this partition as follows:

$$B\boldsymbol{x}^B + R\boldsymbol{x}^R = \boldsymbol{b}. \tag{A.5}$$

By multiplying the above equation by B^{-1}, we have:

$$\boldsymbol{x}^B = B^{-1}\boldsymbol{b} - B^{-1}R\boldsymbol{x}^R. \tag{A.6}$$

Thus, the basic vector $\boldsymbol{x}^B$ is expressed in terms of the nonbasic vector $\boldsymbol{x}^R$. By substituting this expression into the objective function in (A.1), we have:

$$z = \boldsymbol{c}^B B^{-1}\boldsymbol{b} - (\boldsymbol{c}^B B^{-1}R - \boldsymbol{c}^R)\boldsymbol{x}^R. \tag{A.7}$$

Now, we define a *simplex multiplier* $\boldsymbol{\pi} \in R^m$ and *simplex criterion* $\boldsymbol{p} \in R^{n-m}$ by

$$\boldsymbol{\pi} = \boldsymbol{c}^B B^{-1} \tag{A.8}$$

$$\boldsymbol{p} = \boldsymbol{\pi} R - \boldsymbol{c}^R, \tag{A.9}$$

where $\boldsymbol{\pi}$ and $\boldsymbol{p}$ are row vectors. The following vectors are called the *basic solution* corresponding to the basis B:

$$\bar{\boldsymbol{x}}^B = B^{-1}\boldsymbol{b} \tag{A.10}$$

$$\bar{\boldsymbol{x}}^R = \boldsymbol{0}. \tag{A.11}$$

Obviously the basic solution is feasible for (A.2) and (A.3).

A.3 OPTIMAL BASIC SOLUTIONS

We call a basis B *optimal* if it satisfies:

$$\bar{\boldsymbol{x}}^B = B^{-1}\boldsymbol{b} \geq \boldsymbol{0} \tag{A.12}$$

$$\boldsymbol{p} = \boldsymbol{\pi} R - \boldsymbol{c}^R \leq \boldsymbol{0}. \tag{A.13}$$

Theorem A.1 *The basic solution corresponding to an optimal basis is the optimal solution of linear programming* (P).

Proof. It is easy to see that $(\bar{\boldsymbol{x}}^B = B^{-1}\boldsymbol{b},\ \bar{\boldsymbol{x}}^R = \boldsymbol{0})$ is a feasible solution to (P). Furthermore,

$$z = \boldsymbol{c}^B \bar{\boldsymbol{x}}^B - \boldsymbol{p}\boldsymbol{x}^R. \tag{A.14}$$

Hence, by considering $\boldsymbol{p} \leq \boldsymbol{0}$, we find that z attains its minimum when $\boldsymbol{x}^R = \boldsymbol{0}$. □

The *simplex method* for linear programming starts from a basis, reduces the objective function monotonically by changing bases and finally attains an optimal basis.

A.4 DUAL PROBLEM

Given the linear programming (P) (called the *primal problem*), there corresponds the following *dual problem* with the row vector of variables $\boldsymbol{y} \in R^m$.

$$(D) \quad \max\ w = \boldsymbol{y}\boldsymbol{b} \tag{A.15}$$
$$\text{subject to} \quad \boldsymbol{y}A \leq \mathbf{c}, \tag{A.16}$$

and $\boldsymbol{y}$ not otherwise constrained.

Theorem A.2 *For each primal feasible solution $\boldsymbol{x}$ and each dual feasible solution $\boldsymbol{y}$,*

$$\boldsymbol{c}\boldsymbol{x} \geq \boldsymbol{y}\boldsymbol{b}. \tag{A.17}$$

That is, the objective function value of the dual maximizing problem never exceeds that of the primal minimizing problem.

Proof. By multiplying (A.2) from the left by $\boldsymbol{y}$, we have

$$\boldsymbol{y}A\boldsymbol{x} = \boldsymbol{y}\boldsymbol{b}. \tag{A.18}$$

By multiplying (A.16) from the right by $\boldsymbol{x}$ and noting $\boldsymbol{x} \geq \mathbf{0}$, we have:

$$\boldsymbol{y}A\boldsymbol{x} \leq \boldsymbol{c}\boldsymbol{x}. \tag{A.19}$$

Comparing (A.18) and (A.19),

$$\boldsymbol{c}\boldsymbol{x} \geq \boldsymbol{y}A\boldsymbol{x} = \boldsymbol{y}\boldsymbol{b}. \tag{A.20}$$

□

Corollary A.1 *If a primal feasible $\boldsymbol{x}^0$ and a dual feasible $\boldsymbol{y}^0$ satisfy*

$$\boldsymbol{c}\boldsymbol{x}^0 = \boldsymbol{y}^0\boldsymbol{b}, \tag{A.21}$$

then $\boldsymbol{x}^0$ is optimal for the primal and $\boldsymbol{y}^0$ is optimal for its dual.

Theorem A.3 (Duality Theorem) *(i) In a primal-dual pair of linear programs, if either the primal or the dual problem has an optimal solution, then the other does also, and the two optimal objective values are equal.*
(ii) If either the primal or the dual problem has an unbounded solution, then the other has no feasible solution. (iii) If either problem has no solution then the other problem either has no solution or its solution is unbounded.

Proof. (*i*) Suppose that the primal problem has an optimal solution. Then there exists an optimal basis B and $\boldsymbol{p} = \boldsymbol{\pi}R - \boldsymbol{c}^R \leq \mathbf{0}$ as in (A.13). Thus,

$$\boldsymbol{\pi}R \leq \mathbf{c}^R. \tag{A.22}$$

However, multiplying (A.8) on the right by B,

$$\boldsymbol{\pi}B = \mathbf{c}^B. \tag{A.23}$$

Hence,

$$\pi A = \pi \left[B|R\right] \leq \left[\mathbf{c}^B|\mathbf{c}^R\right] = \mathbf{c}. \tag{A.24}$$

Consequently,

$$\pi A \leq \mathbf{c}. \tag{A.25}$$

This shows that the simplex multiplier π for an optimal basis to the primal is feasible for the dual problem. Furthermore, it can be shown that π is optimal to the dual problem as follows: The basic solution ($\bar{\boldsymbol{x}}^B = B^{-1}\boldsymbol{b},\ \bar{\boldsymbol{x}}^R = \mathbf{0}$) for the primal basis B has the objective value $z = \mathbf{c}^B B^{-1}\boldsymbol{b} = \pi\boldsymbol{b}$, while π has the dual objective value $w = \pi\boldsymbol{b}$. Hence, by Corollary A.1, π is optimal for the dual problem. Conversely, it can be demonstrated that if the dual problem has an optimal solution, then the primal problem does also and the two objective values are equal, by transforming the dual to the primal form and by observing its dual. (See Gass, *Linear Programming*, pp. 158-162, for details).

(*ii*) (a) If the objective function value of the primal problem is unbounded below and the dual problem has a feasible solution, then by Theorem A.2,

$$w = \boldsymbol{y}\boldsymbol{b} \leq -\infty. \tag{A.26}$$

Thus, we have a contradiction. Hence, the dual has no feasible solution.

(b) On the other hand, if the objective function value of the dual problem is unbounded upward, it can be shown by similar reasoning that the primal problem is not feasible.

(*iii*) To demonstrate (*iii*), it is sufficient to show the following as an example in which both primal and dual problems have no solution.

<Primal>	min	$-x$	<Dual>	max	y
	subject to	$0 \times x = 1$		subject to	$y \times 0 \leq -1$
		$x \geq 0$			

where x and y are scalar variables. □

A.5 SYMMETRIC DUAL PROBLEMS

The following two LPs, $(P1)$ and $(D1)$, are mutually dual.

$$\begin{aligned} (P1) \quad \min\ z &= \boldsymbol{c}\boldsymbol{x} \\ \text{subject to}\quad A\boldsymbol{x} &\geq \boldsymbol{b} \\ \boldsymbol{x} &\geq \mathbf{0}. \end{aligned} \tag{A.27}$$

$$\begin{aligned} (D1) \quad \max\ w &= \boldsymbol{y}\boldsymbol{b} \\ \text{subject to}\quad \boldsymbol{y}A &\leq \boldsymbol{c} \qquad (A.28) \\ \boldsymbol{y} &\geq \mathbf{0}. \qquad (A.29) \end{aligned}$$

The reason is that, by introducing a nonnegative slack $\boldsymbol{\lambda} \in R^m$, $(P1)$ can be rewritten as $(P1')$ below and its dual turns out to be equivalent to $(D1)$.

$$\begin{array}{rrcl} (P1') & \min\ z & = & \boldsymbol{cx} \\ \text{subject to} & A\boldsymbol{x} - \boldsymbol{\lambda} & = & \boldsymbol{b} \\ & \boldsymbol{x} & \geq & \boldsymbol{0}, \quad \boldsymbol{\lambda} \geq \boldsymbol{0}. \end{array} \tag{A.30}$$

This form of mutually dual problems can be depicted as Table A.1, which is expressed verbally as follows:

For the inequality $\geq$ ($\leq$) constraints of the primal (dual) problem, the corresponding dual (primal) variables must be nonnegative. The constraints of the dual (primal) problem are bound to inequality $\leq$ ($\geq$). The objective function is to be maximized (minimized).

This pair of LPs are called symmetric primal-dual problems. The duality theorem above holds for this pair, too.

Table A.1. Symmetric Primal-Dual Problem

		x_1	...	...	x_n	≥ 0	
	y_1	a_{11}	...	...	a_{1n}		b_1
	.	...	...	...	...		.
$0 \leq$	.	...	...	...	...	$\geq$	.
	.	...	...	...	...		.
	y_m	a_{m1}	...	...	a_{mn}		b_m
			$\leq$				
		c_1	...	...	c_n		

A.6 COMPLEMENTARITY THEOREM

Let us transform the symmetric primal-dual problems into equality constraints by introducing nonnegative slack variables $\boldsymbol{\lambda} \in R^m$ and $\boldsymbol{\mu} \in R^n$, respectively.

$$\begin{array}{rrcl} (P1') & \min\ z & = & \boldsymbol{cx} \\ \text{subject to} & A\boldsymbol{x} - \boldsymbol{\lambda} & = & \boldsymbol{b} \\ & \boldsymbol{x} & \geq & \boldsymbol{0}, \quad \boldsymbol{\lambda} \geq \boldsymbol{0}. \end{array}$$

$$\begin{array}{rrcl} (D1') & \max\ w & = & \boldsymbol{yb} \\ \text{subject to} & \boldsymbol{y}A + \boldsymbol{\mu} & = & \boldsymbol{c} \qquad \text{(A.31)} \\ & \boldsymbol{y} & \geq & \boldsymbol{0}, \quad \boldsymbol{\mu} \geq \boldsymbol{0}. \qquad \text{(A.32)} \end{array}$$

Then, the optimality condition in Duality Theorem A.3 can be stated as follows:

Theorem A.4 (Complementarity Theorem) *Let $(\boldsymbol{x},\ \boldsymbol{\lambda})$ and $(\boldsymbol{y},\ \boldsymbol{\mu})$ be feasible to $(P1')$ and $(D1')$, respectively. Then, $(\boldsymbol{x},\ \boldsymbol{\lambda})$ and $(\boldsymbol{y},\ \boldsymbol{\mu})$ are optimal* *to $(P1')$ and $(D1')$ if and only if it holds:*

$$\boldsymbol{\mu x} = \boldsymbol{y\lambda} = 0. \tag{A.33}$$

Proof.

$$\text{From } \boldsymbol{\mu x} = 0, \text{ we have } (\boldsymbol{c} - \boldsymbol{y}A)\boldsymbol{x} = 0 \Rightarrow \boldsymbol{cx} = \boldsymbol{y}A\boldsymbol{x}. \tag{A.34}$$

$$\text{From } \boldsymbol{y\lambda} = 0, \text{ we have } \boldsymbol{y}(A\boldsymbol{x} - \boldsymbol{b}) = 0 \Rightarrow \boldsymbol{y}A\boldsymbol{x} = \boldsymbol{yb}. \tag{A.35}$$

Thus, $\boldsymbol{cx} = \boldsymbol{yb}$. By the duality theorem, $\boldsymbol{x}$ and $\boldsymbol{y}$ are optimal for the primal and the dual, respectively. □

By (A.33), we have

$$\sum_{j=1}^{n} \mu_j x_j = 0, \quad \sum_{i=1}^{m} y_i \lambda_i = 0. \tag{A.36}$$

By nonnegativity of each term in these two expressions,

$$\mu_j x_j = 0 \quad (j = 1, \ldots, n) \tag{A.37}$$

$$y_i \lambda_i = 0. \quad (i = 1, \ldots, m) \tag{A.38}$$

Thus, either the variable μ_j or the variable x_j must be zero for each j and either y_i or λ_i must be zero for each i. We called this property *complementarity.*

A.7 FARKAS' LEMMA AND THEOREM OF THE ALTERNATIVE

Theorem A.5 (Farkas' Lemma, Theorem of the Alternative) *For each $(m \times n)$ matrix A and each vector $\boldsymbol{b} \in R^m$, either*

$$(I) \quad A\boldsymbol{x} = \boldsymbol{b} \quad \boldsymbol{x} \geq \boldsymbol{0}$$

has a solution $\boldsymbol{x} \in R^n$ or

$$(II) \quad \boldsymbol{y}A \leq \boldsymbol{0} \quad \boldsymbol{yb} > 0$$

has a solution $\boldsymbol{y} \in R^m$ but never both.

Proof. For (I), we consider the following the primal-dual pair of LPs:

$$(P2) \qquad \begin{aligned} &\min \ z = \boldsymbol{0x} \\ &A\boldsymbol{x} = \boldsymbol{b} \\ &\boldsymbol{x} \geq \boldsymbol{0} \end{aligned}$$

$$(D2) \qquad \begin{aligned} &\max \ w = \boldsymbol{yb} \\ &\boldsymbol{y}A \leq \boldsymbol{0}. \end{aligned}$$

If (I) has a feasible solution, then it is optimal for $(P2)$ and hence, by the duality theorem, the optimal objective value of $(D2)$ is 0. Therefore, (II) has no solution.

On the other hand, $(D2)$ has a feasible solution $\boldsymbol{y} = \mathbf{0}$ and is not infeasible. Hence, if $(P2)$ is infeasible, $(D2)$ is unbounded upward. Thus, (II) has a solution.

□

A.8 STRONG THEOREM OF COMPLEMENTARITY

Theorem A.6 *For each skew-symmetric matrix $K\ (= -K^T) \in R^{n \times n}$, the inequality*

$$K\boldsymbol{x} \geq \mathbf{0}, \quad \boldsymbol{x} \geq \mathbf{0} \tag{A.39}$$

has a solution $\bar{\boldsymbol{x}}$ such that

$$K\bar{\boldsymbol{x}} + \bar{\boldsymbol{x}} > \mathbf{0}. \tag{A.40}$$

Proof. Let $\boldsymbol{e}_j \quad (j = 1, \ldots, n)$ be the j-th unit vector and the system (P_j) be,

$$(P_j) \qquad \begin{aligned} &K\boldsymbol{x} \geq \mathbf{0} \\ &\boldsymbol{x} \geq \mathbf{0}, \quad \boldsymbol{e}_j\boldsymbol{x} > 0. \end{aligned}$$

If (P_j) has a solution $\bar{\boldsymbol{x}}^j \in R^n$, then we have:

$$(K\bar{\boldsymbol{x}}^j)_j \geq \mathbf{0}, \quad \bar{\boldsymbol{x}}^j \geq \mathbf{0}, \quad \boldsymbol{e}_j\bar{\boldsymbol{x}}^j = (\bar{\boldsymbol{x}}^j)_j > 0$$

and hence

$$(K\bar{\boldsymbol{x}}^j)_j + (\bar{\boldsymbol{x}}^j)_j > 0.$$

If (P_j) has no solution, then by Farkas' lemma the following system has a solution $\bar{\boldsymbol{v}}^j \in R^n$, $\bar{\boldsymbol{w}}^j \in R^n$.

$$(D_j) \qquad \begin{aligned} &K\boldsymbol{v} = \boldsymbol{e}_j + \boldsymbol{w} \\ &\boldsymbol{v} \geq \mathbf{0}, \quad \boldsymbol{w} \geq \mathbf{0}. \end{aligned}$$

This solution satisfies:

$$(K\bar{\boldsymbol{v}}^j)_j = 1 + (\bar{\boldsymbol{w}}^j)_j > 0$$

and hence

$$(K\bar{\boldsymbol{v}}^j)_j + (\bar{\boldsymbol{v}}^j)_j > 0.$$

Since, for each $j = 1, \ldots, n$, either $\bar{\boldsymbol{x}}^j$ or $\bar{\boldsymbol{v}}^j$ exists, we can define a vector $\bar{\boldsymbol{x}}$ by summing over j. Then $\bar{\boldsymbol{x}}$ satisfies:

$$K\bar{\boldsymbol{x}} + \bar{\boldsymbol{x}} > \mathbf{0}.$$

□

Let a primal-dual pair of LPs with the coefficient $A \in R^{m \times n}$, $\boldsymbol{b} \in R^m$ and $\boldsymbol{c} \in R^n$ be $(P1')$ and $(D1')$ in Section A.6. Suppose they have optimal solutions $(\bar{\boldsymbol{x}}, \bar{\boldsymbol{\lambda}})$ for the primal and $(\bar{\boldsymbol{y}}, \bar{\boldsymbol{\mu}})$ for the dual, respectively. Then, by the complementarity condition in Theorem A.4, we have:

$$\bar{\mu}_j \bar{x}_j = 0 \quad (j = 1, \ldots, n) \tag{A.41}$$

$$\bar{y}_i \bar{\lambda}_i = 0 \quad (i = 1, \ldots, m) \tag{A.42}$$

However, a stronger theorem holds:

Theorem A.7 (Strong Theorem of Complementarity) *The primal-dual pair of LPs (P1′) and (D1′) have optimal solutions such that, in the complementarity condition (A.41) and (A.42), if one member of the pair is 0, then the other is positive.*

Proof. Observe the system:

$$\begin{aligned} A\boldsymbol{x} - r\boldsymbol{b} &\geq \boldsymbol{0} \\ -\boldsymbol{y}A + r\boldsymbol{c} &\geq \boldsymbol{0} \\ \boldsymbol{y}\boldsymbol{b} - \boldsymbol{c}\boldsymbol{x} &\geq 0 \\ \boldsymbol{x} \geq \boldsymbol{0}, \ \boldsymbol{y} &\geq \boldsymbol{0}, \ r \geq 0. \end{aligned}$$

We define a matrix K and a vector $\boldsymbol{w}$ by:

$$K = \begin{pmatrix} O & A & -\boldsymbol{b} \\ -A^T & O & \boldsymbol{c}^T \\ \boldsymbol{b}^T & -\boldsymbol{c} & 0 \end{pmatrix}$$

$$\boldsymbol{w}^T = \left(\boldsymbol{y}^T, \ \boldsymbol{x}, \ r\right)^T.$$

Then, by Theorem A.6, the system

$$K\boldsymbol{w} \geq \boldsymbol{0}, \quad \boldsymbol{w} \geq \boldsymbol{0}$$

has a solution $\widetilde{\boldsymbol{w}}^T = \left(\widetilde{\boldsymbol{y}}^T, \ \widetilde{\boldsymbol{x}}, \ \widetilde{r}\right)^T$ such that

$$K\widetilde{\boldsymbol{w}} + \widetilde{\boldsymbol{w}} > \boldsymbol{0}.$$

This results in the following inequalities:

$$A\widetilde{\boldsymbol{x}} - \widetilde{r}\boldsymbol{b} + \widetilde{\boldsymbol{y}}^T > \boldsymbol{0} \tag{A.43}$$

$$-\widetilde{\boldsymbol{y}}A + \widetilde{r}\boldsymbol{c} + \widetilde{\boldsymbol{x}}^T > \boldsymbol{0} \tag{A.44}$$

$$\widetilde{\boldsymbol{y}}\boldsymbol{b} - \boldsymbol{c}\widetilde{\boldsymbol{x}} + \widetilde{r} > 0 \tag{A.45}$$

We have two cases for $\widetilde{r}$.
(i) If $\widetilde{r} > 0$, we define $\bar{\boldsymbol{x}}$ and $\bar{\boldsymbol{\lambda}}$ by

$$\bar{\boldsymbol{x}} = \widetilde{\boldsymbol{x}}/\widetilde{r}, \quad \bar{\boldsymbol{y}} = \widetilde{\boldsymbol{y}}/\widetilde{r} \tag{A.46}$$

$$\bar{\boldsymbol{\lambda}} = A\bar{\boldsymbol{x}} - \boldsymbol{b}, \quad \bar{\boldsymbol{\mu}} = \boldsymbol{c} - \bar{\boldsymbol{y}}A. \tag{A.47}$$

Then, $(\bar{\boldsymbol{x}},\ \bar{\boldsymbol{\lambda}})$ is a feasible solution of $(P1')$ and $(\bar{\boldsymbol{y}},\ \bar{\boldsymbol{\mu}})$ is a feasible solution of $(D1')$. Furthermore, $\bar{\boldsymbol{y}}\boldsymbol{b} \geq \boldsymbol{c}\bar{\boldsymbol{x}}$. Hence, these solutions are optimal for the primal-dual pair LPs. In this case, (A.43) and (A.44) result in

$$\begin{aligned} \bar{\boldsymbol{\lambda}} + \bar{\boldsymbol{y}} &> \boldsymbol{0} && \text{(A.48)} \\ \bar{\boldsymbol{\mu}} + \bar{\boldsymbol{x}} &> \boldsymbol{0}. && \text{(A.49)} \end{aligned}$$

Thus, strong complementarity holds as asserted in the theorem.
(ii) If $\widetilde{r} = 0$, it cannot occur that both $(P1')$ and $(D1')$ have feasible solutions. The reason is: if they have feasible solutions $\boldsymbol{x}^*$ and $\boldsymbol{y}^*$, then

$$A\boldsymbol{x}^* \geq \boldsymbol{b},\ \ \boldsymbol{x}^* \geq \boldsymbol{0},\ \ \boldsymbol{y}^* A \leq \boldsymbol{c},\ \ \boldsymbol{y}^* \geq \boldsymbol{0}. \qquad \text{(A.50)}$$

Hence, we have:

$$\boldsymbol{c}\widetilde{\boldsymbol{x}} \geq \boldsymbol{y}^* A\widetilde{\boldsymbol{x}} \geq 0 \geq \widetilde{\boldsymbol{y}} A\boldsymbol{x}^* \geq \widetilde{\boldsymbol{y}}\boldsymbol{b}. \qquad \text{(A.51)}$$

This contradicts (A.45) in the case $\widetilde{r} = 0$. Thus, the case $\widetilde{r} = 0$ cannot occur.

□

A.9 LINEAR PROGRAMMING AND DUALITY IN GENERAL FORM

As a more general LP, we consider the case when there are both nonnegative variables $\boldsymbol{x}^1 \in R^k$ and sign-free variables $\boldsymbol{x}^2 \in R^{n-k}$ and both inequality and equality constraints are to be satisfied as follows:

$$\begin{array}{lll} (LP) & \min & z = \boldsymbol{c}^1\boldsymbol{x}^1 + \boldsymbol{c}^2\boldsymbol{x}^2 \\ & \text{subject to} & A_{11}\boldsymbol{x}^1 + A_{12}\boldsymbol{x}^2 \geq \boldsymbol{b}^1 \\ & & A_{21}\boldsymbol{x}^1 + A_{22}\boldsymbol{x}^2 = \boldsymbol{b}^2 \\ & & \boldsymbol{x}^1 \geq \boldsymbol{0} \\ & & \boldsymbol{x}^2 \ \text{free}, \end{array} \qquad \text{(A.52)}$$

where $A_{11} \in R^{\,l\times k}$, $A_{12} \in R^{\,l\times(n-k)}$, $A_{21} \in R^{(m-l)\times k}$ and $A_{22} \in R^{(m-l)\times(n-k)}$. The corresponding dual problem is expressed as follows, with variables $\boldsymbol{y}^1 \in R^l$ and $\boldsymbol{y}^2 \in R^{m-l}$.

$$\begin{array}{lll} (DP) & \max & w = \boldsymbol{y}^1\boldsymbol{b}^1 + \boldsymbol{y}^2\boldsymbol{b}^2 \\ & \text{subject to} & \boldsymbol{y}^1 A_{11} + \boldsymbol{y}^2 A_{21} \leq \boldsymbol{c}^1 \\ & & \boldsymbol{y}^1 A_{12} + \boldsymbol{y}^2 A_{22} = \boldsymbol{c}^2 \\ & & \boldsymbol{y}^1 \geq \boldsymbol{0} \\ & & \boldsymbol{y}^2 \ \text{free}. \end{array} \qquad \text{(A.53)}$$

It can be easily demonstrated that the two problems are mutually primal-dual and the duality theorem holds between them. Table A.2 depicts the general form of the duality relation of Linear Progrmming.

Table A.2. General Form of Duality Relation

		≥ 0	free		
		$\boldsymbol{x}^1$	$\boldsymbol{x}^2$		
$0 \leq$	$\boldsymbol{y}^1$	A_{11}	A_{12}	$\geq$	$\boldsymbol{b}^1$
free	$\boldsymbol{y}^2$	A_{21}	A_{22}	$=$	$\boldsymbol{b}^2$
		$\leq$	$=$		
		$\boldsymbol{c}^1$	$\boldsymbol{c}^2$		

Now, we introduce slack variables $\boldsymbol{\lambda}^1 \in R^{\,l}$ and $\boldsymbol{\mu}^1 \in R^k$ to (LP) and (DP) and rewrite them as (LP') and (DP') below:

$$
\begin{aligned}
(LP') \quad \min \quad & z = \boldsymbol{c}^1\boldsymbol{x}^1 + \boldsymbol{c}^2\boldsymbol{x}^2 \\
\text{subject to} \quad & A_{11}\boldsymbol{x}^1 + A_{12}\boldsymbol{x}^2 - \boldsymbol{\lambda}^1 = \boldsymbol{b}^1 \\
& A_{21}\boldsymbol{x}^1 + A_{22}\boldsymbol{x}^2 = \boldsymbol{b}^2 \\
& \boldsymbol{x}^1 \geq \boldsymbol{0} \\
& \boldsymbol{x}^2 \ \text{free} \\
& \boldsymbol{\lambda}^1 \geq \boldsymbol{0}.
\end{aligned}
\tag{A.54}
$$

$$
\begin{aligned}
(DP') \quad \max \quad & w = \boldsymbol{y}^1\boldsymbol{b}^1 + \boldsymbol{y}^2\boldsymbol{b}^2 \\
\text{subject to} \quad & \boldsymbol{y}^1 A_{11} + \boldsymbol{y}^2 A_{21} + \boldsymbol{\mu}^1 = \boldsymbol{c}^1 \\
& \boldsymbol{y}^1 A_{12} + \boldsymbol{y}^2 A_{22} = \boldsymbol{c}^2 \\
& \boldsymbol{y}^1 \geq \boldsymbol{0} \\
& \boldsymbol{y}^2 \ \text{free} \\
& \boldsymbol{\mu}^1 \geq \boldsymbol{0}.
\end{aligned}
\tag{A.55}
$$

We then have the following complementarity theorem:

Corollary A.2 (Complementarity Theorem in General Form)
Let $(\boldsymbol{x}^1,\ \boldsymbol{x}^2,\ \boldsymbol{\lambda}^1)$ and $(\boldsymbol{y}^1,\ \boldsymbol{y}^2,\ \boldsymbol{\mu}^1)$ be feasible to (LP') and (DP'), respectively. Then they are optimal to (LP') and (DP') if and only if the relation below holds.

$$
\boldsymbol{\mu}^1\boldsymbol{x}^1 = \boldsymbol{y}^1\boldsymbol{\lambda}^1 = 0. \tag{A.56}
$$

Also, there exist optimal solutions that satisfy the following strong complementarity.

Corollary A.3 (Strong Theorem of Complementarity) *In the optimal solutions to the primal-dual pair LPs, (LP') and (DP'), there exist ones such that, in the complementarity condition* (A.56), *if one of the pair is* 0, *then the other is positive.*

Notes

1. S.I. Gass (1985), *Linear Programming,* 5th ed., McGraw-Hill.
2. A. Charnes and W.W. Cooper (1961), *Management Models and Industrial Applications of Linear Programming,* (Volume 1 & 2), John Wiley & Sons.
3. O.L. Mangasarian (1969), *Nonlinear Programming,* McGraw-Hill.
4. K. Tone (1978), *Mathematical Programming,* (in Japanese) Asakura, Tokyo.
5. G.B. Dantzig (1963), *Linear Programming and Extensions* (Princeton: Princeton University Press).
6. W.A. Spivey and R.M. Thrall (1970), *Linear Optimization* (New York: Holt, Rinehart and Winston).
7. E.D. Nering and A.W. Tucker (1993), *Linear Programming and Related Problems* (New York: Academic Press).

Appendix B
Introduction to DEA-Solver

This is an introduction and manual for the attached DEA-Solver. There are two versions of DEA-Solver, the "Learning Version" (called **DEA-Solver-LV**, in the attached CD) and the "Professional Version" (called **DEA-Solver-PRO**: visit the DEA-Solver website at: **http://www.saitech-inc.com/** for further information). This manual serves both versions. DEA-Solver was developed by Kaoru Tone. All responsibility is attributed to Tone, but not to Cooper and Seiford in any dimension.

B.1 PLATFORM

The platform for this software is Microsoft Excel 97/2000 or later (a trademark of Microsoft Corporation).

B.2 INSTALLATION OF DEA-SOLVER

The accompanying installer will install DEA-Solver and sample problems in the attached CD-ROM to the hard disk (C:) of your PC. Click Setup.EXE in the folder "DEA-Solver" in the CD-ROM. Just follow the instruction on the screen. The folder in the hard disk is "C:\DEA-Solver" which includes the code DEA-Solver-LV(V3).xls and another folder "Samples(LV3)." A shortcut to DEA-Solver.xls will be automatically put on the Desktop. If you want to install "DEA-Solver" to another drive or to other folder (not to "C:\DEA-Solver"), just copy it to the disk or to the folder you designate. For the "Professional Version" an installer will automatically install "DEA-Solver."

B.3 NOTATION OF DEA MODELS

DEA-Solver applies the following notation for describing DEA models.

<Model Name> - <I or O> - <C or V or GRS>

where I or O corresponds to "Input"- or "Output"-orientation and C, V or GRS to "Constant", "Variable" or "General" returns to scale. For example, "AR-I-C" means the Input oriented Assurance Region model under Constant returns-to-scale assumption. In some cases, "I or O" and/or "C or V" are omitted. For example, "CCR-I" indicates the Input oriented CCR model which is naturally under constant returns-to-scale. "FDH" (= Free Disposal Hull) has no extensions. The abbreviated model names correspond to the following models,

1. CCR = Charnes-Cooper-Rhodes model (Chapters 2, 3)
2. BCC = Banker-Charnes-Cooper model (Chapters 4, 5)
3. IRS = Increasing Returns-to-Scale model (Chapter 5)
4. DRS = Decreasing Returns-to-Scale model (Chapter 5)

5. GRS = Generalized Returns-to-Scale model (Chapter 5)
6. AR = Assurance Region model (Chapter 6)
7. ARG = Assurance Region Global model (Chapter 6)
8. NCN = Non-controllable variable model (Chapter 7)
9. NDSC = Non-discretionary variable model (Chapter 7)
10. BND = Bounded variable model (Chapter 7)
11. CAT = Categorical variable model (Chapter 7)
12. SYS = Different Systems model (Chapter 7)
13. SBM = Slacks-Based Measure model (Chapter 4)
14. Weighted SBM = Weighted Slacks-Based Measure model (Chapter 4)
15. Cost = Cost efficiency model (Chapter 8)
16. New-Cost = New-Cost efficiency model (Chapter 8)
17. Revenue = Revenue efficiency model (Chapter 8)
18. New-Revenue = New-Revenue efficiency model (Chapter 8)
19. Profit = Profit efficiency model (Chapter 8)
20. New-Profit = New-Profit efficiency model (Chapter 8)
21. Ratio = Ratio efficiency model (Chapter 8)
22. Bilateral = Bilateral comparison model (Chapter 7)
23. FDH = Free Disposal Hull model (Chapter 4)
24. Window = Window Analysis (Chapter 9)
25. Super-efficiency = Super-efficiency model (Chapter 10)

B.4 INCLUDED DEA MODELS

The "Learning Version" includes all models and can solve problems with up to 50 DMUs; The "Professional Version" includes *Malmquist, Scale elasticity, Congestion* and *Undesirable output* models in addition to the above models and can deal with large-scale problems within the capacity of Excel worksheet.

B.5 PREPARATION OF THE DATA FILE

The data file should be prepared in an Excel Workbook prior to execution of DEA-Solver. The formats are as follows:

B.5.1 *The CCR, BCC, IRS, DRS, GRS, SBM, Super-Efficiency and FDH Models*

Figure B.1 shows an example of data file for these models.

1. **The first row (Row 1)**
   The first row (Row 1) contains Names of Problem and Input/Output Items, i.e.,
   Cell A1 = Problem Name
   Cell B1, C1, ...= Names of I/O items.
   The heading (I) or (O), showing them as being input or output should head the names of I/O items. The items without an (I) or (O) heading will not be considered as inputs and outputs. The ordering of (I) and (O) items is arbitrary.

2. **The second row and after**
   The second row contains the name of the first DMU and I/O values for the corresponding I/O items. This continues up to the last DMU.

3. **The scope of data domain**
   A data set should be bordered by at least one blank column at right and at least one blank row at bottom. This is a necessity for knowing the scope of the data domain. The data set should start from the top-left cell (A1).

4. **Data sheet name**
   A preferable sheet name is "DAT" (not "Sheet 1"). Never use names "Score", "Rank", "Projection", "Weight", "WeightedData", "Slack", "RTS", "Window", "Graph1" and "Graph2" for data sheet. These are reserved for this software.

The sample problem "Hospital(CCR)" in Figure B.1 has 12 DMUs with two inputs "(I)Doctor" and "(I)Nurse" and two outputs "(O)Outpatient" and "(O)Inpatient". The data set is bordered by one blank column (F) and by one blank row (14). The GRS model has the constraint $L \leq \sum_{j=1}^{n} \lambda_j \leq U$. The values of $L(\leq 1)$ and $U(\geq 1)$ must be supplied through the Message-Box on the display by request. Defaults are $L = 0.8$ and $U = 1.2$.

As noted in **1.** above, items without an (I) or (O) heading will not be considered as inputs or outputs. So, if you delete "(I)" from "(I)Nurse" to "Nurse," then "Nurse" will not be accounted for in this efficiency evaluation. Thus you can add (delete) items freely to (from) inputs and outputs without changing your data set.

	A	B	C	D	E	F
1	Hospital	(I)Doctor	(I)Nurse	(O)Outpatient	(O)Inpatient	
2	A	20	151	100	90	
3	B	19	131	150	50	
4	C	25	160	160	55	
5	D	27	168	180	72	
6	E	22	158	94	66	
7	F	55	255	230	90	
8	G	33	235	220	88	
9	H	31	206	152	80	
10	I	30	244	190	100	
11	J	50	268	250	100	
12	K	53	306	260	147	
13	L	38	284	250	120	
14						

Figure B.1. Sample.xls in Excel Sheet

B.5.2 The AR Model

Figure B.2 exhibits an example of data for the AR (Assurance Region) model. This problem has the same inputs and outputs as in Figure B.1. The constraints for the assurance region are described in rows 15 and 16 after "one blank row" at 14. This blank row is necessary for separating the data set and the assurance region constraints. These rows read as follows: the ratio of weights "(I)Doctor" vs. "(I)Nurse" is not less than 1 and not greater than 5 and that for "(O)Outpatient" vs. "(O)Inpatient" is not greater than 0.2 and not less than 0.5. Let the weights for Doctor and Nurse be $v(1)$ and $v(2)$, respectively. Then the first constraint implies

$$1 \leq v(1)/v(2) \leq 5.$$

Similarly, the second constraint means that the weights $u(1)$ (for Outpatient) and $u(2)$ (for Inpatient) satisfies the relationship

$$0.2 \leq u(1)/u(2) \leq 0.5.$$

Notice that the weights constraint can be applied between inputs and outputs as well.

	A	B	C	D	E	F
1	Hospital	(I)Doctor	(I)Nurse	(O)Outpatient	(O)Inpatient	
2	A	20	151	100	90	
3	B	19	131	150	50	
4	C	25	160	160	55	
5	D	27	168	180	72	
6	E	22	158	94	66	
7	F	55	255	230	90	
8	G	33	235	220	88	
9	H	31	206	152	80	
10	I	30	244	190	100	
11	J	50	268	250	100	
12	K	53	306	260	147	
13	L	38	284	250	120	
14						
15	1	(I)Doctor	(I)Nurse	5		
16	0.2	(O)Outpatient	(O)Inpatient	0.5		
17						

Figure B.2. Sample-AR.xls in Excel Sheet

B.5.3 The ARG Model

Instead of restricting ratios of virtual multipliers, this model imposes bounds on the virtual input (output) relative to the total virtual input (output). For example, in the above hospital case, the virtual input of Doctor is expressed by v(1) × (Number of) Doctor and the total virtual input is denoted by v(1) × (Number of) Doctor + v(2) × (Number of) Nurse, where v(1) and v(2) are weights to Doctor and Nurse, respectively. We impose lower and upper bounds, L and U, to the ratio of these two factors. Thus, we have constraints as expressed below.

$$L \leq \frac{v(1) \times \text{Doctor}}{v(1) \times \text{Doctor} + v(2) \times \text{Nurse}} \leq U.$$

In the Excel worksheet, we designate L and U along with the input (output) name as exhibited in Figure B.3. This means that $L = 0.5$ and $U = 0.8$ for Doctor in the above expression. See Section 6.3 in Chapter 6.

	A	B	C	D	E	F
1	Sample-ARG	(I)Doctor	(I)Nurse	(O)Outpatien	(O)Inpatient	
2	A	20	151	100	90	
3	B	19	131	150	50	
4	C	25	160	160	55	
5	D	27	168	180	72	
6	E	22	158	94	66	
7	F	55	255	230	90	
8	G	33	235	220	88	
9	H	31	206	152	80	
10	I	30	244	190	100	
11	J	50	268	250	100	
12	K	53	306	260	147	
13	L	38	284	250	120	
14						
15	0.5	(I)Doctor	0.8			
16	0.2	(I)Nurse	0.3			
17	0.2	(O)Outpatien	0.5			
18	0.4	(O)Inpatient	0.8			
19						

Figure B.3. Sample-ARG.xls in Excel Sheet

B.5.4 The NCN and NDSC Models

The non-controllable and non-discretionary models have basically the same data format as the CCR model. However, the uncontrollable inputs or outputs must have the headings (IN) or (ON), respectively. Figure B.4 exhibits the case where 'Doctor' is an uncontrollable (i.e., "non-discretionary" or "exogenously fixed") input and 'Inpatient' is an uncontrollable output.

	A	B	C	D	E	F
1	Hospital	(IN)Doctor	(I)Nurse	(O)Outpatient	(ON)Inpatient	
2	A	20	151	100	90	
3	B	19	131	150	50	
4	C	25	160	160	55	
5	D	27	168	180	72	
6	E	22	158	94	66	
7	F	55	255	230	90	
8	G	33	235	220	88	
9	H	31	206	152	80	
10	I	30	244	190	100	
11	J	50	268	250	100	
12	K	53	306	260	147	
13	L	38	284	250	120	
14						

Figure B.4. Sample-NCN (NDSC).xls in Excel Sheet

B.5.5 The BND Model

The bounded inputs or outputs must have the headings (IB) or (OB). Their lower and upper bounds should be designated by the columns headed by (LB) and (UB), respectively. These (LB) and (UB) columns must be inserted immediately after the corresponding (IB) or (OB) column. Figure B.5 implies that 'Doctor' and 'Inpatient' are bounded variables and their lower and upper bounds are given by the columns (LB)Doc., (UB)Doc., (LB)Inpat., and (UB)Inpat, respectively.

	A	B	C	D	E	F	G	H	I
1	Hospital	(IB)Doc.	(LB)Doc.	(UB)Doc.	(I)Nurse	(O)Outpat	(OB)Inpat.	(LB)Inpat.	(UB)Inpat.
2	A	20	15	22	151	100	90	80	100
3	B	19	15	23	131	150	50	45	55
4	C	25	20	25	160	160	55	50	60
5	D	27	21	27	168	180	72	70	76
6	E	22	20	25	158	94	66	60	80
7	F	55	45	56	255	230	90	80	100
8	G	33	31	36	235	220	88	80	95
9	H	31	29	33	206	152	80	70	90
10	I	30	28	31	244	190	100	90	110
11	J	50	45	50	268	250	100	90	120
12	K	53	45	54	306	260	147	130	160
13	L	38	30	40	284	250	120	110	130
14									

Figure B.5. Sample-BND.xls in Excel Sheet

B.5.6 The CAT, SYS and Bilateral Models

These models have basically the same data format as the CCR model. However, in the last column they must have an integer showing their category, system or bilateral group, as follows.

For the CAT model, the number starts from 1 (DMUs under the most difficult environment or with the most severe competition), 2 (in the second group of difficulty) and so on. It is recommended that the numbers be continuously assigned starting from 1.

For the SYS model, DMUs in the same system should have the same integer starting from 1.

For the Bilateral model, DMUs must be divided into two groups, denoted by 1 or 2.

Figure B.6 exhibits a sample data format for the CAT model.

	A	B	C	D	E	F
1	Hospital	(I)Doctor	(I)Nurse	(O)Outpatient	(O)Inpatient	Cat.
2	A	20	151	100	90	1
3	B	19	131	150	50	2
4	C	25	160	160	55	2
5	D	27	168	180	72	2
6	E	22	158	94	66	1
7	F	55	255	230	90	1
8	G	33	235	220	88	2
9	H	31	206	152	80	1
10	I	30	244	190	100	1
11	J	50	268	250	100	2
12	K	53	306	260	147	2
13	L	38	284	250	120	2
14						

Figure B.6. Sample-CAT.xls in Excel Sheet

B.5.7 The Cost and New-Cost Models

The unit cost columns must have the heading (C) followed by the *input* name. The ordering of columns is arbitrary. If an input has no cost column, its cost is regarded as zero. Figure B.7 is a sample.

	A	B	C	D	E	F	G	H
1	Hospital	(I)Doctor	(C)Doctor	(I)Nurse	(C)Nurse	(O)Outpat.	(O)Inpat.	
2	A	20	500	151	100	100	90	
3	B	19	350	131	80	150	50	
4	C	25	450	160	90	160	55	
5	D	27	600	168	120	180	72	
6	E	22	300	158	70	94	66	
7	F	55	450	255	80	230	90	
8	G	33	500	235	100	220	88	
9	H	31	450	206	85	152	80	
10	I	30	380	244	76	190	100	
11	J	50	410	268	75	250	100	
12	K	53	440	306	80	260	147	
13	L	38	400	284	70	250	120	
14								

Figure B.7. Sample-Cost(New-Cost).xls in Excel Sheet

B.5.8 The Revenue and New-Revenue Models

The unit price columns must have the heading (P) followed by the *output* name. The ordering of columns is arbitrary. If an output has no price column, its price is regarded as zero. See Figure B.8 for an example.

	A	B	C	D	E	F	G	H
1	Hospital	(I)Doctor	(I)Nurse	(O)Outpat.	(P)Outpat.	(O)Inpat.	(P)Inpat.	
2	A	20	151	100	550	90	2010	
3	B	19	131	150	400	50	1800	
4	C	25	160	160	480	55	2200	
5	D	27	168	180	600	72	3500	
6	E	22	158	94	400	66	3050	
7	F	55	255	230	430	90	3900	
8	G	33	235	220	540	88	3300	
9	H	31	206	152	420	80	3500	
10	I	30	244	190	350	100	2900	
11	J	50	268	250	410	100	2600	
12	K	53	306	260	540	147	2450	
13	L	38	284	250	295	120	3000	
14								

Figure B.8. Sample-Revenue(New-Revenue).xls in Excel Sheet

B.5.9 The Profit, New-Profit and Ratio Models

As a combination of *Cost* and *Revenue* models, these models have cost columns headed by (C) for inputs and price columns headed by (P) for outputs.

B.5.10 The Window Models

Figure B.9 exhibits an example of data format for a Window Analysis model. Top-left corner (A1) contains the problem name, e.g., "Car" in this example. The next right cell (B1) must include the first time period, e.g., "89." The second row beginning from the B column exhibits "I/O items", e.g., "(I)Sales" and "(O)Profit." The name of DMUs appears from the third row in the column A. The contents (observed data) follow in the third row and after. This style is repeated until the last time period. Notice that each time period is placed at the top-left corner of the corresponding frame and (I)/(O) items have the same names throughout the time period. It is not necessary to insert headings (I)/(O) to the I/O names of the second time period and after. I/O items are determined as designated in the first time period. Figure B.9 demonstrates performance of four car-manufacturers, i.e., Toyota, Nissan, Honda and Mitsubishi, during five time periods, i.e., from (19)89 to (19)93, in terms of the input "Sales" and the output "Profit."

A	B	C	D	E	F	G
Car	89		90		91	
DMU	(I)Sales	(O)Profit	Sales	Profit	Sales	Profit
Toyota	719	400	800	539	850	339
Nissan	358	92	401	139	418	120
Honda	264	74	275	100	280	65
Mitsubishi	190	44	203	49	231	66

H	I	J	K
92		93	
Sales	Profit	Sales	Profit
894	125	903	103
427	34	390	0
291	54	269	33
255	56	262	57

Figure B.9.   Sample-Window.xls in Excel Sheet

B.5.11   Weighted SBM Model

This model requires weights to inputs/outputs as data. They should be given
at the rows below the main body of data set with one inserted blank row. See
Figure B.10. The first column (A) has **WeightI** or **WeightO** designating input
or output, respectively, and the weights to inputs or outputs follow consecu-
tively in the order of input (output) items recorded at the top row. The values
are relative, since the software normalizes them properly. Refer to (4.81)-(4.83)
in Chapter 4. If they are vacant, weights are regarded as even. Figure B.10
designates that weights to Doctor and Nurse are 10:1 and those to Outpatient

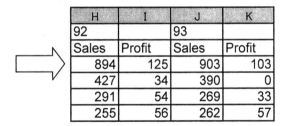

and Inpatient are 1:5.

	A	B	C	D	E	F
1	WSBM	(I)Doctor	(I)Nurs	(O)Outpatie	(O)Inpatient	
2	A	20	151	100	90	
3	B	19	131	150	50	
4	C	25	160	160	55	
5	D	27	168	180	72	
6	E	22	158	94	66	
7	F	55	255	230	90	
8	G	33	235	220	88	
9						
10	WeightI	10	1			
11	WeightO	1	5			
12						

Figure B.10. Sample-Weighted SBM.xls in Excel Sheet

B.6 STARTING DEA-SOLVER

After completion of the data file in an Excel sheet on an Excel book as mentioned above, close the data file and click either the icon or the file "DEA-Solver" in Explorer. This starts DEA-Solver. First, click "Enable Macros" and then follow the instructions on the display.

Otherwise if the file "DEA-Solver" is already open (loaded), click "Tools" on the *Menu Bar*, then select "Macro" and click "Macros." Finally, click "Run" on the Macro.

This Solver proceeds as follows,

1. Selection of a DEA model

2. Selection of a data set in Excel Worksheet

3. Selection of a Workbook for saving the results of computation and

4. DEA computation

B.7 RESULTS

The results of computation are stored in the selected Excel workbook. The following worksheets contain the results, although some models lack some of them.

1. **Worksheet "Summary"**
   This worksheet shows statistics on data and a summary report of results obtained.

2. **Worksheet "Score"**
This worksheet contains the DEA-score, reference set, λ-value for each DMU in the reference set and ranking in input and in the descending order of efficiency scores. A part of a sample Worksheet "Score" is displayed in Figure B.11, where it is shown that DMUs A, B and D are efficient (Score=1) and DMU C is inefficient (Score=0.882708) with the reference set composed of B ($\lambda_B = 0.9$) and D ($\lambda_D = 0.13889$) and so on.

No.	DMU	Score	Rank	Reference set (lambda)					
1	A	1	1	A	1				
2	B	1	1	B	1				
3	C	0.8827083	8	B	0.9	D	0.13888889		
4	D	1	1	D	1				
5	E	0.7634995	12	A	0.5794409	B	5.72E-02	D	0.1526401
6	F	0.8347712	10	B	0.2	D	1.11111111		
7	G	0.9019608	7	A	0.2588235	B	1.29411765		
8	H	0.7963338	11	A	0.3866921	B	1.35E-02	D	0.6183983
9	I	0.9603922	4	A	0.6470588	B	0.83529412		
10	J	0.8706468	9	D	1.3888889				
11	K	0.955098	6	A	0.86	D	0.96666667		
12	L	0.9582043	5	A	0.6470588	B	1.23529412		

Figure B.11. A Sample Score Sheet

3. **Worksheet "Rank"**
This worksheet contains the ranking of DMUs in the descending order of efficiency scores.

4. **Worksheet "Projection"**
This worksheet contains projections of each DMU onto the efficient frontier by the chosen model.

5. **Worksheet "Weight"**
Optimal weights $v(i)$ and $u(i)$ for inputs and outputs are exhibited in this worksheet. $v(0)$ corresponds to the constraints $\sum_j^n \lambda_j \geq l$ and $u(0)$ to $\sum_j^n \lambda_j \leq u$. In the BCC model where $l = u = 1$ holds, $u(0)$ stands for the value of the dual variable for this constraint.

6. **Worksheet "WeightedData"**
This worksheet shows the optimal weighted I/O values, $x_{ij}v(i)$ and $y_{rj}u(r)$ for each DMU$_j$ (for $j = 1, \ldots, n$).

7. **Worksheet "Slack"**
This worksheet contains the input excesses s^- and output shortfalls s^+

for each DMU. In the radial models, e.g., CCR and BCC, s^- and s^+ are calculated by using the formula (3.10) for the input-oriented case. Hence, notice that the (total) input-slacks are obtained as $s^- + (1-\theta)x_o$. In the non-radial models, e.g., SBM (Slacks-based measure), s^- and s^+ are defined via (4.48), and they indicate the total slacks of the concerned DMU.

8. **Worksheet "RTS"**
   In case of the BCC, AR-I-V and AR-O-V models, the returns-to-scale characteristics are recorded in this worksheet. For inefficient DMUs, returns-to-scale characteristics are those of the (input- or output-oriented) projected DMUs on the frontier.

9. **Graphsheet "Graph1"**
   The bar chart of the DEA scores is exhibited in this graphsheet. This graph can be redesigned using the Graph functions of Excel.

10. **Graphsheet "Graph2"**
    The bar chart of the DEA scores in the ascending order is exhibited in this graphsheet. A sample of Graph2 is exhibited in Figure B.12.

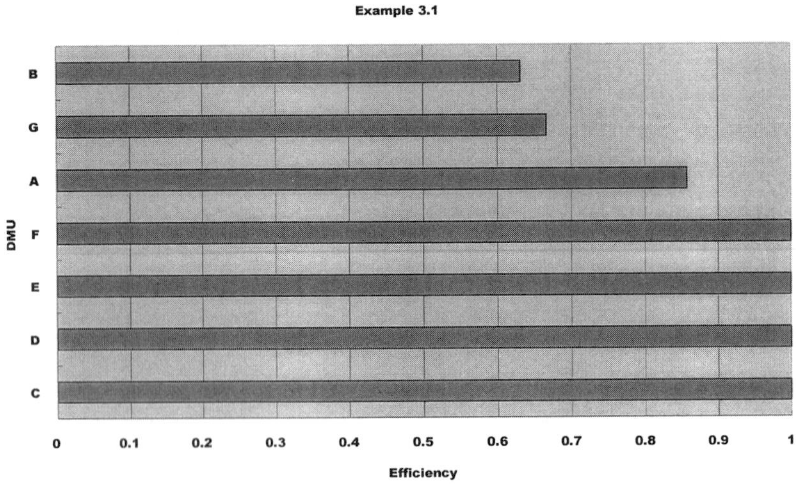

Figure B.12. A Sample Graph2

11. **Worksheets "Window*k*"**
    These sheets are only for Window models and k ranges from 1 to L (the length of time periods in the data). The contents are similar to Table 11.1 in Chapter 11. They also include two graphs, 'Variations through Window' and 'Variations by Term'. We will illustrate them in the case of Sample-Window.xsl in Figure B.9. Let $k = 3$ (so we deal with three adjacent years, for example). The results of computation in the case of "Window-I-C" are summarized in Table B.1.

Table B.1. Window Analysis by Three Adjacent Years

Maker	89	90	91	92	93	Average	C Average
Toyota	0.826	1	0.592			0.806	
		1	0.592	0.208		0.600	
			1	0.351	0.286	0.546	0.651
Nissan	0.381	0.515	0.426			0.441	
		0.515	0.426	0.118		0.353	
			0.720	0.200	0	0.307	0.367
Honda	0.416	0.540	0.345			0.434	
		0.540	0.345	0.275		0.387	
			0.582	0.465	0.308	0.452	0.424
Mitsubishi	0.344	0.358	0.424			0.375	
		0.358	0.424	0.326		0.369	
			0.716	0.551	0.545	0.604	0.449

From this table we can see row-wise averages of scores for each maker, which we call "Average through Window." The graph "Variations through Window" exhibits these averages. See Figure B.13.

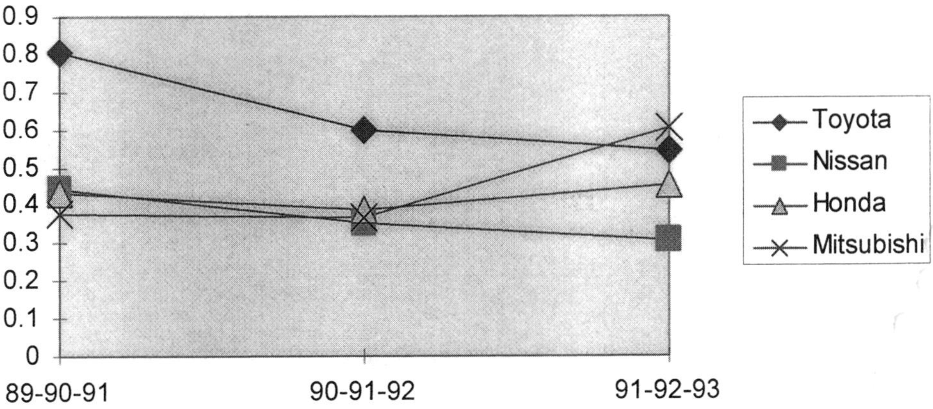

Figure B.13. Variations through Window

We can also evaluate column-wise averages of scores for each maker, which we call "Average by Term." The graph "Variations by Term" exhibits these averages. See Figure B.14.

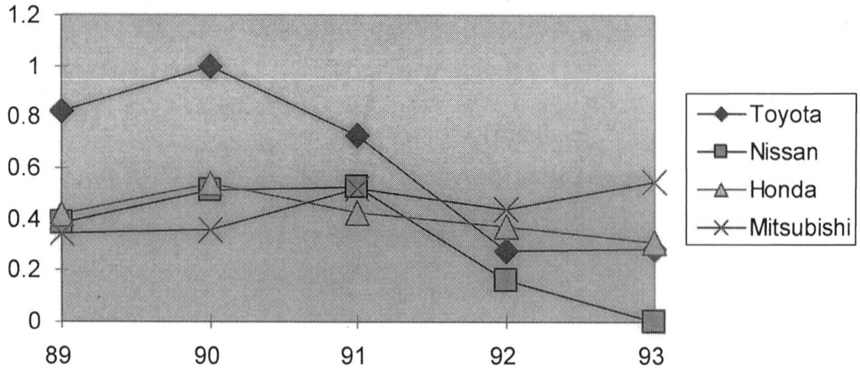

Figure B.14. Variations by Term

Note. The BCC, AR-I-V and AR-O-V models contain all the worksheets except "Window." The CCR, IRS, DRS, GRS, AR-I-C, AR-O-C and SBM models contain all sheets except "RTS" and "Window." The NCN, BND, CAT, SYS, Cost, Revenue, Profit, Ratio and FDH models produce "Summary," "Rank," "Score," "Projection," "Graph1" and "Graph2." The Bilateral model shows "Summary," "Score" and "Rank" sheets. The Window models return only "Window" and "Summary" sheets.

B.8 DATA LIMITATIONS

B.8.1 Problem Size

The "Learning Version" in the attached CD can solve problems with up to 50 DMUs. For the "Professional Version," the problem size is unlimited in terms of the number of DMUs and I/O items within the capacity of an Excel worksheet and the main memory of PC. More concretely, the data limitations for the "Professional Version" are as follows;

1. No. of DMUs must be less than 60000.
2. If No. of DMUs $\times$ (No. of Inputs + No. of Outputs + 2) $\geq$ 60000, then the "Projection" sheet will not be provided.

B.8.2 Inappropriate Data for Each Model

DMUs with the following irregular data are excluded from the comparison group as "inappropriate" DMUs. They are listed in the Worksheet "Summary." We will adopt the following notations for this purpose.

$xmax$ ($xmin$) = the max (min) input value of the DMU concerned *ymax* (*ymin*) = the max (min) output value of the DMU concerned

costmax (*costmin*) = the max (min) unit cost of the DMU concerned

pricemax (*pricemin*) = the max (min) unit price of the DMU concerned

1. For the *CCR, BCC-I, IRS, DRS, GRS, CAT* and *SYS* models, a DMU with no positive value in inputs, i.e., $xmax \leq 0$, will be excluded from computation. Zero or minus values are permitted if there is at least one positive value in the inputs of the DMU concerned.

   For the *BCC-O* model, DMUs with no positive value in outputs, i.e., $ymax \leq 0$, will be excluded from computation.

2. For the *AR* model, i.e., *AR-I-C, AR-I-V, AR-O-C* and *AR-O-V*, DMUs with $xmin < 0$, $xmax \leq 0$ or $ymax \leq 0$ will be excluded from the comparison group.

3. For the *FDH* model, DMUs with no positive input value, i.e., $xmax \leq 0$, or a negative input value, i.e., $xmin < 0$, will be excluded from computation.

4. For the *Cost* model, DMUs with $xmax \leq 0$, $xmin < 0$, $costmax \leq 0$, or $costmin < 0$ are excluded. DMUs with the current input cost ≤ 0 will also be excluded.

5. For the *Revenue*, *Profit* and *Ratio* models, DMUs with no positive input value, i.e., $xmax \leq 0$, no positive output value, i.e., $ymax \leq 0$, or with a negative output value, i.e., $ymin < 0$, will be excluded from computation. Furthermore, in the *Revenue* model, DMUs with $pricemax \leq 0$, or $pricemin < 0$ will be excluded from the comparison group. In the *Profit* model DMUs with $costmax \leq 0$ or $costmin < 0$ will be excluded. Finally, in the *Ratio* model, DMUs with $pricemax \leq 0$, $pricemin < 0$, $costmax \leq 0$ or $costmin < 0$ will be excluded.

6. For the *NCN* and *BND* models, negative input and output values are automatically set to zero by the program. DMUs with $xmax \leq 0$ in the controllable (discretionary) input variables will be excluded from the comparison group as "inappropriate" DMUs. In the *BND* model, the lower bound and the upper bound must enclose the given (observed) value, otherwise these values will be adjusted to the given value.

7. For the *Window-I-C* and *Window-I-V* models, no restriction exists for output data, i.e., positive, zero or negative values for outputs are permitted. However, DMUs with $xmax \leq 0$ will be characterized as being zero efficiency. This is for purpose of completing the score matrix. So, care is needed for interpreting the results in this case. If the number of DMUs per one period (term) exceeds 255, no graph will be produced.

8. For the *SBM* model, nonpositive inputs or outputs are replaced by a small positive value.

9. For the *Bilateral* model, we cannot compare two groups if some inputs are zero for one group while the other group has all positive values for the corresponding input item.

B.9 SAMPLE PROBLEMS AND RESULTS

The attached "DEA-Solver LV (learning version)" includes the sample problems and results for all models in the folder "Samples".

The "Professional Version" is available via http://www.saitech-inc.com.

B.10 SUMMARY OF HEADINGS TO INPUTS/OUTPUTS

Table B.2 exhibits headings to input/output and samples.

Table B.2. Headings to Inputs/Outputs

Heading	Description	Example	Models employed
(I)	Input	(I)Employee	All models
(O)	Output	(O)Sales	All models
(IN)	Non-controllable or Non-discretionary input	(IN)Population	NCN (Non-controllable) NDSC (Non-discretionary)
(ON)	Non-controllable or Non-discretionary output	(ON)Area	As above
(IB)	Bounded input	(IB)Doctor	BND (Bounded variable)
(OB)	Bounded output	(OB)Attendance	As above
(LB)	Lower bound of bounded variable	(LB)Doctor	As above
(UB)	Upper bound of bounded variable	(UB)Doctor	As above
(C)	Unit cost of input	(C)Manager	Cost, New-Cost, Profit, New-Profit, Ratio
(P)	Unit price of output	(P)Laptop	Revenue, New-Revenue, Profit, New-Profit, Ratio

Appendix C
Bibliography

Comprehensive bibliography of 2800 DEA references is available in the attached CD-ROM.

Author Index

Topic Index